The Paradoxes
of the
American Presidency

The Paradoxes of the American Presidency

Thomas E. Cronin
Michael A. Genovese

New York   Oxford
OXFORD UNIVERSITY PRESS
1998

Oxford University Press

Oxford   New York
Athens   Auckland   Bangkok   Bogota   Bombay
Buenos Aires   Calcutta   Cape Town   Dar es Salaam
Delhi   Florence   Hong Kong   Istanbul   Karachi
Kuala Lampur   Madras   Madrid   Melbourne
Mexico City   Nairobi   Paris   Singapore
Taipei   Tokyo   Toronto   Warsaw

and associated companies in
Berlin   Ibadan

Library of Congress Cataloging-in-Publication Data

Cronin, Thomas E.
    The paradoxes of the American presidency / Thomas E. Cronin,
Michael A. Genovese.
        p.   cm.
    Includes bibliographical references and index.
    ISBN 0-19-511692-5 (cl). — ISBN 0-19-511693-3   (pb)
    1. Presidents—United States.   I. Genovese, Michael A.
II. Title.
JK516.C73   1997
352.23′0973—dc32   97-29534
                                                        CIP

1 3 5 7 9 8 6 4 2

Printed in the United States of America
on acid-free paper

Contents

11
"If Men Were Angels . . .": Presidential Leadership
and Accountability *349*

Preface

Alexander Hamilton persuaded us that an American presidency was *necessary*. Many others have now demonstrated that the American presidency is *unique* and always *potentially dangerous*. Yet beyond this consensus—that the presidency is unique, necessary, and potentially dangerous—the American presidency remains a challenging institution to explain.

One sign of a discipline's maturity is the ability of those practicing in the field to agree on a "unifying theory" that both explains and predicts behavior. Presidency researchers have long lamented the absence of such theories.

What is it about the American presidency that defies theoretical precision? Why can't we devise propositions that predict the behavior of presidents and explain presidential leadership?

Much of the difficulty stems from the unusual character of the institution. The presidency is both unique and evolving. It defies simple explanations. It is dynamic, variable, and often a contradictory office. Lacking a precise constitutional delineation of powers, the American presidency is elastic and changing. Different occupants at different times may mold the institution to the nation's needs, or to suit their own needs; at other times the office and the U.S. system of separated and shared powers limit and constrain a president.

This book is an effort to understand the American presidency by viewing it through the lens of a series of paradoxes that shape and define the office. Our goal is to convey the complexity, the many-sidedness, and the contrarian aspects of the office.

A paradox is a sentiment or statement that is seemingly contradictory or opposed to common sense and yet may nonetheless be true. We use the term in a general sense, that we often hold clashing or contradictory notions of what a leader should do. A more rigorous definition is as follows:

> A logical paradox consists of two contrary, or even contradictory propositions to which we are led by apparently sound arguments. The arguments are sound because when used in other contexts they do not seem to invite any difficulty. It is only in the particular combination in which the paradox occurs that the arguments lead to a troublesome conclusion. In

its most extreme form a paradox consists in the apparent equivalence of two propositions, one of which is the negative of the other.

John van Heijenort, "Logical Paradoxes," in Paul Edwards, ed., *The Encyclopedia of Philosophy* (New York: Macmillan Co. and Free Press, 1967), vol. 5, p. 45.

In many ways paradoxes define the presidential office, and it is the paradoxical nature of the institution that makes the search for a unified theory elusive.

This book is an interpretation and a synthesis. It is an interpretative treatment in that it offers a fresh look at this complicated institution and at the even more complicated series of leaders who have served as president.

It is also a synthesis of presidential scholarship done by able journalists, biographers, historians, and political scientists over the past fifty years. We borrow and quote liberally from the rich body of literature that has developed on the American presidency. Our bibliography acknowledges the work of those who have helped us to understand this institution.

This book is informed too by the extensive interviewing both of us have done over the past thirty years. One of us has had the opportunity to work in Congress and in the White House, and to serve on the staff at the Brookings Institution; these experiences have proved invaluable.

Finally, this book builds on and is an extension of our earlier works on the presidency, especially Cronin's *The State of the Presidency* (1980), *Rethinking the Presidency* (1982), and *The Inventing of the American Presidency* (1989); and Genovese's *The Presidential Dilemma* (1995), *The Nixon Presidency* (1990), and *The Supreme Court, the Constitution and the Presidential Power* (1980).

We thank Pauline Batrikian, Clare Clamico, Kent Jancarik, Adrianne Ralph, Ken Singer, and Mahira Vallin, for thoughtful research assistance. We are grateful too for the expert secretarial assistance provided by Bernadette Bernard, JoAnn Collins, Margaret Edwards, and Donna Jones. Special thanks to Tania Cronin and Vicki Graf for their support and patience, our professional colleagues for their encouragement and counsel, and Whitman College and Loyola Marymount University for their institutional support. Tom Cronin also thanks the Luce Foundation and the Jerry McHugh family of Denver for their support and encouragement of his research. Nancy Lane, Karen Shapiro, and Gioia Stevens, our editors at Oxford University Press, provided splendid encouragement.

T.E.C. and M.G.
Autumn, 1997

The Paradoxes
of the
American Presidency

CHAPTER   **1**
_____

Presidential Paradoxes

He must have "common opinions." But it is equally
imperative that he be an "uncommon man." The
public must see themselves in him, but they must, at
the same time, be confident that he is something
bigger than themselves.

> Harold J. Laski, *The American Presidency: An
> Interpretation* (Harper & Brothers, 1940), p. 38

To become president, Lincoln had had to talk more
radically on occasion than he actually felt; to be an
effective president he was compelled to act more
conservatively than he wanted.

> Richard Hofstadter, *The American Political Tradition*
> (Vintage, 1948), p. 128

A law of opposites frequently influences the American
Presidency. Once in office, Presidents are seen to do
things least expected of them, often things they had
explicitly promised not to do. Previous commitments
or perceived inclinations act as a kind of insurance
that protects against any great loss if a President
behaves contrary to expectation.

> Daniel P. Moynihan, *The New Republic,* December 31,
>
> 1983, p. 18

The mind searches for answers to the complexities of life. We often gravitate toward simple explanations for the world's mysteries. This is a natural

1

way to try and make sense out of a world that seems to defy understanding. We are uncomfortable with contradictions so we reduce reality to understandable simplifications. And yet, contradictions and clashing expectations are part of life. "No aspect of society, no habit, custom, movement, development, is without cross-currents," says historian Barbara Tuchman. "Starving peasants in hovels live alongside prosperous landlords in featherbeds. Children are neglected and children are loved."[1] In life we are confronted with paradoxes for which we seek meaning. The same is true for the American presidency. We admire presidential power, yet fear it. We yearn for the heroic, yet are also inherently suspicious of it. We demand dynamic leadership, yet grant only limited powers to the president. We want presidents to be dispassionate analysts and listeners, yet they must also be decisive. We are impressed with presidents who have great self-confidence, yet we dislike arrogance and respect those who express reasonable self-doubt.

How then are we to make sense of the presidency? This complex, multidimensional, even contradictory institution is vital to the American system of government. The physical and political laws that seem to constrain one president, liberate another. What proves successful in one, leads to failure in another. Rather than seeking one unifying theory of presidential politics that answers all our questions, we believe that the American presidency might be better understood as a series of paradoxes, clashing expectations and contradictions.

Leaders live with contradictions. Presidents, more than most people, learn to take advantage of contrary or divergent forces. Leadership situations commonly require successive displays of contrasting characteristics. Living with, even embracing, contradictions is a sign of political and personal maturity.

The effective leader understands the presence of opposites. The aware leader, much like a first-rate conductor, knows when to bring in various sections, knows when and how to turn the volume up and down, and learns how to balance opposing sections to achieve desired results. Effective presidents learn how to manage these contradictions and give meaning and purpose to confusing and often clashing expectations. The novelist F. Scott Fitzgerald once suggested that, "The test of a first-rate intelligence is the ability to hold two opposed ideas in the mind at the same time."[2] Casey Stengel, long-time New York Yankee manager and occasional (if accidental) Zen philosopher, captured the essence of the paradox when he noted, "Good pitching will always stop good hitting, and vice versa."

Our expectations of, and demands on, the president are frequently so contradictory as to invite two-faced behavior by our presidents. Presidential powers are often not as great as many of us believe, and the president gets unjustly condemned as ineffective. Or a president will overreach or resort to unfair play while trying to live up to our demands.

The Constitution is of little help. The founders purposely left the presidency imprecisely defined. This was due in part to their fears of both the monarchy and the masses, and in part to their hopes that future presidents would create a more powerful office than the framers were able to do at the time. They knew that at times the president would have to move swiftly and effectively, yet they went to considerable lengths to avoid enumerating specific powers and duties in order to calm the then widespread fear of monarchy. After all, the nation had just fought a war against executive tyranny. Thus the paradox of the invention of the presidency: To get the presidency approved in 1787 and 1788, the framers had to leave several silences and ambiguities for fear of portraying the office as an overly centralized leadership institution. Yet when we need central leadership we turn to the president and read into Article II of the Constitution various prerogatives or inherent powers that allow the president to perform as an effective national leader.

Today the informal and symbolic powers of the presidency account for as much as the formal, stated ones. Presidential powers expand and contract in response to varying situational and technological changes. The powers of the presidency are thus interpreted so differently that they sometimes seem to be those of different offices. In some ways the modern presidency has virtually unlimited authority for almost anything its occupant chooses to do with it. In other ways, a president seems hopelessly ensnarled in a web of checks and balances.

Presidents and presidential candidates must constantly balance conflicting demands, cross pressures, and contradictions. It is characteristic of the American mind to hold contradictory ideas without bothering to resolve the conflicts between them. Perhaps some contradictions are best left unresolved, especially as ours is an imperfect world and our political system is a complicated one, held together by countless compromises. We may not be able to resolve many of these clashing expectations. Some of the inconsistencies in our judgments about presidents doubtless stem from the many ironies and paradoxes of the human condition. While difficult, at the least we should develop a better understanding of what it is we ask of our presidents, thereby increasing our sensitivity to the limits and possibilities of what a president can achieve. This might free presidents to lead and administer more effectively in those critical times when the nation has no choice but to turn to them. Whether we like it or not, the vitality of our democracy depends in large measure upon the sensitive interaction of presidential leadership with an understanding public willing to listen and willing to provide support. Carefully planned innovation is nearly impossible without the kind of leadership a competent and fair-minded president can provide.

The following are some of the paradoxes of the presidency. Some are cases of confused expectations. Some are cases of wanting one kind of presidential behavior at one time, and another kind later. Still others stem from the contradiction inherent in the concept of democratic leadership, which on the surface at least, appears to set up "democratic" and "leadership" as warring concepts. Whatever the source, each has implications for presidential performance and for how Americans judge presidential success and failure.

Paradox #1. Americans demand powerful, popular presidential leadership that solves the nation's problems. Yet we are inherently suspicious of strong centralized leadership and the abuse of power. Thus we place significant limits on the president's powers.

Paradox #2. We yearn for the democratic "common person" and simultaneously a leader who is uncommon, charismatic, heroic, and visionary.

Paradox #3. We want a decent, just, caring, and compassionate president, yet we also admire a cunning, guileful, and, on occasions that warrant it, even a ruthless, manipulative president.

Paradox #4. We admire the "above politics" nonpartisan or bipartisan approach, and yet the presidency is perhaps the most political office in the American system, which requires a creative entrepreneurial master politician.

Paradox #5. We want a president who can unify diverse people and interests; however, the job requires taking firm stands, making unpopular or controversial decisions that necessarily upset and divide.

Paradox #6. We expect our presidents to provide bold, visionary, innovative, *programmatic* leadership, and at the same time to respond *pragmatically* to the will of public opinion majorities. That is to say, we expect presidents to lead and to follow, and to exercise "democratic leadership."

Paradox #7. Americans want powerful, self-confident presidential leadership. Yet we are inherently suspicious of leaders who view themselves as infallible and above criticism.

Paradox #8. What it takes to become president may not be what is needed to govern the nation.

Paradox #9. The presidency is sometimes too strong yet at other times too weak.

To govern successfully, presidents must manage these paradoxes, and must balance a variety of competing demands and expectations.

Paradox #1. Americans demand powerful, popular presidential leadership that solves the nation's problems. Yet we are inherently suspicious of strong centralized leadership and especially the abuse of power and therefore we place significant limits on the president's powers.

We admire power but fear it. We love to unload responsibilities on our leaders, yet we intensely dislike being bossed around. We expect impressive leadership from presidents, and we simultaneously impose constitutional, cultural, and political restrictions on them. These restrictions often prevent presidents from living up to our expectations.

Our ambivalence toward executive power is hardly new. The founders knew the new republic needed more leadership, yet they feared the development of a popular leadership institution that might incite the people and yield factious or demagogic government. Thus, the early conception of the

American president was of an informed, virtuous statesman whose detached judgment and competence would enable him to work well with Congress and other leaders in making and implementing national public policy. This early presidency, as envisaged by the founders, did not encourage a popularly elected leader who would seek to directly shape and respond to the public's views. On the contrary, popular leadership grounded in the will of the people was "a synonym for demagoguery which, combined with the possibility of majority tyranny, was regarded as the peculiar vice to which democracies were susceptible."[3] The founders' goal was to provide some distance between the public and national leaders, especially the president, that distance to be used to refine the popular view, to allow for leadership and statesmanship rather than to do what the people wanted done.[4] To be sure, the people should be represented through Congress, yet they were not to be embodied in the executive. The president was to exercise wisdom, not responsiveness; judgment, not followership.

But the presidency of 1787 is not the presidency of today. Today it is a larger office, structurally similar to the original design but politically more powerful and more closely connected to popular passions. With the evolution of the nation has come an alternative conception of the presidency—the president is, in essence, the one truly national voice and representative of the American people. New arrangements for nominating and electing presidents have reinforced this conception, as has access to the magic of television. Today's president is seen as an agent of the people who should provide popular leadership.

The demand for a responsive president often conflicts with the demand for an informed, judicious statesman. The claims of politics and popular leadership have altered and undermined the early notions of presidential behavior. And as the presidency has become the lightning rod for nearly all of society's discontents, so also have presidents sometimes sought to be all things to all people.[5]

Still, while looking for strong, popular, presidential leadership, Americans also remain profoundly cautious about concentrating power in any one person's hands. Schooled in the tradition of negative government, Americans have historically feared the abuse of leadership power. Thus we have a system designed to thwart presidential leadership, save in times of crisis.

Presidents are supposed to follow the laws and respect the constitutional procedures that were designed to restrict their power, yet still they must be powerful and effective when action is needed. For example, we approve of presidential military initiatives and covert operations when they work out well, but we criticize presidents and insist they work more closely with Congress when the initiatives fail. We recognize the need for secrecy in certain government actions, but we resent being deceived and left in the dark—again, especially when things go wrong, as in Reagan's Iranian arms sale diversions to the Contras.

Although we sometimes do not approve of the way a president acts, we often approve of the end results. Thus Lincoln is often criticized for acting outside the limits of the Constitution, but at the same time he is forgiven due to the obvious necessity for him to violate certain constitutional principles in order to preserve the Union. FDR was often flagrantly deceptive and manipulative not only of his political opponents but also of his staff and allies. FDR even relished pushing people around and toying with them. But leadership effectiveness in the end often comes down to whether a person acts in terms of the highest interests of the nation. Most historians conclude Lincoln and Roosevelt were responsible in the use of presidential power, to preserve the Union, to fight the depression and nazism. Historians also conclude that Nixon was wrong for acting beyond the law in pursuit of personal power.

In short, we expect presidents, as the most visible representatives of big government and as popular leaders, to solve the entire scope of our problems by mustering all the powers and strengths that office now appears to confer, and then some. We are, however, unwilling to allow presidents to infringe on our rights in any significant way. In a self-help, individualistic, do-it-yourself society (which we still like to think we are), we remain ambivalent about the role of centralized presidential popular leadership because, as a people, we are somewhat ambivalent about the ends of government. We need government, but we resist its powers when we can, and we dislike admitting our growing dependence on it. It is also true that while we want strong leadership, if such leadership comes in the wrong form, we diminish our enthusiasm for it. This results in a dramatic roller coaster ride of strong support for the heroic presidency model followed by equally strong condemnations of presidential power.

Paradox #2. We yearn for the democratic "common person" and also for the uncommon, charismatic, heroic, visionary performance.

We want our presidents to be like us, but better than us. We like to think America is the land where the common sense of the common person reigns. Nourished on a diet of Frank Capra's "common-man-as-hero" movies, and the literary celebration of the average citizen by authors such as Emerson, Whitman, and Thoreau, we prize the common touch. The plain-speaking Harry Truman, the up-from-the-log cabin "man or woman of the people," is enticing. Few of us, however, settle for anything but the best; we want presidents to succeed and we hunger for brilliant, uncommon, and semiregal performances from presidents.

Thus, while we fought a revolution to depose royalty, part of us yearns for the majesty and symbolism of a royal family.

> From the beginning, American presidential style has had to lean simultaneously in two different directions; the character in the White House should behave with becoming democratic modesty. . . . But he must also express—in word, deed, bearing, personality . . . the greatness of the idea and the people he represents.[6]

King *and* commoner, we yearn for and demand both. The human heart, we are told, secretly and ceaselessly reinvents royalty and quests for the heroic. At the same time, we are told the hero is the individual the democratic nation must guard itself against. "Pity the nation that needs heroes," goes an old proverb. Why is this a haunting warning? Strong leaders, it is believed, can sap, diminish, and possibly even destroy the very wellsprings of self-government. Hence the old saying: Strong leaders make for a weak people.

Cicero wasn't sure whether great oratory was a national asset or a national peril. Presidents have to be persuasive, but if they are too convincing they may con us. Sometimes we are so mesmerized by the effective communicator that we fail to object, criticize, or resist: We fail to maintain the healthy skepticism so necessary in a political democracy.

The Field Research Corporation, a California public opinion organization, once asked a cross section of Californians to describe in their own words the qualities a presidential candidate should have. *Honesty* and *trustworthiness* topped the list. But one intriguing finding was that most (72 percent) prefer someone with *plain and simple tastes*. On the other hand, they expressed a strong preference (66 percent) for a spellbinding speaker or great communicator, someone like JFK or Ronald Reagan who inspires the nation.[7]

An old Greek saying describes two leadership "types": hedgehogs and foxes. The hedgehog knows only one or two things but is dogged in pursuit of these goals. Foxes, on the other hand, know very many things and they go about pursuing these multiple goals. Lincoln, FDR, and Reagan are the archetype hedgehogs. They were determined to win the war or cut taxes. And they regularly simplified the goals to dramatize the need. They were wholesalers. Hoover, Carter, and Bush are viewed as foxes who got caught up with so many less clear undertakings that they eventually became buried under the load.

Leaders must be hedgehogs *and* foxes, wholesalers *as well as* retailers. They must have macro and micro agendas. They should be willing and able to dramatize their missions, but they also must know that personal heroics need to be counterbalanced by reflective leadership.

It is said the American people crave to be governed by a president who is greater than anyone else yet not better than themselves. We are inconsistent; we want our president to be one of the folks yet also something special. If presidents get too special, however, they get criticized and roasted. If they try to be too folksy, people get bored. We cherish the myth that anyone can grow up to be president, that there are no barriers and no elite qualifications, but we don't want someone who is too ordinary. Would-be presidents have to prove their special qualifications—their excellence, their stamina, and their capacity for uncommon leadership. Fellow commoner, Truman, rose to the demands of the job and became an apparently gifted decision maker, or so his admirers would have us believe.

In 1976 Governor Jimmy Carter seemed to grasp this conflict and he ran as local, down-home, farm-boy-next-door makes good. The image of the peanut farmer turned gifted governor contributed greatly to Carter's success as a national candidate and he used it with consummate skill. Early in his presidential bid, Carter enjoyed introducing himself as peanut farmer *and* nuclear physicist, once again suggesting he was down to earth but cerebral as well.

Ronald Reagan illustrated another aspect of this paradox. He was a representative all-American—small-town, midwestern, and also a rich celebrity of stage, screen, and television. He boasted of having been a Democrat, yet campaigned as a Republican. A veritable Mr. Smith goes to Washington, he also had uncommon star quality. Bill Clinton liked us to view him as both a Rhodes scholar and an ordinary saxophone-playing member of the high school band from Hope, Arkansas; as a John Kennedy and even an Elvis figure; and also as just another jogger who would stop by for a Big Mac on the way home from a run in the neighborhood.

A president or would-be president must be bright but not too bright, warm and accessible but not too folksy, down to earth but not pedestrian. Democratic party presidential candidate in 1952 and 1956, Adlai Stevenson was witty and clever, talents that seldom pay in politics. Voters apparently prefer plainness and solemn platitudes, but these too can be overdone; witness Gerald Ford's talks, no matter what the occasion, that dulled our senses with the banal. People also joked that when Jimmy Carter gave a fireside chat even the fire fell asleep. The "catch 22" here is that an uncommon performance puts distance between a president and the truly common man. We persist, however, in wanting an uncommon individual as president.

We are often torn between demanding dynamic, charismatic leadership from our presidents and wanting presidents to heed our views. The preeminently successful presidents radiated courage and hope and stirred the hearts and minds of Americans with an almost demagogic ability to simplify and convince. "We need leaders of inspired idealism," said Theodore Roosevelt, "leaders to whom are granted great visions, who dream greatly and strive to make their dreams come true, who can kindle the people with the fire from their own burning souls."[8]

Do we pay a price when we get, permit, or encourage heroic popular leadership in the White House? Does it diminish the vitality of nongovernmental America? Does it possibly dissipate citizen and civic participation and responsibility? We haven't studied these effects carefully enough to know. We do know, however, that those heroic, larger-than-life presidents sometimes have inadvertently weakened the office for their successors. The grand performances of Jefferson, Jackson, Lincoln, Wilson, and FDR made it difficult for their successors to lead. The stretching of the powers and the strenuous performance often invited backlash. It was as if the presidency had roamed too far from the normal and acceptable. Invariably, a leveling of their successors seemed to take place.

There is another related problem with the notion of heroic presidential leadership. Most of the time those who wait around for heroic leaders in the White House are disappointed. This is because most of the time presidents do not provide galvanizing, brilliant policy leadership. In practice, the people make policy more often than presidents do; solutions percolate up rather than being imposed from the top down. Indeed, on many of the more important issues the people generally have to wait for presidents to catch up. In the overall scheme of the untidy policymaking process the public often is out in front, as they were in the move to get out of Vietnam, as they were in civil rights and in demanding better and more sustained arms control negotiations. Thus the old question of who leads whom needs to be addressed. Presidents, much of the time, are shrewd followers; they are not allowed to be heroic, pace-setting, or in-advance-of-their-times leaders.

Paradox #3. We want a decent, just, caring, and compassionate president, yet we admire a cunning, guileful, and, on occasions that warrant it, even a ruthless, manipulative president.

There is always a fine line between boldness and recklessness, between strong self-confidence and what the Greeks called "hubris," between dogged determination and pigheaded stubbornness. Opinion polls indicate people want a just, decent, and intellectually honest individual as our chief executive. Almost as strongly, however, the public also demands the quality of toughness.

We may admire modesty, humility, and a sense of proportion, but most of our great leaders have been vain and crafty. After all, you don't get to the White House by being a wallflower. Most have aggressively sought power and were rarely preoccupied with metaphysical inquiry or ethical considerations.

Franklin Roosevelt's biographers, while emphasizing his compassion for the average American, also agree he was vain, devious, and manipulative and had a passion for secrecy. These, they note, are often the standard weaknesses of great leaders. Significant social and political advances are made by those with drive, ambition, and a certain amount of brash, irrational self-confidence.

One psychologist believes Americans insist that our country's leaders display moral judgment, "but nothing dismays them as much as an officeholder who appears publicly anguished or in conflict over the ambiguities of an issue."[9]

George McGovern and Gerald Ford were criticized for being "too nice," "too decent." Being a "nice guy" is sometimes equated with being soft and afraid of power. The public dislikes the idea that a chief executive might be spineless or sentimental. Gerald Ford, Jimmy Carter, and Bill Clinton were all criticized by their own staffs for failing to "crack the whip" and discipline their advisers.

Would-be presidents simultaneously have to win our trust by displays of integrity while possessing the calculation, single-mindedness, and pragmatism of a jungle fighter.

Ronald Reagan seemed to understand these conflicting signals. He talked often of his concern for the truly needy and he portrayed himself as a devout believer in traditional values. This was manifest in his unusually strong pro-school prayer, pro-life/anti-abortion, and pro-family values positions. At the same time, his family relationships were hardly ideal.

Perhaps Dwight Eisenhower reconciled these clashing expectations better than recent presidents. Blessed with a wonderfully seductive, benign smile and a reserved, calming disposition, he was also the disciplined, strong, no-nonsense five-star general with all the medals and victories to go along with it. His ultimate resource as president was this reconciliation of decency and proven toughness, likability alongside demonstrated valor. Some of his biographers suggest his success was at least partly due to his uncanny ability to appear guileless to the public yet act with ample cunning in private.

While Americans want a president who is somewhat religious, they are wary of one who is too much so. Presidents often go out of their way to be photographed going to church or in the presence of noted religious leaders. Nixon and Reagan, who were not religious in the usual sense, held highly publicized Sunday services in the White House and spoke at fundamentalist religious gatherings.

This country was founded by refugees fleeing from religious persecutions. Americans want moral leadership on occasion, but they do not want a president who will inhibit religious freedom and diversity. Nor do they want preachiness or public policies guided solely by moral principles as opposed to considerations of power and national interest.

One of the ironies of the American presidency is that those characteristics we condemn in one president, we look for in another. Thus a supporter of Jimmy Carter's once suggested that Sunday school teacher Carter wasn't "rotten enough," "a wheeler-dealer," "an s.o.b."—precisely the virtues (if they can be called that) that Lyndon Johnson was most criticized for a decade earlier. President Clinton was viewed as both a gifted Southern Baptist–style preacher by some of his followers and a man who was character challenged, by opponents.

We appear to demand a double-faced personality. We demand the sinister as well as the sincere, the cunning as well as the compassionate, President Mean and President Nice, the president as Clint Eastwood and the president as Mr. Rogers—tough and hard enough to stand up to Khrushchev and North Korea, to Saddam Hussein and the Ayatollah, or to press the nuclear button, yet compassionate enough to care for the ill fed, ill clad and ill housed. In this case, the public seems to want a kindhearted s.o.b. or a clean wheeler-dealer, hard roles to cast and an even harder role to perform over eight years.

Former President Nixon, in writing about leaders he worked with, said a modern-day leader has to employ a variety of unattractive qualities on occasion in order to be effective, or at least to appear effective. Nixon may have carried these practices too far when he was in office, but his retirement writings are instructive nonetheless:

> In evaluating a leader, the key question about his behavioral traits is not whether they are attractive or unattractive, but whether they are useful. Guile, vanity, dissembling—in other circumstances these might be unattractive habits, but to the leader they can be essential. He needs guile in order to hold together the shifting coalitions of often bitterly opposed interest groups that governing requires. He needs a certain measure of vanity in order to create the right kind of public impression. He sometimes has to dissemble in order to prevail on crucial issues.[10]

Nixon should know. Despite other failings, he was often an effective foreign affairs president.

We want decency and compassion at home, but demand toughness and guile when presidents have to deal with our adversaries. We want presidents to be fierce or compassionate, nice or mean, sensitive or ruthless depending on what we want done, on the situation, and, to some extent, on the role models of the recent past. But woe to a president who is too much or too little possessed of these characteristics!

Americans severely criticize would-be leaders who are viewed as soft or afraid to make decisions, use power, or fire anyone. Jimmy Carter, Gerald Ford, and Bill Clinton, among others, were faulted for indecision, timidity, or failure to be pragmatic. Journalists said they merely didn't know how to play "hardball," or were unwilling to display it.

"He must know when to dissemble, when to be frank. He must pose as a servant of the public in order to become its master," wrote Charles de Gaulle in his book on leadership, *The Edge of the Sword.* De Gaulle also said leaders need strong doses of egotism, pride, and hardness.[11]

This paradox is sometimes wrongly perceived as a tension between male and female characteristics. But differences within genders are often larger than those between them. Everyone—male, female, or androgenous, East or West—has to blend competitiveness with cooperation, assertiveness with concern, toughness with tenderness.

Leaders vitally need to balance their competing impulses. Because leadership is wholly contextual, what works in one setting may fail in another. Thus, to be effective across issues and time, leaders must be flexible in matching style to circumstance.

In other words, leaders have to be uncommonly active, attentive listeners. They must squint with their ears. Along with listening we expect leaders to decide, and make judgments. Leaders such as Hamlet wait too long to act. Others, like King Lear or Sophocles' King Creon of Thebes, listen little and act in foolish haste.

Plainly, however, ambition is essential if a leader is to make a major difference. And, to gain power and retain it, one must have a love of power, and this love is often incompatible with moral goodness. In fact, it is often more linked to qualities of pride, duplicity, and cruelty. It raises all the worst fears associated with some of Machiavelli's slash-and-burn tactics.

Too, the intentional use of coercion, force, and even killing may, under certain circumstances, be morally and legally justifiable. The moral dilemma often becomes a choice between two competing evils. The key, usually, is whether the leader is seeking only self-aggrandizement. Leadership divorced from worthy purposes is merely manipulation or deception and, in the extreme, repressive tyranny. We insist that leaders, whatever their style, act with integrity and remain committed to larger purposes and commonly held goals.

Still, a paradox remains. "Power, or organized energy, may be a man-killing explosive or a life-saving drug," wrote Saul Alinsky. "The power of a gun may be used to enforce slavery, or to achieve freedom."[12] And so it is with leadership. Good leaders respect the preciousness of human life. Elements of calculation, abrasiveness, manipulation, and egoism are endemic in leadership. But a president must also be able to consider people in all of their relationships, in the wholeness of their lives, and not just as a means for enhancing profit or productivity.

A person who permits a cynical shell to harden around the heart will not long be able to exercise creative leadership. "The exercise of the heart is that of experiencing, thinking critically, willing, and acting, so as to overcome egocentrism and to share passion with other people . . . and to respond to their needs with the help one can give," writes Michael Maccoby.[13] Presidents who are hell bent on success rarely ponder whether the ends justify their means. Yet we ask that they raise questions of integrity and motivation with themselves.

Abraham Lincoln said that few things are wholly good or wholly evil. Most public policies or ideological choices are an indivisible compound of the two. Thus, our best judgment of the balance between them is continually demanded. The best of presidents are balanced individuals; they are sure of themselves, not dogmatic; they are self-confident, yet always willing to learn from their mistakes.

Paradox #4. We admire the "above politics" nonpartisan or bipartisan approach, yet the presidency is perhaps the most political office in the American system, a system in which we need a creative entrepreneurial master politician.

The public yearns for a statesman in the White House, for a George Washington or a second "era of good feelings"—anything that might prevent partisanship or politics as usual in the White House. Former French President Charles de Gaulle once said, "I'm neither of the left nor of the right nor of the center, but above." In fact, however, the job of president demands that the officeholder be a gifted political broker, ever attentive to changing political moods and coalitions.

Many of our early presidents unambiguously condemned parties while blatantly reaching out for party support when they needed to get their programs through Congress. "It is one of the paradoxes of the office," wrote historian Robert J. Morgan, "that a President must seek to balance his position as chief of his party with an equal need for support of his policies from all quarters of the nation regardless of partisan lines. He owes his office to the efforts of the party which put him there and, yet, once in power, his success as a leader rests in no little measure upon his securing a broad base of popular approval."[14]

Franklin Roosevelt illustrates this well. Appearing so remarkably non-partisan while addressing the nation, he was in practice one of the craftiest manipulators and political-coalition builders to occupy the White House.[15] He could punish friends and reward enemies when needed, or vice versa. He did not always succeed—for example when he tried to "pack the Court" in 1937 and to purge some Democratic members of Congress in 1938.

Presidents are often expected to be above politics in some respects while being highly political in others. Presidents are never supposed to act with their eyes on the next election, yet their power position demands they must. They are neither supposed to favor any particular group or party nor wheel and deal and twist too many arms. That's politics and that's bad! Instead, a president is supposed to be "president of all the people," above politics. A president is also asked to lead a party, to help fellow party members get elected or reelected, to deal firmly with party barons, interest group chieftains, and congressional political brokers. His ability to gain legislative victories depends on his skills at party leadership and on the size of his party's congressional membership. Jimmy Carter once lamented that "It's very difficult for someone to serve in this office and meet the difficult issues in a proper and courageous way and still maintain a combination of interest-group approval that will provide a clear majority at election time."[16]

To take the president out of politics is to assume, incorrectly, that a president will be generally right and the public generally wrong, that a president must be protected from the push and shove of political pressures. But what president has always been right? Over the years, public opinion has usually been as sober a guide as anything else on the political waterfront. And, lest we forget, having a president constrained and informed by public opinion is what democracy is all about.

The fallacy of antipolitics presidencies is that only one view of the national interest is tenable, and a president may pursue that view only by ignoring political conflict and pressure. Politics, properly conceived, is the art of accommodating the diversity and variety of public opinion to meet public goals. Politics is the task of building durable coalitions and majorities. It isn't always pretty. "The process isn't immaculate and cannot always be kid-gloved. A president and his men must reward loyalty and punish opposition; it is the only way."[17]

A president has to be far more than a party leader; but if he is not at least that he usually will fail—fail to accomplish his goals, however worthy, and

fail to perpetuate the influence of the forces and considerations that
elected him. Governing takes politics; it also takes persistence.[18]

Factions and partisan coalitions are a reality in American politics—
even if many of our founders tried to wish them away. Paradoxically, men
like Alexander Hamilton, John Adams, and Thomas Jefferson, all of whom
had condemned political parties in the 1780s, vigorously reached out for
party support only a few years later. Jefferson, the man who unambiguously proclaimed that if he "could go to heaven but with a party, I would
not go there at all," became one of the most astute party leaders in American presidential history.

No president can escape politics. Presidents must first be elected by
their party. And presidents who are effective must rely on party leaders in
Congress and the states.

In their attempts to be unifying leaders, presidents often avoid polariz-
ing conflicts. One of the lessons of history, however, is that early confrontation of controversial and divisive issues may avoid later violence. Further, sharpening conflict is often an important leadership responsibility.
After he had left the White House, Harry Truman said a president "who
is any damn good at all makes enemies. I even made a few myself when I
was in the White House, and I wouldn't have been without them." The
desire for a nonpolitical presidency sometimes is a staunchly status quo
desire to curb presidents, like Harry Truman and Franklin Roosevelt, from
becoming advocates and coalition builders for humanitarian progress
through political reform.

If past is prologue, presidents will go to considerable lengths to portray themselves as unconcerned with their own political future. They will
do so in large part because the public applauds a divorce between the
presidency and politics. People naively think that we can somehow turn
the job of president into that of a managerial or strictly executive post.
(The six-year, single-term proposal reflects this thinking.) Not so. The
presidency is a highly political office, and it cannot be otherwise. Moreover, its political character is for the most part desirable. A president separated from, or somehow above, politics might easily become a president
who doesn't listen to the people, doesn't respond to majority sentiment or
pay attention to views that may be diverse, intense, and at odds with his
own. Presidents may not always wish to obey the will of the majority—in
fact, leadership sometimes requires them to publicly argue against majority sentiment—but they cannot be unmindful of the will of the people. A
president immunized to politics would be a president who would become
too easily isolated from the processes of government and from the
thoughts and aspirations of the people.

This clashing expectation will endure. A standard diagnosis of what's
gone wrong in an administration will be that the presidency has become
too politicized. But it will be futile to try to take the president out of politics. A more useful approach is to realize that certain presidents try too

hard to hold themselves above politics, or at least to give that appearance, rather than engage in it deeply, openly, and creatively. A democratic president has to act politically regarding controversial issues if any semblance of government by consent is to be achieved.

Paradox #5. We want a president who can unify us, yet the job requires taking firm stands, making unpopular or controversial decisions that necessarily upset and divide us.

Closely related to paradox #4, paradox #5 holds that we ask the president to be a national unifier and a *harmonizer* while at the same time the job requires priority setting and *advocacy* leadership. The tasks are near opposites.

It is widely held that presidents must pull us together. Presidents must build coalitions and seek consensus. Presidents must not be too far ahead of their times if they are to be successful.

> He must see what he sees with the eyes of the multitude upon whose shoulders he stands. To get anywhere he must win understanding; to win it, the policy he pursues must never be so remote from the view about him that he cannot get that understanding. At bottom, his real power is in the popular support he can rally. . . .[19]

Our nation is one of the few in the world that calls on its chief executive to serve as its symbolic, ceremonial head of state *and* as its political head of government. Elsewhere, these tasks are spread around. In some nations there is a monarch and a prime minister; in others there are three visible national leaders—a head of state, a premier, and a powerful party chief.

In the absence of an alternative office or institution, we demand that our president act as a unifying force in our lives. Perhaps it all began with George Washington, who so artfully performed this function. At least for a while he truly was above politics, a unique symbol of our new nation. He was a healer, a unifier, and an extraordinary man for several seasons. Today we ask no less of our presidents than that they should do as Washington did, and more.

We have designed a presidential job description, however, that often forces our contemporary presidents to act as national dividers. Presidents must necessarily divide when they act as the leaders of their political parties, when they set priorities to the advantage of certain goals and groups at the expense of others, when they forge and lead political coalitions, when they move out ahead of public opinion and assume the role of national educators, when they choose one set of advisers over another. A president, as a creative executive leader, cannot help but offend certain interests. When Franklin Roosevelt was running for a second term, some garment workers unfolded a great sign that said, "We love him for the enemies he has made." Such is the fate of a president on an everyday basis; if presidents choose to use power they will lose the goodwill of those who preferred inaction.

The opposite is, of course, also true. Presidents obsessed with striving to protect their power and popularity by choosing to act can run into difficulties as well. Presidents may avoid divisive conflict in favor of short-run, low-risk policymaking. Such a president may seem effective but may in reality undermine the confidence of the American people and the credibility of the presidency in the long run.

Look at it from another angle. The nation is torn between the view that a president should primarily preside over the nation and merely serve as a referee among the various powerful interests that actually control who gets what, when, and how and a second position, which holds that a president should gain control of government processes and powers so as to use them for the purpose of furthering public, as opposed to private, interests.

Harry S. Truman noted that fourteen or fifteen million Americans had the resources to have representatives in Washington to protect their interests, and that the interests of the great mass of other people were the responsibility of the president of the United States.

The president is sometimes seen as the great defender of the people, the ombudsman or advocate general of "public interests." But this should be viewed as merely a claim, for many presidents have acted otherwise, even antagonistically, to mass or popular preferences.

Hence, we alternately want our presidents to stir things up or calm things down, to inject conflict or to lessen conflict. Much depends, too, on how you define leadership. And only the narrowest conception of leadership would call solely for the avoidance of conflict.

The public demands results, yet often judges presidents on the basis of means. Whereas the public will usually approve of the ends of certain policies, they may disapprove of the means. Means often require sacrifice, and this can turn the public against a president.

Leaders are forever having to decide how much harmony or stirring up is needed to achieve objectives. All organizations need action and decisiveness, but they also need shared values, community, and integration. The truly effective president understands how to use selective conflict for revitalization.

Reasonable people, George Bernard Shaw and others have observed, adjust to reality and cope with what they find. Unreasonable people dream of a different, better world and try to adapt the world to themselves. This discontent or unreasonableness is often the first step in the progress of an organization or a nation-state.

But everyone has to be aware that stirrer-uppers and conflict-polarizers are threatening people. In the kingdom of the blind, the one-eyed man is king. This may be as the proverb has it, but in the kingdom of the one-eyed person, the two-eyed person is looked on with profound suspicion and is commonly feared. Leaders are often those two-eyed, or "twice-born types," to borrow William James' phrase, who have different ideas about what should be. And thus they challenge the status quo, they war on complacency, they try to become "movers and shakers"—all of which done too fast or overdone can create major disequilibrium or instability.

Paradox #6. We expect our presidents to provide bold, visionary, innovative, *programmatic* leadership and at the same time to *pragmatically* respond to the will of public opinion majorities; that is to say, we expect presidents to lead and to follow, to exercise "democratic leadership."

We want both pragmatic and programmatic leadership. We want principled leadership and flexible, adaptable leaders. *Lead us,* but also *listen to us.*

Most people can be led only where they want to go. "Authentic leadership," wrote James MacGregor Burns, "is a collective process." It emerges from a sensitivity or appreciation of the motives and goals of both followers and leaders. The test of leadership, according to Burns, "is the realization of intended, real change that meets people's enduring needs." Thus a key function of leadership is "to engage followers, not merely to activate them, to commingle needs and aspirations and goals in a common enterprise, and in the process to make better citizens of both leaders and followers."[20]

We want our presidents to offer leadership, to be architects of the future and to offer visions, plans, and goals. At the same time we want them to stay in close touch with the sentiments of the people. We want a certain amount of innovation, but we resist being led too far in any one direction.

We expect vigorous, innovative leadership when crises occur. Once a crisis is past, however, we frequently treat presidents as if we didn't need or want them around. We do expect presidents to provide us with bold, creative, and forceful initiatives "to move us ahead," but we resist radical new ideas and changes and usually embrace "new" initiatives only after they have achieved some consensus.

Most of our presidents have been conservatives or at best "pragmatic liberals." They have seldom ventured much beyond the crowd. They have followed public opinion rather than shaped it. John F. Kennedy, the author of the much-acclaimed *Profiles in Courage,* was often criticized for presenting more profile than courage. He avoided political risks where possible. Kennedy was fond of pointing out that he had barely won election in 1960 and that great innovations should not be forced on the public by a leader with such a slender mandate. President Kennedy is often credited with encouraging widespread public participation in politics, but he repeatedly reminded Americans that caution is needed, that the important issues are complicated, technical, and best left to the administrative and political experts. Seldom did Kennedy attempt to change the political context in which he operated. Instead he resisted, "the new form of politics emerging with the civil rights movement: mass action, argument on social fundamentals, appeals to considerations of justice and morality. Moving the

American political system in such a direction would necessarily have been long range, requiring arduous educational work and promising substantial political risk."[21]

Kennedy, the pragmatist, shied away from such an unpragmatic undertaking.

Leadership requires radiating hope, grounded in reality. Be dedicated to "visions" and be passionately committed to carrying them out. Be flexible enough to change direction quickly. This is just one of the many paradoxical demands we place on all our modern leaders. The art of the possible is wanted, the challenges of bringing about the doable and the desirable, yet creative dreaming also is required. A constant balance, reconciling dreams and reality, intuition and logic, is needed.

We occasionally look to idealists and prophets because we know that major breakthroughs and reforms often come only from the principled few, not the conventional political or organizational leaders. This is true, in part, because majority opinion often stifles new approaches. As every politician knows: if you oppose your constituents too directly on an issue too close to their hearts, you are not going to get elected, or re-elected.

Presidents can get caught coming and going. The public wants them to be both leaders of the country and representatives of the people. We want them to be decisive and exercise their own judgment; we want them to be responsive to public opinion, especially to the "common sense" of *our* opinions. It was perhaps with this in mind that many have defined the ideal democratic leader as an "uncommon man of common opinions."

Americans may admire consistency in the abstract, but in politics consistency has its costs. The late Everett Dirksen, a popular Republican senator from Illinois, used to say, "I'm a man of fixed and unbending principle—but my first fixed and unbending principle is to be flexible at all times."

Franklin D. Roosevelt proclaimed that the presidency is preeminently a place for moral leadership. But he was also an opportunist and pragmatist just as Lincoln and Theodore Roosevelt were before and as Kennedy, Reagan, and Clinton would be after him. Former California governor and erstwhile presidential candidate Jerry Brown once told an interviewer that "a little vagueness goes a long way in this business."

Those men knew that political leadership responsibilities in America meant sometimes being detached, vague, and uncommitted, while at other times taking a stand and being passionately committed. Timing often determined their success or failure. But perhaps more important, the lesson for those who want political power in America is that they must narrow their range of ideas and ideals to conform to the prevailing political and economic climate of opinion.

Our paradoxical desire for programmatic but pragmatic leadership is compounded by the fact that what we find desirable in one president we condemn in another. This factor often renders the lessons of history immaterial, and makes such a thing as standard pragmatic or programmatic leadership difficult, if not impossible.

A president who becomes too committed risks being called rigid; a president who becomes too pragmatic risks being called wishy-washy. Some, like Clinton, get accused of both. Some, like Bush, counter this dilemma by stressing character, competence, rectitude, and experience and by avoiding strong stands that offend important segments of the population. However, the more a president tries to please everyone and respond only to immediate pressures, the more that president is likely to be drawn away from the more significant remote problems of the day. The more pragmatic presidents become, the more they are shaped by events rather than being shapers of them.

Jimmy Carter was criticized by the press and others for avoiding commitments and stressing his "flexibility" on the issues. This prompted a major discussion of what came to be called the "fuzziness issue." In fact, he took stands on most issues, but being a centrist or a pragmatic moderate, his stands were often either unpopular or dismissed as nonstands by most liberals and conservatives, especially purists. Bill Clinton suffers a similar fate. What strikes one person as fuzziness or even duplicity appeals to another person as remarkable political skill.

Most candidates view a campaign as a fight to win office, not an occasion for adult education. Barry Goldwater in 1964 ran with the slogan "We offer a choice not an echo," referring to his unusually thematic strategy, but Republican party regulars who yearned to win the election preferred "a chance, not a choice." Once in office, presidents often operate the same way—the electoral connection looms large as an issue-avoiding, controversy-ducking political incentive. Most presidents strive to maximize their options, and hence leave matters up in the air or delay choices. JFK mastered this strategy. LBJ, on the other hand, permitted himself to be trapped in a corner that seemed to allow no escape on Vietnam. His options had swiftly dissolved.

We want idealism *and* realism, optimism *and* levelheadedness. Be inspirational, we tell the president, but also be realistic—don't promise more than you can deliver. We ask our presidents to stir our blood, giving us a sense of glory about ourselves, but also to appeal to our reason. Too much inspiration will invariably lead to dashed hopes, disillusionment, and cynicism. The best leaders often suffer from one of their chief virtues—an instinctive tendency to raise aspirations, to summon us to transcend personal interests and subordinate ourselves to dreaming dreams of a bolder, more majestic national community.

Most Americans want to be inspired. We savor the upbeat rhetoric and promises of a brighter tomorrow. We genuinely want to hear about New Nationalism, New Deals, New Frontiers, Great Societies, New American Revolutions, and New Covenants; we want our fears to be assuaged during a "fireside chat" or a "conversation with the President"; we want to be told "the torch has been passed to a new generation of Americans . . .

and the glow from that fire can truly light the world." We want something to believe. We want fearless leaders to tell us that the "only fear we have to fear is fear itself," that "we are Number One," that a recession has "bottomed out," or that "America is back standing tall again." To understate the state of the nation is to seem unpresidential.

These promises, while they make electoral and psychological sense, often backfire. As candidates promise too much and build expectations too high, the public is disappointed when reality fails to match promise. This may compel presidents to break their promises, as Bush did when he broke his "Read my lips, no new taxes!" pledge.

Presidents who do not raise hopes are criticized for letting events shape their presidency rather than making things happen. A president who eschewed inspiration of any kind would be rejected as un-American. For people everywhere America has been the land of promise, of possibilities, of dreams. No president can stand in the way of this truth, regardless of the current dissatisfaction about the size of big government in Washington and its incapacity to deliver the services it promises.

Do presidents overpromise because presidential aspirants are congenital optimists, or because they are pushed by the demanding public? Or are competing candidates engaged in an escalating spiral of promise heaped upon promise as they try to outbid each other for votes? Surely the public demands it, but only self-confident optimists need apply in the first place. Whatever the source, few presidents have been able to keep all their major promises and fulfill their intentions. Poverty was not ended; a Great Society was not realized. Vietnam dragged on and on. Watergate outraged a public that had been promised an open presidency. Energy independence remains an illusion just as crime in the street continues.

Charlie Brown of Peanuts cartoon fame once said, "I have very strong opinions, but they don't last long." So also American public opinion shifts—sometimes quickly, sometimes slowly. There are times when we want presidents to be engaged actively as innovators, and on other occasions we would like to see them sit back and let things run their course.

Historically, conservatives have been more opposed to a strong presidency than liberals. This has changed, especially when conservatives occupy the White House. The conservatives may fear a powerful planning presidency at home, but they want a powerful presidency for national security leadership. Liberals, of course, have generally applauded a strong presidency when it has been in the hands of Roosevelt, Truman, or Kennedy. Overall, in the debate over the proper range of presidential power our attitudes are shaped considerably by the party and by the kind of leader in the White House at the time. Conservatives and liberals have generally favored or opposed a powerful presidency depending on whether or not it suited their immediate purposes.

American historian Arthur M. Schlesinger, Sr., when writing about public mood shifts in America, once claimed that "a period of concern for the rights of the few has been followed by one of concern for the wrongs of the many. . . . An era of quietude has been succeeded by one of rapid movement."[22] A period of affirmative government is followed by backlash and a period of lull.

This examination of shifting public mood is of interest because a president is constrained by the dominant national mood or climate of expectations. Sometimes the coming of a new president can help to recast or shift the national mood. More often, however, the national mood responds to major events, or challenges. Admittedly, full accounting for these shifts remains somewhat elusive. Still, Carter's difficulties arose in no small measure because he sought to lead a nation that had turned inward, introspective, and perhaps somewhat narcissistic. Likewise, Bill Clinton is an activist in an age of anxiety; a progressive in an age of retrenchment.

"We call them stiff-necked and inflexible when they won't revise a position, fuzzy and opportunistic when they do. . . . We urge them, once nominated, to move to the center, the mainstream . . . and then bash them for fakery and cynicism when they so move."[23] Should it be considered so intolerable for presidents to admit they have changed, grown, seen something differently, yielded to the arguments or evidence or even pleas of others? George Bush suffered immeasurably when he did an about face on his vivid pledge not to raise taxes. And Bill Clinton was not forgiven for abandoning his middle-class tax cut pledge.[24]

Paradox #7. Americans want powerful, self-confident presidential leadership. Yet we are inherently suspicious of leaders who are arrogant, infallible, and above criticism.

We unquestionably cherish our three branches of government with their checks and balances and theories of dispersed and separated powers. We want our presidents to be successful and to share their power with their cabinets, Congress, and other "responsible" national leaders. In theory, we oppose the concentration of power, we dislike secrecy, and we resent depending on any one person to provide all of our leadership.

But Americans also yearn for dynamic, aggressive presidents—even if they do cut some corners. We celebrate the gutsy presidents who make a practice of manipulating and pushing Congress. We perceive the great presidents to be those who stretched their legal authority and dominated the other branches of government. It is still Jefferson, Jackson, Lincoln, and the Roosevelts who get top billing. Whatever may have been the framers' intentions for the three branches, most experts now agree that most of the time, especially in crises, our system works best when the presidency is strong and when we have a self-confident, assertive president.

There is, of course, a fine line between confidence and arrogance, between firmness and inflexibility. We want presidents who are not afraid to exert their will, but at what point does this become antidemocratic, even authoritarian?

We want presidents to consult widely and use the advice of cabinet members and other top advisers. We like the idea of collegial leadership and shared responsibility. But do we want presidents to sacrifice their own ideas and priorities to those of their cabinet officers? No. We elect the president, not the advisers. While we want presidents to be open minded, we also admire the occasional "profile in courage" type of decision. One of the most fondly remembered Lincoln stories underscores this point. President Lincoln supposedly took a vote at a cabinet meeting and it went entirely against him. He announced it this way: "Seven nays and one aye, the ayes have it." But most of the time, Abraham Lincoln followed the leadership of Congress, his advisers, and the general public.[25]

> A president who is believed not to make up his own mind rapidly loses the power to maintain the hold. The need to dramatize his position by insistence upon his undoubted supremacy is inherent in the office as history has shaped it. A masterful man in the White House will, under all circumstances, be more to the liking of the multitude than one who is thought to be swayed by his colleagues.[26]

We want simultaneously a secular leader and a civil religious mentor; we praise our three-branched system, but we place capacious hopes upon and thus elevate the presidential branch. Many people believe only the president can give us sustained leadership, can dramatize and symbolize our highest expectations of ourselves as an almost "chosen" people with a unique mission.

Another aspect of this dilemma is that we want to be reassured, but we rally to and admire crisis leadership. Although a president is expected to exude hope, reassurance, and an "I'm OK we're OK" sense of confidence, the public nevertheless likes to see presidents visibly wrestling with crises. We admire the solitary executive embattled in conflict sitting alone, making the tough decision, exerting will. Presidents learn that more political support is available when the United States is at odds with an enemy. It is almost easier to go to the brink of war than to defend advances in detente or new arms agreements. President Kennedy found "the politics of confrontation" with Cuba as well as with Khrushchev were politically helpful. President Reagan benefited from a military mission with Grenada, and President Bush benefited, at least for a time, from the 1991 Gulf War to liberate Kuwait.

Presidents are simultaneously asked to build a lasting peace and to maintain United States superiority as the number one superpower. Promote peace, yes. But don't yield anything to the enemy! Moreover, presidents who wish to appear presidential must accentuate the nation's sense of being in a severe predicament. They perceive that a sense of heightened crisis must be created. FDR did it, as did Kennedy, especially in the summer of 1961 when he launched his "everybody build a fallout shelter" program. Ronald Reagan's claims of a "window of vulnerability" and the "evil empire" of the Soviet Union may have been partly the product of a desire to alarm the public into action.

The danger of falling under the "leadership" of one who lacks will or confidence to lead persists. Humility is admirable, while excessive humility paralyzes. Significant advances in the world have often been made by confident innovators.

Leaders must face all the complexities of the situation but they must also act. Leaders are conscious of the flaws in their own organization, yet minimize the risks. They know their followers often want and need relentless optimism about impending battles. Shakespeare's *Henry V* provided this almost irrational, gamble-all, whatever-the-odds call to arms. Winston Churchill in the midst of World War II also personified this dimension.

"Any self-doubts the leader may have, especially in the battlefield, must be concealed at all costs," wrote military historian John Keegan. "The leader of men in warfare can show himself to his followers only through a mask, . . . made in such form as will mark him to men of his time and place as the leader they want and need."[27]

Rare is the great commander, the truly successful executive, or the politician who is not self-centered and conceited. Similarly, the politician who does not exude the desire for leadership is not likely to be effective or remembered. A leader must be conceited enough to believe that he is indispensable.

Untempered confidence, however, is dangerous. Hitler oozed it. So did Melville's mad Captain Ahab. Both had vision, purpose, and enormous drive.

The question always is whether these large egos are subject to reasonable self-control. Self-discipline is key. An unrestrained, ego that constantly needs to be fed and isn't placed in disciplined service to worthy ends is an ego that invariably corrupts.

Leaders, of course, must believe in themselves, but they cannot afford to discredit the ideas, plans, counsel, or criticism of others. Leaders who encourage thoughtful dissent in their organizations are, according to several studies, likely to produce better organizational decision-making. Effective presidents encourage and reward criticism without retaliating against the critics. Hitler eliminated his critics. Ahab ignored his. In *Antigone*, Sophocles' King Creon listened almost entirely to himself, which proved fatal. His son, Haemon, chided him in vain, saying, "Let not your first thought be your only thought. Think if there cannot be some other way. Surely, to think you own the only wisdom and yours the only word, the only will, betrays a shallow spirit, an empty heart."

But Creon dismisses his son's advice, saying, "Indeed, am I to take lessons at my time of life from a fellow of his age?" He ignores everyone else as well until it is too late.

We are all a little like Creon. A fine line separates self-confidence from pigheaded pride, boldness from recklessness, mindless adherence to the course from reevaluation and redirection. The challenge is how to blend the competing impulses and combine them effectively in particular situations.

Paradox #8. What it takes to become president may not be what is needed to govern the nation.

To win a presidential election takes ambition, money, luck, and masterful public relations strategies. It requires the formation of an electoral coalition. To govern a democracy requires much more. It requires the formation of a *governing* coalition, and the ability to compromise and bargain.

"People who win primaries may become good presidents—but 'it ain't necessarily so'" wrote columnist David Broder. "Organizing well is important in governing just as it is in winning primaries. But the Nixon years should teach us that good advance men do not necessarily make trustworthy White House aides. Establishing a government is a little more complicated than having the motorcade run on time."[28]

Ambition (in heavy doses) and stiff-necked determination are essential for a presidential candidate, yet too much of either can be dangerous. A candidate must be bold and energetic, but in excess these characteristics can produce a cold, frenetic candidate. To win the presidency obviously requires a single-mindedness, yet our presidents must also have a sense of proportion, be well-rounded, have a sense of humor, be able to take a joke, and have hobbies and interests outside the realm of politics.

To win the presidency many of our candidates (Lincoln, Kennedy, and Clinton come to mind) had to pose as being more progressive or even populist than they actually felt; to be effective in the job they are compelled to appear more cautious and conservative than they often want to be. One of Carter's political strategists said, "Jimmy campaigned liberal, but governed conservative." And as Bill Clinton pointed out toward the end of his first year in office, "We've all become Eisenhower Republicans."[29]

Another aspect of campaigning for the White House is the ambiguous position candidates take on issues in order to increase their appeal to the large bulk of centrist and independent voters. The following is a typical view replete with paradox: "I want my presidential candidate to have clearcut policies, to be as clear and precise as possible on his positions, not hazy and ambiguous—to run a campaign that educates people and persuades them to adopt the candidate's positions. But I also want my candidate to win—I realize they have to be ambiguous to build the winning coalitions."[30]

Policy positions are seldom comprehensively outlined; bumper-sticker slogans and entertaining TV ads are designed to please everyone and offend no one. Such presidential pledges as LBJ's "We will not send American boys to fight the war that Asian boys should be fighting," Richard Nixon's "open presidency" (1968) and "peace is at hand" (1972), Carter *and* Reagan's "I will balance the budget," and Bush's "no new taxes" are illustrative.

One of the most difficult aspects of campaigning is to win without proving you are unworthy of the job you are seeking. Thus it is a common temptation for all candidates, including some incumbents, to run a "bureaucrats-are-bums," anti-Washington outsider, kind of campaign. There is something more than a little deceitful, and certainly a lot that is absurd, in a presidential candidate who is trying to get to Washington by saying that he is running "against Washington," and one hoping to be elected to the most powerful office in the world by proclaiming that he is against big government.[31] One irony of the 1980s is that Reagan campaigned against big government but left bloated deficits, a proliferation of defense programs, and a bigger government than anyone ever would have imagined in 1980.

We expect a president to be able to work closely and effectively with Congress and civil servants. Candidates who bad-mouth Washington officials will breed resentment; if they get to the White House, they will have a difficult time winning sustained cooperation from these same officials. Candidates who get too pious about how ethical standards have to be raised in Washington will be called hypocrites if their own top aides are caught in compromised conflict-of-interest positions. Or if, as is the case with Clinton, they become embroiled in scandals.

We often also want both a "fresh face," an outsider, as a presidential candidate *and* a seasoned, mature, experienced veteran who knows the corridors of power and the back alleyways of Washington. That's why Colin Powell fascinated so many people. Frustration with past presidential performances leads us to turn to a "fresh new face" uncorrupted by Washington's politics and its "buddy system" (Carter, Reagan, Clinton). But inexperience, especially in foreign affairs, has sometimes led to blunders by the outsiders.

New nominating rules combine with the requirements of the media and this "high tech" campaign age to sharpen the clash between what is required of a successful candidate and the successful president.

To be a winning candidate, the would-be president must put together an electoral coalition involving a majority of voters advantageously distributed across the country. The candidate must thus appeal to all regions and interest groups and cultivate the appearance of honesty, sincerity, and experience. Note that this is not all bad. It's good to travel around the country, meeting people, learning about their problems and testing ideas on diverse audiences. As Jimmy Carter put it, "I was a hell of a better American because I had to go around the country and learn about the different regions and people. . . . You have to be able to stand up under pressure and have the physical endurance and ability to express yourself . . . and let the public learn about you as you learn about them."[32]

Once elected, however, the electoral coalition has served much of its purpose and a governing coalition is the order of the day.

Candidates depend on television to transform candidacy into incumbency. Studies find that candidates spend well over half their funds on radio and television broadcasting. Moreover, this is how most people "learn" about the candidates. Approximately two-thirds of the American public report television is the best way for them to follow candidates, and about half of the public acknowledges that their best understanding of the candidates and issues is derived from television coverage.

While very important, television has mixed and confusing effects for the presidency. Faults are exaggerated and "looking good" often becomes as important as doing well.

Thus, what it takes to become president may differ from what it takes to *be* president. To become president takes a determined, and even a driven, person, a master fund-raiser, a person who is glib, dynamic, charming on television, and hazy on the issues. But once president, the person must be well rounded, careful in reasoning, clear and specific in communications, and not excessively ambitious. It may well be that our existing primary-convention system adds up to an effective obstacle course for testing would-be presidents. But with the experiences of the past generation in mind we have some reason for asking whether our system of electing presidents is not at odds with what is required to end up with a president who is competent, fair-minded, and emotionally healthy.

Paradox #9. The presidency is sometimes too strong, yet other times too weak.

Presidents are granted wide latitude in dealing with events abroad. At times, presidents can act unilaterally, without the express consent of Congress. While the constitutional grounds for such action may be dubious, the climate of expectations allows presidents to act decisively abroad. This being the case, the public comes to think the president can do the same at home. But this is usually not the case. A clashing expectation is built into the presidency when strength in some areas is matched with weakness in other areas.[33]

It often seems that our presidency is *always too strong* and *always too weak*. Always too powerful given our worst fears of tyranny and our ideals of a "government by the people." Always too strong, as well, because it now possesses the capacity to wage nuclear war (a capacity that doesn't permit much in the way of checks and balances and deliberative, participatory government). But always too weak when we remember nuclear proliferation, the rising national debt, the budget deficit, lingering discrimination, poverty, and the clutch of other fundamental problems yet to be solved.

The presidency is always too strong when we dislike the incumbent. Its limitations are bemoaned, however, when we believe the incumbent is striving valiantly to serve the public interest as we define it. The Johnson presidency vividly captured this paradox: many who believed he was too strong in Vietnam also believed he was too weak to wage his War on Poverty. Others believed just the opposite.

Like everyone else, presidents have their good days and their bad days, their creative leadership periods and their periods of isolation, their times of imperiousness and of ineptitude. On their good days we want the presidency to be stronger. On their bad days we want all the checks and balances that can be mustered. The dilemma of the presidency today is that we can't have it both ways. Since President Washington took office we have vastly multiplied the requirements for presidential leadership and made them increasingly difficult to fulfill. Students of the presidency usually conclude that more power, not less, will be needed if presidents are to get the job done. There are now just too many constraints on governmental action—when presidents are at their best.

But if the presidency is to be given more power, should it not also be subject to more controls? Perhaps so. But what controls will curb the power of a president who abuses the public trust and at the same time not undermine the capacity of a fair-minded president to serve the public interest? The riddle here arises because we sometimes disagree over which days are good and which are bad. Thus Ford's extensive use of the veto power was applauded by some and condemned by others, as was his pardon of Nixon. Likewise, Clinton's efforts to end the ban on gays in the military was seen by some as a strengthening of human rights and equality, by others as condoning or encouraging what they saw as immoral behavior.

CONCLUSION

Leadership means many things to many people. For some it has a rich, positive meaning. For others it connotes manipulation, deception, or even oppression.

Neither presidents nor the general public should be relieved of their respective responsibilities for trying to fashion a more effective and fair-minded leadership system simply because the presidency is an office full of clashing expectations. We cannot throw up our hands and say, "Well, no one makes a person run for that crazy job in the first place."

All leaders face countervailing pressures that pull them in different directions. Sisyphus of Greek mythology was a man condemned forever to roll a rock up a hill only to have it roll down before reaching the top. As Nelson Mandela said, "I have discovered the secret that after climbing a great hill, one only finds that there are many more hills to climb."[34]

We shall have to select presidents who understand these dilemmas and have a gift for the improvisation necessitated by their contrary demands. Too often we have selected persons who have simply learned to manipulate these contending expectations for short-term electoral purposes.

We need to probe the origins and to assess the consequences of the clashing expectations and to learn how presidents and we as the public can better coexist with them. It is apparent that they often serve to confuse us in our evaluation of presidents, and they create tremendous ambiguity for presidents.

When Americans realize that the presidency is incapable of dealing with everything well and that democratic politics, in general, is not suited to provide quick answers to every social and economic malaise, then we will be more responsible in the way we judge presidents.

CHAPTER 2

Electing the American President

He shall hold his Office during the Term of four Years.
. . . No person except a natural born Citizen, or a
Citizen of the United States at the time of the
Adoption of this Constitution, shall be eligible to the
Office of President; neither shall any person be
eligible to that Office who shall not have attained to
the Age of thirty-five years, and been fourteen Years a
Resident within the United States.

U.S. Constitution, Article II

If Jefferson is elected, the Bible will be burned, the
French "Marseillaise" will be sung in Christian
churches, and we may see our wives and daughters
the victims of legal prostitution; soberly dishonored;
speciously polluted.

Reverend Timothy Dwight, President, Yale University,
quoted in D. C. Coyle, Order of the Presidency
(1960), p. 43

The way things now stand, in [the 1980] election a
decision of major national importance will again be
entrusted to the outdated, flawed mechanism of the
electoral college. Most importantly, the Nation will
again run the serious risk that due to the workings of
this system the candidate obtaining the most popular
votes might not be selected as President.

Hon. Jonathan B. Bingham (D-NY), Congressional Record,
February 26, 1979, p. H.882

Can anyone remember when a kind word was said about the way we elect our presidents? Rarely have Americans felt they had a choice between first-rate candidates. Although Americans may love their country, celebrate democracy, and prize the institution of the presidency, they are far less confident about the procedures and machinations of the presidential selection process.

Most Americans see the selection process as an increasingly self-defeating system that offers a menu of unappetizing choices while taking the best candidates out of the race. Political journalist Elizabeth Drew said, "The process by which we choose our presidential candidates more resembles a demolition derby than a rational procedure. It's an elimination contest, offering us the last man—or men—standing at the end of a long, gruelling, and expensive series of matches, in some strange arenas. Meanwhile, what this is supposed to be about is who is best fit to lead our country. It should be no surprise that people seem to be increasingly unhappy about how we choose our president."[1]

The selection process itself is so full of paradoxes, ironies, and unintended consequences that it is widely held that "elections are rarely our finest hour." Indeed, one of the glaring paradoxes of the presidency, already mentioned in chapter 1, is that the qualities it takes to get elected are often different from, and even contradictory to, those qualities a president needs in order to govern.

This chapter explores and explains how we elect our presidents. The following questions will be discussed: Who becomes president? What do we look for in prospective presidents? What must a person do to be taken seriously as a presidential candidate? How important are the primaries? Are the national conventions still needed? What about the so-called advantages of incumbency? What has been the effect of the Twenty-second Amendment? Why do people vote or not vote for a president? What about the controversial electoral college and its role in presidential elections? And finally, does the selection process do what we need in democratic and quality-of-outcome terms? After all, a presidential election is about more than merely determining who will assume office. It is also, or should be, the great conversation a democracy has concerning its future.

WHO BECOMES PRESIDENT?

Our presidents are usually middle-aged white male Protestant lawyers of European lineage from the larger states. Thirty-six of our forty-two presidents trace ancestry to the British Isles, three were Dutch, and two were German. All but one, John Kennedy, were Protestants. About half have been lawyers and served in Congress. Most others have held state or community elective office.

Gone are the days when party leaders could select a presidential candidate. Today, with the proliferation of primary elections, candidates who are well financed and can devote themselves to full-time campaigning often have an advantage over public figures who occupy office. Candidates such as Jimmy Carter and Ronald Reagan had the advantage over most of their rivals. Wealthy individuals such as H. Ross Perot and Steve Forbes can "buy" their way into a race.

In a sense, there is an "on-deck" circle of at most forty individuals in any given presidential election year: governors, prominent U.S. senators, a few members of the House of Representatives, and a handful of recent governors or vice presidents who have successfully kept themselves in the news media. These men, and an occasional woman, form the core of electables whom "The Great Mentioner," that mythic conglomeration of prominent media figures, is willing to anoint as serious candidates.

Americans often gripe about the small number of able people who vie for the presidency. Is this the best we can do? With our large and highly educated population, why can't we produce more candidates like Washington, Jefferson, and Lincoln?

No doubt there are talented business entrepreneurs, educators, or other outstanding individuals who might make as capable a president as any of the activist politicians who inevitably become the serious candidates. Seldom, however, are these nonprofessional politicians willing to enter the political thicket at age 56 or 61. Billionaire H. Ross Perot, who won 19 percent of the popular vote in 1992 and 8 percent in 1996, is the exception to the rule. Even he could barely stomach the process. Some, perhaps many, do not want to spend two or three years on the campaign trail or raise the thirty million dollars needed to run. Others may be reluctant to disclose their finances and subject their families and themselves to the brutal public scrutiny involved in a presidential race. General Colin Powell readily acknowledged that his wife's objection to his running for president was a major factor in his decision not to seek the presidency in 1996. He also stated vaguely that he hadn't quite felt the call to political service. The demanding presidential campaign, especially with its more than thirty primary elections, is unappealing even to seasoned politicians.

The general public harbors a disdain for politicians. When asked about their career aspirations for their children, parents rate "politician" or even "president" low on their list. A 1945 Gallup poll found that only two people in ten said they wanted their son to go into politics. In 1955, three out of ten felt that politics was a good profession for their sons. A similar finding occurred in 1965. In 1991 Gallup questions included aspirations for daughters, and the results were the same as for sons. A recent Gallup poll found that only 8 percent of teenagers said public office was an attractive career goal.

If career questions focus on the presidency, the results are similarly disappointing. A 1996 Princeton Survey Research Associates poll revealed that 63 percent said they would not want a child to seek the presidency. And a 1996 *Time*/CNN poll found 61 percent saying they did not want their child to run for president. A 1992 Yankelovich Clancy Shulman poll asked, "Which one of these would you most like to see one of your children grow up to be?": 38 percent said university president; 28 percent chose head of a large corporation; 11 percent said a sports star; only 7 percent said president of the United States.[2]

As noted, our candidates are drawn from a narrow range of backgrounds. Eight of the eighteen twentieth-century presidents were attorneys, eight were members of Congress, eight served as governors, and seven had been vice president. Only George Bush was elected president while running as an incumbent vice president; the other six became president on the death or resignation of the incumbent president, or were elected after leaving the vice presidency.

Usually by about January of election year one or two front-runners have emerged from the pack to go on to the semifinal—the nominating convention. Both parties and the public are usually presented with a virtual *fait accompli.* In practice, name familiarity, access to large financial resources, and the substantial spare time necessary to prepare for the primaries give certain candidates advantages over others.

A presidential candidate must be at least 35 years old, must have lived within the United States for fourteen years, and must be a natural born citizen. Whether a person born abroad of American parents is qualified to serve as president has never been decided. Many scholars believe that such a person would be considered "natural born" even if not native born.

The three overriding factors that influence how people vote in presidential elections are their *party orientation,* their *public policy preferences,* and the way they perceive the *integrity, character, and judgment* of the candidates.

Parties may be decaying as organized, vital institutions in America, but they are still important factors when the votes are counted. Americans are not as partisan as they were in the early twentieth century; they now are better educated, have multiple sources of political information, and are no longer dependent on party leaders for welfare, patronage, or certain other intermediary services. Still, roughly two-thirds of Americans identify themselves as Democrats or Republicans. Most of the time, Republicans vote Republican, Democrats vote Democrat. Republicans, however, are a more cohesive party in the general election. More than 80 percent of Republicans voted for Barry Goldwater in 1964, 95 percent voted for Nixon in 1972, 89 percent for Ford in 1976, 86 percent for Reagan in 1980, 92 percent for Reagan in 1984, 91 percent for Bush in 1988. Only 73 percent of Republicans voted for Bush in 1992 (Ross Perot drew votes away from Bush that year). Democrats, for most of the post–World War II era, are more numerous, less unified, and less predictable than Republicans. Carter got more than 80 percent of the Democratic vote in 1976, but George McGovern got only 67 percent in 1972. A third of the people calling themselves Democrats voted for Richard Nixon that year. In 1980, Carter's percentage of the Democrat vote dropped to 67. In 1984, Walter Mondale received 73 percent, and in 1988 Dukakis got 82 percent. In 1992, Bill Clinton won only 77 percent of the Democrat vote (again, Ross Perot drew some votes away from Clinton).

The core of party loyalists upon whom candidates have traditionally relied has shrunk. The public is less attached to parties and more dependent on television for voting cues, which forces presidents to "go public" in an effort to woo the independent-minded voters. Thus, candidates often feel the need to reinvent themselves at each campaign to better conform to what they perceive voters' preferences to be, and to tailor their message to the current fashion. While it is important for leaders to be flexible, they must also have character. There is a difference between a candidate willing to bend, and one who will break.

Issues or public policy preferences play an even more complicated role in how people cast their presidential vote. In certain years, like 1800, 1860, 1896, 1936, 1964, and 1992, issues played a larger, more clear-cut role. In other years, 1956, 1960, and 1984, issues weren't as important. Sometimes both major candidates take the same position on a major issue, as in 1968 when both Nixon and Hubert Humphrey said they favored ending the war, honorably, in Vietnam.

Issues seem to be less important in many elections because of the tendency for both parties to offer candidates who take moderate policy stands to appease widespread public opinion. Thus, both candidates often echo the public mood rather closely. Moreover, candidates' policy preferences are often deliberately highly ambiguous. Clear-cut policy stands are infrequent and inconspicuous as candidates devote much more attention to their concerns about general goals, problems, and past performance.

While candidates often try to avoid specific controversial issues, such issues do serve a broader and in many ways more important function: They allow candidates to simplify complexity for voters. Effective candidates provide meaning, purpose, and hope in a confusing world. They help make sense of the times. A candidate, or president, who can give comfort, assurance, and hope amid confusion can generate loyal and wide support.

With the collapse of the Soviet Union and the end of the Cold War, the focus of ideological conflict has changed in American electoral politics. From World War II to the late 1980s, Cold War issues such as "Can candidate X be trusted with his finger on the nuclear trigger?" weighed heavily in voters' decisions. This helped Lyndon Johnson defeat Barry Goldwater in 1964—Goldwater was viewed as too bellicose. In 1972, the same question helped Richard Nixon defeat George McGovern—as McGovern was generally viewed as too soft.

Honesty and integrity top the list of desired characteristics. We are keenly aware, more so because of Watergate, the Iran-Contra scandal, and Whitewater, of the need for a president to set a tone, and to serve as an example, of credibility. Dishonesty and duplicity are qualities we dislike.

One of President Clinton's biggest problems is that even after a fairly impressive list of accomplishments, the public still does not trust him. In a *Los Angeles Times* poll, when voters were asked for the "Top reasons they would vote against Clinton," 18 percent answered that he was "Not fulfilling campaign promises," and 14 percent answered, "Don't trust him," the two highest responses in the survey.[3] There are times, paradoxically, especially when the president deals with foreign adversaries, that we want the president to use guile, cunning, even dishonesty, if it serves the nation's interest.

Intelligence; a capacity to clarify, communicate, and mobilize; as well as flexibility, compassion, and open-mindedness are also leading characteristics sought in presidential candidates.

Most voters look for a moderate candidate. They generally vote against extremists of any kind and for the middle-of-the-road, don't-rock-the-boat candidates. This has led to the aphorism that "the only extreme in American politics that wins is the extreme middle." Ronald Reagan is somewhat of an exception to this rule, but the public seldom saw Reagan as an extremist; his rhetoric was often comforting and nurturing to the point that he appeared to be mainstream.

In the post–Cold War era, questions of nuclear arms and foreign policy have been less meaningful to voters. In 1992, Americans elected a president who had no foreign policy experience and had avoided the draft during the Vietnam War. In recent years, the terms of debate have changed as voters turned inward, and the ideological cleavages have been less over foreign affairs than over domestic issues such as the economy, crime, education, the environment.[4] The irony here is that regardless of the lack of electoral importance for presidential campaigning, foreign affairs soon come to dominate a president's agenda. Bill Clinton, who ran on the promise that he would place attention on the economy, education, the environment, health care, and other issues close to home, soon found himself distracted by Haiti, Somalia, Russia, Bosnia, and other international hot spots.

Personality and character also count. In local elections people often just rely on party labels. But with governors, U.S. senators, and especially presidents, and with the availability of extensive television coverage, people plainly intrude their own personal judgment about who is most fit to serve in office.

Many people deplore the fact that a candidate's personality and style are evaluated as equal to or more important than issues and substance. But a candidate's personality is a perfectly legitimate and, indeed, a proper subject for voters to weigh. There is little doubt that a candidate's sense of self-confidence and his personal style of conduct can and usually do affect how he would behave in office. "Presidents do far more than respond to the issue preferences of the voters. They have enormous discretionary power, and their personalities can importantly affect the way they handle issues and decide public policy."[5]

The basic insecurities of certain presidents have also led to failure in the White House. We have reason to be alert to whether a candidate can accept criticism and whether he demands absolute loyalty from subordinates. "The great danger is that a President who feels threatened by events or harassed by 'enemies' will precipitate a crisis in order to shore up his own inner doubts and confound his opposition."[6]

What then do we look for in our presidents? We want persons who are self-confident and self-aware, persons who are not rigid, defensive, compulsive, self-striving, and torn by self-doubt and self-pity or who blame problems on "enemies" and punish those on their "enemies list."

Political scientist James David Barber, a pioneer in studying presidential personality, urges us to look first at a candidate's character. The issues of an election year will change, but the character of the president will last. Politics, he says, is politicians. There is no way to understand the former without understanding the latter. We need to look closely at the rhetoric, skill, and world view of presidential candidates. Barber believes Wilson, Hoover, Johnson, and Nixon displayed tendencies toward compulsiveness and rigidity and were not as desirable as FDR, Truman, and JFK. These latter candidates combined a high volume and fast tempo of activity with marked enjoyment of politics and people. These "activist-positives," as Barber calls them, manifested strong self-esteem and distinct success in relating to their environment.[7] Barber's analysis has had an influence on opinion shapers—members of the press who cover national elections—and even on some presidential candidates. Although his book has raised almost as many questions as it has answered, Barber has correctly identified the importance of evaluating character and personality in presidential elections.

In recent years, questions of personal character have loomed large in presidential politics.[8] During the 1988 primaries, Democratic candidate Gary Hart was forced to withdraw from the race when he was discovered on a pleasure yacht, the "Monkey Business," with a woman other than his wife. During the 1992 primaries, candidate Bill Clinton had to defend himself against charges by a woman that she had had a long-term affair with the then governor. As president, Clinton had to fend off legal charges of improper behavior allegedly committed while he was governor of Arkansas. Polls indicate that the public has a much higher regard for Clinton's policies than for Clinton as a person.[9] Columnist David Broder scolded his colleagues for their tendency to "swoop down" on candidates in search of the "sin of the week."[10] And a *Times-Mirror* poll indicated that personal character issues ranked low in the minds of voters. A mere 14 percent said candidate character was a top priority, while 49 percent said that "ability to accomplish things" was most important, followed by "stand on issues" at 33 percent.[11] Thus, character is important, but other factors may be more important.

In 1996, Americans reelected a president about whom they had serious character-related reservations. The voters did not ignore or dismiss the character question when they voted. But since all elections are judgments about past performance, an incumbent president is judged mostly on the quality of the job done. Character matters, yet performance generally matters more.

To what extent are questions of private moral character relevant to presidential politics? While the president is, in many ways, a moral or symbolic spokesperson for the nation, must the person who fills the office be personally "pure" to be a good president? It doesn't seem so. The public may want its presidents to be of the highest character, but a look at the private lives of our great presidents reveals an array of personal foibles.

The only modern president who might have passed today's "character test" is Richard Nixon. In his private life Nixon was seemingly spotless. In his public life, however, Nixon left much to be desired. In a test of character "which Nixon passes and FDR fails, something is evidently amiss with our current prejudices about the kind of character we desire in political leaders."[12]

Precisely what do we mean by character? Some people with roguish private lives exhibit a tremendous amount of public integrity. Others of spotless private life exhibit public qualities of treachery and duplicity. Today, journalists and rival candidates search the backgrounds of candidates, looking for any indiscretion. If found, that candidate is savaged and usually discarded from the race. Is it fair to base our judgments of character on isolated events that may have occurred years ago, and from which the candidate may have drawn useful lessons?[13]

Questions about presidential character are as old as the office itself. In 1800, ministers denounced Thomas Jefferson from their pulpits as "godless," and Andrew Jackson was pilloried as a barbarian and adulterer.

The election of 1884 provides a fascinating case study of character in politics. In that election, Democratic candidate Grover Cleveland was charged with fathering a child out of wedlock. Cleveland took responsibility, and agreed to pay for the child's upbringing. Understandably, this became a prime issue for his opponent James G. Blaine. The dilemma was that while Cleveland had character questions about his private life, his public or political reputation was spotless. Blaine, on the other hand had an "upright" private life, but was generally thought of as a public crook. The voters selected Cleveland.

With the depoliticization of politics as witnessed in the decline in party loyalty and in issue importance in voting for presidents, personal qualities loom larger in voting decisions. Today we see a person's private life as a barometer that reveals the truth about the person's public life. But a review of history suggests that this is not so. Presidents, like the rest of us, are human, they make mistakes. Some learn from their mistakes, others don't. The perfect person does not exist. Our presidents come with a wide array of strengths and weaknesses. To disqualify candidates because it is revealed they made mistakes years ago seems shortsighted. Does Bill

Clinton's early experimentation with pot disqualify him for the presidency?

We must also remember that on occasion, we punish presidents for doing what they believe is right. Few presidents exhibit a "profiles in courage" type of leadership because it is electorally dangerous. President Ford's principled stand on the Nixon pardon, and Carter's insistence that we do the honorable thing about the Panama Canal, serve as examples of how politically hazardous speaking truth to power can be.

Can a president be too nice, too honorable? Both Ford and Carter were very decent people. Would they have been more effective if they had been more Machiavellian?

Judging character is especially difficult given the varied motives of those who pursue political careers. Some enter politics for the wrong reasons—to fill a void or need that arises from low self-esteem. Others go into politics for more noble reasons—to accomplish good things for the nation. It is often difficult to discern.

One clue into character may be how a person deals with adversity or defeat. FDR's polio would have overwhelmed most people, but Roosevelt overcame adversity, and in doing so, was even more convincing when he told us we had nothing to fear but fear itself. In Roosevelt's case, adversity made him stronger. His character was forged in the fire of personal crisis.

Yes, presidents lie. But is there a difference between the behavior of FDR prior to U.S. entry into World War II and Nixon's lies about Watergate? The essential difference is that FDR misled with the public good in mind, and Nixon misled to save himself. In the long run, historians judge FDR's methods as questionable but his goals and outcome as honorable. Historians see Nixon's actions as self-serving and dishonorable.

While no single definition of character adequately covers all our needs, the following qualities are certainly admirable in the person who becomes president:

- Courage of conviction
- Internal moral compass
- Respect for others
- Commitment to the public good
- Respect for democratic standards
- Trustworthiness
- Generosity of spirit
- Compassion and empathy
- Optimism and hope
- Sense of decency and fair play
- Inner strength and confidence

Is there a useful model for judging character in presidents? The following three tests may serve as a beginning: (1) Does the behavior exceed the boundaries of what most reasonable people would think of as normal? (2) Is there a clear pattern of this behavior, and not an isolated act or event? (3) Does it affect the job performance?

In the end, questions of presidential character serve as yet another example of the paradoxical nature of presidential politics. What we applaud in one president we condemn in another. In the end, the real test of presidential character may well be, "Did this president bring the best out in all of us?"

Rexford Tugwell, a former adviser to FDR, tried his hand at defining the qualities that contribute to successful political careers. He wrote about specific political success stories of his own acquaintance, and his generalizations are instructive. "All were," Tugwell writes, "hearty, full-blooded types, vital, overflowing with energy, restless, driven by ambitions long before their compulsions had any focus:

- They were unintellectual in the scientific sense.
- They were strongly virile and attractive.
- All were extroverts, enjoying sensual pleasures.
- All were superb conversationalists; all knew the uses of parables.
- All were insensitive to others' feelings except as concerned themselves.
- All seemed to have thick skins because they were abused, but this was only seeming; all were hurt and all were unforgiving; and all were anxious for approval.
- All were ruthless in the sense of not reciprocating loyalty; they punished friends and rewarded enemies.
- All had thick armor against probings. Not even those nearest to them knew their minds.
- All were driven by an ambition to attain power in the political hierarchy, and all allowed it to dominate their lives.[14]

However much we seek the well-rounded leader for all seasons, we usually get ambitious, vain, and calculating candidates who do not know what is to be done (though they are willing to try). The men who run and win view presidential campaigns less as dialogues or programs for adult education than as a fight to win office, a fight to get there. Once they get there, they will experiment and see what works. The voters may like a person who knows all the answers—but few candidates will make commitments that are not reversible once in office. One of the first laws of presidential politics seems to be to promise a fresh approach but avoid specificity, to give the appearance of executive ability but not spell out what one intends to execute—at least not in any risky detail.[15]

THE INVISIBLE PRIMARY

Some political analysts used to hold that nothing that happens before the New Hampshire presidential primary (held in February of presidential election years) has any meaning. In times past, it was a confession of weakness for a presidential candidate to get too organized before the New Hampshire primary. Now, however, candidates work for three and even four years before that primary to prepare for their race. Thus Jimmy Carter announced his candidacy for the White House on December 12, 1974, almost two full years before the 1976 election, and he admits he made his decision in 1972, four years before the election. And Senator Bob Dole (R-Kan.) began his bid for the 1996 presidential nomination almost immediately after the 1992 race was over. Hence, "the invisible primary"—the determining events of an election that actually occur a year or more before the New Hampshire primary.[16] Since 1936 the active candidate who ranked as most popular within his own party in the Gallup poll taken one month before New Hampshire's primary has won his party's nomination almost every time. Pre-primary activity is indeed significant. A candidate needs to be convincing on at least several "tracks" before he gets contender status. All of the following are needed:

- To become as well known as possible
- To raise substantial sums of money (approximately $30 million)
- To attract and organize a staff
- To pay numerous visits to key primary states, especially New Hampshire
- To identify issues and build a supportive constituency
- To become identified as a spokesperson for at least a few key issues
- To devise a "winning" strategy
- To speak at over 200 party functions (the rubber-chicken circuit)
- To devise an effective relationship with the media
- To develop a psychological preparedness and a self-confidence that radiates hope

These do not necessarily occur in this order. This short list does not exhaust the self and organizational testing of this period. However, without these a candidate has little chance of being taken serious.

The first need of a would-be president is to become known. No other effort commands as much time as the battle to gain name recognition. Candidates like Roosevelt, Eisenhower, Kennedy, Reagan, and Dole had a leg up on most others because they had become celebrities or had inherited a well-known name even before they ran for the presidency. Candidates like Carter, Dukakis, Clinton, and Lamar Alexander, on the other hand, had to go out and become known the hard way—by crisscrossing the nation, visiting city after city, and giving unremitting, bone-numbing speeches and interviews.

The second major need for a presidential candidate is money. Large sums are needed to pay for staff, for travel, and later in the campaign for crucial television and radio advertising. Money is convertible into many other resources. It is exceedingly difficult to raise money for a presidential race unless you look like a "winner" or unless you take especially strong positions on controversial issues. A conservative such as Barry Goldwater, George Wallace, Ronald Reagan, or Phil Gramm can more easily raise early money than a moderate such as Howard Baker, Bruce Babbitt, Paul Tsongas, or Richard Lugar. Making matters worse, the national media generally follow the "star system" of giving primary coverage to those who are already well known or to the controversial candidates who command intense followings.

The process of raising money in large sums is compromising and often corrupting.[17] The burden of having to raise millions for a presidential race is at the heart of why many able persons do not consider running for the office. It is also at the heart of why some people are turned off by our political process. Two Republican presidential hopefuls in 1996, former Defense Secretary Dick Cheney and former HUD Secretary Jack Kemp, decided not to seek the office, citing among their chief reasons the distastefulness of raising large sums of money. As Kemp noted, "There are a lot of grotesqueries in politics, not the least of which is the fund-raising side."[18] Several top prospects dropped out of the race or failed to enter because of the difficulty of raising money, while the candidates of considerable wealth such as H. Ross Perot and Steve Forbes could enter almost at will. This raises serious questions about the selection process. When does a political contribution become a bribe? When does systematic campaign soliciting become equivalent to a conspiracy to extort funds? The late Hubert Humphrey, who twice ran for president, put it bluntly:

> Campaign financing is a curse. It's the most disgusting, demeaning, disenchanting, debilitating experience of a politician's life. It's stinky. It's lousy. I just can't tell you how much I hate it. I've had to break off in the middle of trying to make a decent, honorable campaign and go up to somebody's parlor or to a room and say, "Gentlemen, and ladies, I'm desperate. You've got to help me . . .
>
> . . . And you see people—a lot of them you don't want to see. And they look at you, and you sit there and you talk to them and tell them what you're for and you need help and, out of the twenty-five who have gathered, four will contribute. And most likely one of them is in trouble and is somebody you shouldn't have had a contribution from.[19]

Reliance on big money made the presidential selection process vulnerable to charges of corruption. Finally, in the aftermath of the Watergate scandal, the Congress passed legislation for public financing of presidential elections. Enacted in 1974, the campaign finance reform called for public disclosure of all contributions exceeding one hundred dollars, established ceilings on contributions, created a system of federal subsidies, and established spending limits. This law was challenged in court, and in *Buckley v. Valeo* (424 U.S. 1, 1976) the Supreme Court determined that while the Congress could regulate contributions and expenditures of campaign organizations, it could not prevent or limit independent individuals or groups from exercising free speech (and spending) rights.

The Congress then passed the Federal Election Campaign Act (FECA), to revise existing laws to keep them in line with constitutional standards. These new guidelines have been in effect since the 1976 presidential election. Since that election, campaign organizations have discovered various loopholes in the law that allow campaigns to get around the law.[20]

First, the "independent" money exemption allows individuals and organizations to spend money for or against candidates as long as the effort is not conducted with the advice or assistance of a candidate's campaign organization. This has opened the door for independent spending on behalf of a candidate, or, as is more likely, negative campaign ads against a candidate. Second, the spending of what is called "soft money" allows party organizations to raise and spend money for a variety of purposes in support of the campaign. In 1992, the Republicans outspent the Democrats in "soft money" by $15 million.[21] Soft money expenditures for 1996 were estimated at over $600 million.[22] This money was used for voter registration drives, get out the vote efforts, and party building in general.[23]

To run a respectable campaign in 1996, a candidate had to spend around $30 million. Enormous time was spent on fund raising. When asked why he robbed banks, Willie Sutton said, "Because that's where the money is." Where do would-be presidents go for their money? Wealthy individuals, corporate elites, and increasingly to California for Hollywood money (known as "the Beverly Hills Primary").[24] Lamar Alexander attended nearly three hundred fund-raising events in 1995, prompting him to comment, "I feel like a trained dog." Another 1996 Republican aspirant, U.S. Senator Arlen Specter, spent four hours on the phone every morning soliciting contributions.[25]

While it would be unfair to say that money literally buys elections, it is nonetheless clear that money plays a significant role in determining who the eventual winner will be. Almost always, the primary candidate who spends the most money wins. In general elections, while the Republicans usually outspend the Democrats (1992 was an exception, and the Democrats won), the role of money tends to be a bit more overshadowed by other factors.[26]

Campaign contributions are often given to reward and influence candidate positions. Steel magnate Henry C. Frick complained of Teddy Roosevelt, "We bought the son of a bitch and then he did not stay bought."[27] Most campaign contributions are seen as investments in the future, and contributors do expect, and often get, a return on their investments.

Incumbent presidents have always used the lure and trappings of the office to elicit campaign contributions, but no president in the past twenty years has elevated it to the art form practiced by the Clinton White House. In an elaborately orchestrated series of "coffees" and "sleep-overs" (in the Lincoln bedroom), the president invited big donors to the White House in hopes that they would contribute to the Clinton reelection effort. The campaign of 1996 was the costliest in history, with each nominee receiving $61.8 million from the government, and the party national committees raising over $880 million in "soft money." President Clinton's fund-raising efforts were unusual *only* for the brazen manner in which the White House was used to raise money.[28]

Attracting loyal staff, identifying key issues, becoming identified with vital issues, and devising a sensible strategy are all vital to the successful launching of a candidacy. Often underestimated is the capacity to establish a good working relationship with reporters and television interviewers. In an age of candidate-centered campaigns, where political parties are of decreasing importance, the electoral role of the media has greatly increased. Some candidates, great speakers who are superb at raising funds, perform poorly when interviewed by the press. Sometimes, too, the great stump speaker looks foolish and too "hot" on television. Friends of former U.S. Senator and Vice President Hubert Humphrey generally admitted that Humphrey and television were not made for each other. He was an outstanding orator at political rallies, but he talked too fast and too much for the television viewer. Ronald Reagan was masterful at giving a prepared speech but when interviewed one-on-one made errors, and his handlers were forced to restrict press access to their candidate.

What might be called the psychological test—how a candidate reacts to the strain, the temptations, and the intense public scrutiny of the campaign—is one of the most important measures of a candidate. Arthur Hadley asked, "How much does the candidate want the presidency? How much of his private self and belief will be compromised to the public man? To what extent will he abandon family, friends and other normal joys of life, and how does he handle this isolation?"[29]

Some candidates develop self-doubts. Some develop a tendency to tell audiences what they want to hear, and over the course of a few weeks they become inconsistent and look ridiculous. Others, such as Edmund Muskie, become plagued by a need for more sleep. Muskie also found his presidential bid in 1972 exacerbated by his inability to control his temper. In 1988, Gary Hart seemed to challenge reporters to catch him in an act of indiscretion, and they did. The exacting invisible primary period is always an exhausting ordeal and a formidable test, as well, of whether an individual can hold up physically and can control himself emotionally.

PRESIDENTIAL PRIMARIES

Presidential primaries began as an outgrowth of the Progressive movement's efforts in the early twentieth century to eliminate "boss rule" and encourage popular participation in government. Presidential primaries began to take shape after 1905 when the La Follette Republicans in Wisconsin provided a system for the direct election of members of the state's delegation to the national nominating convention. Now thirty-seven, including most of the big, electoral-rich states, are using some type of presidential primary.[30]

The concept of popular participation in the nomination of the presidential party nominees evolved slowly. First we relied on the congressional caucus system, which did not allow for direct popular participation at all. Until 1828 members of Congress from each party met and selected the person they wanted as their nominee. With the growth of democratic sentiment and the coming of the Age of Jackson, the national nominating convention system began to emerge as the replacement. In 1828 state legislatures and state party conventions were relied on to nominate party nominees. After that, national conventions took hold, although it was not until 1840 that national party conventions were accorded full recognition.

Not until 1912 did primaries begin to be used regularly (about twelve states used them that year) in enough places to begin to have a serious impact on the presidential nominations. Many party leaders, however, have never been enthusiastic about primaries, in large part because they believe they undermine the two-party structure by strengthening the hand of candidate loyalists and issue-oriented zealots at the expense of the party regulars. Primaries allow people to vote who may have little or no loyalty to the party and no interest in the party's future. In fact, those voting in party primary races are more extremist in their views than party regulars. Thus primaries favor more ideological candidates (to the left in the Democratic party, to the right in the Republican party). These candidates may do well among committed activists within one wing of the party, but spell disaster in the general election where the electorate is more moderate. The Republicans in 1964 with Barry Goldwater, and the Democrats in 1972 with George McGovern, are examples of what can happen when crusading issue activists capture the nomination only to be trounced in the general election. Thus party regulars generally bemoan the "reforms" of the post-Watergate era as misdirected and counterproductive. The goal of giving "power to the people" has, they argue, led to chaos and confusion. The old pre-primary method in which three-fourths of the delegates were chosen by state party conventions or caucuses plainly permitted party regulars and long-time party professionals to control who would be on the party's ticket.

The importance of primaries has waxed and waned during this century, but they have become increasingly important since the 1950s. Candidates in the past forty years or so have viewed the primaries as an essential test to demonstrate their vote-getting appeal. In 1996, for example, roughly 90 percent of all Democratic convention delegates were chosen in primary elections (See Table 2.1). Two-thirds of Republican delegates were likewise chosen in primaries.

Not all states have adopted the primary system. In 1968, seventeen

TABLE 2.1

Number of Presidential Primaries and Percent of Convention Delegates from Primary States

Year	Republican		Democratic	
	# of primaries	% of delegates	# of primaries	% of delegates
1912	12	32.9	13	41.7
1916	20	53.5	20	58.9
1920	16	44.6	20	57.8
1924	14	35.5	17	45.3
1928	17	42.2	16	44.9
1932	16	40.0	14	37.7
1936	14	36.5	12	37.5
1940	13	35.8	13	38.8
1944	14	36.7	13	38.7
1948	14	36.3	12	36.0
1952	15	38.7	13	39.0
1956	19	42.7	19	44.8
1960	16	38.3	15	38.6
1964	17	45.7	17	45.6
1968	17	37.5	16	34.3
1972	23	60.5	22	52.7
1976	29	72.6	28	67.9
1980	31	71.8	35	76.0
1984	26	62.1	30	71.0
1988	37	66.6	37	76.9
1992	39	70.0	39	84.0
1996	36	67.0	43	90.0

Source: Updated from *Presidential Primaries: Road to the White House,* 2d ed., James W. Davis. Copyright © 1980 by Greenwood Publishing Co. Reproduced with permission of Greenwood Publishing Group, Inc., Westport, CT.

states held primaries. By 1992, thirty-seven states chose delegates by primaries. For 1996, Kansas canceled its presidential primary, but will resume primaries in 2000. Some states still select a certain number of delegates to the national convention through local or district conventions or allow a state committee or convention of party officials to choose the remaining delegates.

The rules for primaries vary from state to state and from party to party. Usually, however, voters elect delegates directly or by showing a preference for a presidential candidate. Some of the early primaries, such as those in New Hampshire and Wisconsin, or "Super Tuesday" in the South, can be important in giving a psychological lift to a front-runner or a new challenger. Later primaries, such as those in California, Ohio, and New Jersey, can be important in giving the final edge to one candidate over others as the front-runner heads into the national convention. Rarely, however, do the later primaries have an impact, since the front-runner usually has a sufficient number of committed delegates before the California primary. This has led California and a number of other states to "frontload" their primaries, moving them earlier in the calendar year. In 1996, primaries were so compressed —that in just forty-four days—from February 12 to March 26 when California held its primary, approximately 70 percent of all Republican delegates were selected. This gives a great advantage to those who are well known, well funded, and well organized early in the process, and inhibits a lesser-known candidate from building momentum over time. This has created a rush to the starting line, as only those candidates who do well early have a chance of winning. It has also scared away capable but lesser-known candidates who might otherwise have run.[31]

In recent years the system of presidential primaries has become one of the most passionately debated aspects of the presidential selection process. Critics say it is a case of "democracy gone mad," and a "very questionable method of selecting presidential candidates." It is especially good at eliminating candidates.

Nearly always, the criticism of the primaries ends up focusing on these alleged flaws: the system takes too long, costs too much, highlights entrepreneurial personalities at the expense of issues, makes pseudoenemies out of true political allies, invites factionalism, undermines parties, often favors colorful ideological candidates over moderates, and frequently does not even affect the outcome of the nomination process. Critics point to the Goldwater and McGovern nominations as prime examples of flaws in the primary system.

Primary voters are older, have higher incomes, are more educated, and are more ideological and politically active than are primary nonvoters. Turnout is relatively low. This leads some analysts to conclude that as a democratic institution designed to stimulate popular participation, the presidential primary has limited effectiveness, and has turned off rather than turned on potential voters.

The frequent and sustained criticism of primaries leads some observers to suggest that we should abolish or at least reform them. A national primary, a one-shot winner-take-all event in August or September of election year, is supported by a majority of adults answering Gallup poll surveys. Others favor regional (multistate) primaries or a return to state conventions as a better means by which to select competent presidential nominees.

The primary system surely has its blemishes, but it has also served us reasonably well. Although "the people" do not fully control the nominating process, primaries have increased the public's potential to influence who will be convention delegates and has opened the process to some candidates and ideas that might otherwise have been excluded.

Primaries have decreased the party leaders' firm control over the nomination process. Students of our party system worry about this. They would prefer a system that sends responsible party regulars of the state and local parties to the national convention, not bound by rigid instructions from a primary verdict but as representatives, free to seek out the national interest according to their best judgment. They believe these party regulars would be delegates concerned with the majority of the party's rank and file and also with the acceptability and electability of a candidate as well as that candidate's ability to serve effectively as president. This view perceives party regulars as those who are most informed and best qualified to select the nominee,[32] and ignores the increasingly large number of independent voters in America.

Another criticism of the primary system stems from the undue weight placed on the first primary in New Hampshire. Does this small state reflect the nation or does the very conservative nature of the state give undue advantage to more conservative candidates? In an effort to offset the power of the New Hampshire primary, many states have "front-loaded" their primaries, that is, moved them up in the calendar so as to have a greater impact. But the probable impact of front-loading has been to give New Hampshire even greater influence, as winning that primary is seen as essential if a candidate is to be regarded as credible in those primaries that follow shortly thereafter. Again, the unintended consequences of reform often do more harm than good.[33]

It is true that primaries do allow, on occasion, for a Goldwater or a McGovern to be nominated. But primaries also allow for fresh faces, and young new blood to emerge, as happened in 1960, 1976, and 1992. The 1976, and 1992 campaigns showed that the Democrats, for example, could emerge from the primaries as a fairly unified party. By giving candidates several opportunities to present themselves to the public, our present procedures make it possible for a candidate to win substantial support during a relatively short time period. Abolishing this aspect of the presidential selection process could limit the infusion of new blood into the presidential race. Further, to return to the choosing of the nominee by a small establishment group is unacceptable to most Americans today who see politics as a dirty word and politicians as the enemy. People are demanding more involvement, not less. That Americans fail to utilize the very thing they demand to have, only reinforces the paradoxical nature of presidential politics.

Modest participation in the primaries prompts reform suggestions. Universal voter registration or the motor-voter bill (allowing people to register to vote at the time they register for a driving license) slightly increases participation. Newspapers and television could do a better job of getting to the heart of candidates' views, exploring inconsistencies, and piecing together the candidates' positions on various policies.

One of the virtues of the primary system is that candidates are required to present themselves to the people. Candidates have to organize their thoughts, to clarify and define key issues. They are required to communicate with all kinds of people and to react under pressure. Most of the time it is an excellent learning experience for both candidates and the public. It also allows room for the people to sharpen and alter their initial views of candidates.

The chances of destroying or eliminating a truly outstanding front-runner in the primaries are slim.

Primaries were designed to give the people the right to be involved in the choice of their party's nominee and should not be abandoned because qualified persons choose not to run. While more must be done to encourage the best people to seek the office, we need not sacrifice the democratic integrity of the process to attain our goals. The convention can still draft a "dark horse" if there is no popular favorite. Primaries may be imperfect, but so are democratic societies. They are an inexact, expensive, and overlong way of coming up with nominees, but the system is not as flawed as critics contend.[34] And many of these flaws can be modified (not necessarily fixed) to make the process more congenial to the integrity of the process.[35]

Return to state conventions or the ancient congressional caucus procedure would serve only to increase the influence of political bosses and special interests, who find it easier to bring pressure to bear on a few individuals in those old "smoke-filled rooms" than on entire electorates. Moving to a national or even a regional primary would lessen contact between candidate and voter, virtually prevent the less well-known candidate from running, and increase reliance on television advertising.

Perhaps it is naive to believe that candidates who have more direct contact with the electorate and are dependent on their votes for nomination as well as election will focus more on the voters' needs than on the needs of special interests. However, this would be the case if we continued to improve our use of state presidential primaries. Constant attention and further improvements are needed in order to obtain the high standards that have been set.

Despite certain flaws, primaries are the most effective way of involving the populace in the nominating process.

NATIONAL CONVENTIONS

For most of the past 150 years the national conventions have performed (or tried to perform) several functions. They have nominated presidential candidates acceptable to most factions within the party. By winning plurality victories in the primaries a candidate can secure the nomination without being acceptable to virtually all elements within a party. It is only the acquiescence of these other interests at the convention that signals to the party's rank and file that the nominee is the legitimate party standard-bearer. Carter and Ford won that legitimization at the 1976 conventions, Bush and Clinton did so in 1992, and Dole did so in 1996. Goldwater and McGovern failed at their conventions in 1964 and 1972, respectively.

It has always been the purpose of a convention to select or ratify nominees who possess a strong likelihood of winning voter support in November.

The goal is to produce a winning ticket. Thus a second function of the convention is to shape the vice presidential choice in such a way as to both strengthen the ticket and reconcile factional cleavages within the party.

A more general function of conventions is trying to unify a party that is not inherently unified. This is a time not to examine differences but to seek unity in the face of disagreements and diversity. It is naturally in the party's interest to build enthusiasm and rally the party faithful to work for the national ticket.

Conventions also hammer out party platforms. Platforms are generally less meaningful than the campaign statements of the presidential candidates, but they are a useful guide to the major concerns of a party. They are often inclusive "something-for-everyone" reports. A platform is invariably a compromise of sectional views, diverse caucuses (women, blacks, etc.), and the policy preferences of dominant party elites. The winning candidate is often willing to concede a plank in the platform to the wishes of one of the runner-ups in the primaries. This can be a quid pro quo offer to a faction or a candidate-based organization within the convention that must be won over to unite the party. Thus Gerald Ford's people had to agree to a more hawkish foreign and defense policy plank to placate the Ronald Reagan forces at the 1976 Republican convention, and George Bush and Bob Dole had to give in to the demands of the right wing of their party in 1992 and 1996.

If the party has an incumbent president, the platform is often drafted in the White House or approved by the president and his top policy advisers. Seldom does the party adopt a platform critical of its incumbent president.

Criticism of the national party nominating conventions has been loud and frequent. President Eisenhower called them a national disgrace, and social critic H. L. Mencken once wrote that "There is something about a national convention that makes it as fascinating as a revival or a hanging." Critics contend conventions are too big, noisy, unwieldy, unrepresentative, irresponsible, and a waste of time. Others say they function with too much concern for selecting a winner rather than the best qualified person. Similarly, critics say vice presidential choices are often made too hastily, with much regard for balancing the ticket and little regard for selecting a person who may eventually become president. They give as examples the selections of Agnew (1968), Eagleton (1972), and Quayle (1988).

Television and the proliferation of presidential primaries have altered the role of modern conventions. Today, conventions are less about selecting candidates and more about political theater and marketing; less about where the party wishes to lead the nation and more about entertainment. Bill Greener, who managed the 1996 Republican convention, argued:

> We don't make the rules to the game. We are dealing with the rules. And
> the rules are: You get an hour and don't forget in that hour we have got
> our commercials to air and our station breaks, and don't forget that we

need time to showcase our talent, and don't forget that even before we think about going to the podium we have got to establish the presence of our anchor and our four floor correspondents, and don't forget that after that we are going to take a commercial break, and then we are going to go back to the anchor, and then we *might* take what you've got going at the podium. And that's just the way it is.[36]

Yet another complaint about national conventions has came about in the wake of the Democratic party's McGovern-Fraser commission reforms of the early 1970s. This commission was established to try to improve the delegate selection processes and to "open up" the Democratic party. The basic goal of the McGovern-Fraser reforms was to end the traditional dominance of regular party leaders and to make the Democratic convention more representative of the party rank and file. Jeane Kirkpatrick's book, *The New Presidential Elite,* offers strong criticism of those reforms and suggests that so long as they are in effect the party conventions will not be able to perform their traditional functions as well as they once did.

The key criticism here is that according to the new rules many, perhaps even a majority, of the delegates to the Democratic convention, are not acting as responsible members of a party, with all the memories, past participation in, and commitment to its future. Instead, they are acting as members of a candidate-centered organization whose loyalties are almost exclusively to specific candidates and their issues. This may lead, some critics charge, to allowing the "zealots" to rule.

This last criticism is a serious charge, but it does not stand up to critical analysis. "Greater participation . . . does not mean that party leaders must automatically lose. All it means is that party leaders must compete with nonleaders for influence."[37] Nor are issue activists inclined to be party wreckers. Issue activists supported JFK in 1960, LBJ in 1964, Carter in 1976, and Clinton in 1992—even though all were moderates. "The groups advantaged by the [new participatory] process are not homogeneous, they are not necessarily more or less representative than the party organization people they have replaced . . . and their participation does not necessarily guarantee continued internal divisions in the Democratic party."[38]

While it is true that party activists and convention delegates are more extremist (in the sense of being more left or right) than the general public, it is not true that this leads them to nominate extremists for the presidency. Of the ten individuals nominated over the last five election cycles, only Ronald Reagan leaned to the political extreme, and he rarely seemed extreme to the voters who elected him twice.

One wag commented that "Democrats have to control their left-wing nuts and Republicans have to worry about their right-wing nuts." And we should not be surprised when committed true believers get active in politics in pursuit of their cherished agenda items. At the 1996 Republican convention, 21 percent of the delegates identified themselves as members of the religious right. Only 15 percent of the public has such an affiliation.

Seventy percent of the delegates said they were conservative. Only 34 percent of the public so identifies itself. Did these "zealots" nominate a right-wing radical? No. They chose Bob Dole (or, more accurately, the primary voters did). On the Democratic side, 43 percent of the delegates said they were liberal, yet only 16 percent of the voters and 21 percent of Democrats so identify themselves. Did they nominate a leftist? No. Bill Clinton was their nominee.[39]

On balance, the national conventions have served us rather well, if imperfectly.[40] New rule changes, the far greater role of television, and the reality that nominees are more and more "selected" and "nominated" in the state primaries require us to reexamine the traditional functions of the convention. Although substantial change has taken place, the conventions still serve many intended goals. Despite their deficiencies, they have nominated some capable leaders, Lincoln, the Roosevelts, Wilson, FDR, Eisenhower, and JFK among them. Conventions have had many triumphs and only a few outright failures.

Conventions remain a uniquely American pageant. Where most Western democracies select their leader (prime minister) from a closed party caucus, away from the hustle and bustle of mass politics, in the United States selecting a president is the people's job, as messy and chaotic as this process may be.

Many of the criticisms of the convention process are really criticisms of elections in general. Elections may not be our finest hour, they may not always address our better selves, they may not be the ideal way to select leaders, but they remain the most reliable of the known available devices.

INCUMBENCY: ADVANTAGE OR DISADVANTAGE?

The advantages of incumbency for a president seeking reelection are traditionally revered as a political article of faith. For a sitting president, the benefits of incumbency are easily distinguishable: instant recognition; full access to government research resources; ability to dominate events and make news and to attract constant media exposure; party organizational structure at one's disposal; ability to dispense government contracts; and some ability to manipulate the economy.

Regardless of a president's record of accomplishment, an incumbent president is supposed to benefit from a public relations machine that shows the president as a person of action, a commander in chief, a traveling statesman, and a strategic crisis manager. Not the least of a president's assets is a loyal White House staff that, in the unavoidable blurring of presidential and political functions, performs a myriad of services. A president and the national party can also raise money far more easily than the opposition, and the resources of the office provide millions of dollars' worth of publicity.

A rival for the presidency is generally a political candidate and little more. He is a seeker whose motives are unclear, a pursuer with a feverish gleam in his eye. He covets what his rival already possesses. The most he can give is promises.

Perhaps the most important asset for the incumbent is the selective or manipulative use of government contracts, patronage, and other political controls over the economy. President Nixon's manipulation of milk prices is alleged to have aided his campaign treasury in the 1972 election. In what was called "the incumbency-responsiveness program," Nixon aides sought to maximize their control over the federal government's enormous resources to their best advantage. Federal grants were evaluated according to political benefits. Political appointments and ambassadorships were sometimes promised in exchange for large campaign contributions. And corporations were "encouraged" to contribute to the upcoming Nixon campaign in a near-extortionist manner.

But is incumbency really that much of an advantage? Political scientist Edward Tufte, who studied the relationship of the economy's performance and election outcomes, concluded that short-run economic performance has a good deal to do with ensuring an incumbent's reelection. Tufte finds positive support going back to 1918 for the not surprising hypothesis "that an incumbent administration, while operating within political and economic constraints and limited by the usual uncertainties in successfully implementing economic policy, may manipulate the short-run course of the national economy in order to improve its party's standing in upcoming elections and to repay past political debts."[41] Today, however, presidents have less opportunity to manipulate the economy in this manner. With an enormous national debt and budget challenges, presidents have less discretionary money and fewer resources to throw at the voters in the months before an election. Thus, where Richard Nixon and the Democrats who controlled Congress could, prior to the 1972 election, hand out large increases in Social Security benefits, Presidents Bush and Clinton were forced to find ways to limit such spending and seldom could use the largess of the federal government to their advantage.

A discussion of the advantages of incumbency certainly suggests that the odds of unseating an incumbent president are formidable. Still, the record is clear—incumbents can and often have lost. Indeed, incumbency is sometimes as much a burden as a benefit. Since the Jacksonian era, which inspired the rise of mass political parties and the party convention system, five presidents have been denied party renomination: Tyler, Fillmore, Pierce, Andrew Johnson, and Arthur. Another nine, John Adams, Van Buren, Cleveland, Benjamin Harrison, Taft, Hoover, Ford, Carter, and Bush, were defeated for a second term after winning their party's nomination. Against these fourteen failures, incumbent presidents have been successful in winning both the nomination and the election only sixteen times, including three reelection triumphs by Franklin Roosevelt.

One of the paradoxes of recent years can be seen in the incumbency advantage being transformed into the incumbency disadvantage. In 1968, incumbent Lyndon Johnson was compelled not to seek renomination when faced with mass protests over his Vietnam policy. His successor, Richard Nixon was forced to resign from office one step ahead of impeachment. Nixon's successor, Gerald Ford, nearly lost his nomination bid and ended up losing the election in 1976, and the man who defeated him, Jimmy Carter, also faced a serious nomination challenge and lost in his 1980 bid for reelection. Reagan's successor George Bush lost in his 1992 reelection bid to Bill Clinton. Plainly, being the incumbent can be dangerous to the president's political health. With a strong anti-government mood, in an age of cynicism spawned by Vietnam, Watergate, the Iran-Contra scandal, and other events, we blame presidents for much that goes wrong.

Incumbency, then, is a double-edged sword.[42] It can help a president who presides over a period of prosperity and peace and projects an image of being in charge of events. But it can just as readily act against a president associated with troubled, perplexing times, who does not seem to be in full possession of his office. If prosperity favors the incumbents, depression favors the challengers. Distrust of politicians and low morale in the nation favor the challengers; strong confidence in the national government favors the incumbent.

An incumbent is necessarily on the defensive: His record is under detailed scrutiny, his administration's every flaw and unfulfilled promise is exposed to microscopic examination. The American people can conveniently, if often unfairly, blame a whole range of problems on the president, whereas the astute challenger presents a smaller target. The challenger can talk about secret plans and make a rash of promises without being specific about what they would cost and who would wind up paying for them. A challenger is also in a far better position to present illusory problems as real ones and to play upon public fear about such things as missile gaps, inflation, immigrants, and crime in the streets in ways that may distort reality while conveniently winning headlines. The challenger can take the offensive and try to convince the voters that a new president would do better.

The incumbent is often judged against the idealized model of the perfect president, and not unnaturally, is found wanting. Paradoxically, the incumbent may at times be the symbol of the nation's pride, but just as readily may be the nation's most convenient scapegoat.

People forget what you have done for them and remember only what you did *to* them. Thus, if taxes and inflation go down, people will ascribe the situation to American ingenuity and the success of the free enterprise system. But an increase in taxes and inflation gets blamed on a failure of presidential leadership. In the past generation the presidency has gradually acquired more responsibility for peace, prosperity, and improvement in the quality of life than it has the authority to implement. It is a case of demanding more of the presidency, even when the capacity to act isn't possible. The "what have you done for us lately" refrain is even more recurrent in an age of heightened single-issue, narrow special-interest-group politics.

In sum, while incumbency on balance is usually an advantage, especially in the hands of an astute politician, a variety of burdens go along with it. In practice, for various reasons, "American presidential tenure experience comes closer to being a one-term tradition."[43]

THE GENERAL ELECTION: WHAT MATTERS?

Presidential candidates are not free agents who can choose among strategies at will. Their strategy is seldom based on mere choice, it is usually forced by circumstance. For example, about every eight or twelve years there is a strong underlying desire to throw out the party in office. (In 1920, 1932, 1952, 1960, 1968, 1980, and 1992 voters seemed in part to be punishing those in office who were unable to improve things.) Slogans such as "It's time for a change" or "Throw the rascals out" are a familiar refrain as voters turn incumbents out of office with an almost predictable alternation.

Between 65 and 75 percent of the voters usually have made up their minds as to how they will vote by the end of the national conventions (late August of election year, a good eight or ten weeks before election day). The basic organizational effort of a candidate must be aimed at stirring up the support of voters at the grass roots. The strategy that makes the most sense is to get out all possible supporters and potential supporters and independents and to target only secondary resources at converting the opposition. Supporters need general reassurance on both substantive and stylistic matters, but opponents want to know specific policy plans and program ideas.

A lot depends on how candidates conduct themselves. Our election process usually, although not always, excludes broad policy questions. Of necessity, then, in addition to observing the candidates' party, the people assess the contenders not so much on what they believe but on how good a job they might do.[44] This is why we carefully watch how they answer tricky questions and whether they keep their cool with hecklers. Looking "presidential" is an important part of the voter's equation when deciding on a preferred candidate.

Public financing of the general election lessens the contender's dependence on wealthy special interests. However, candidates still need to build a coalition of well-organized interest groups to get out the vote. Candidates cannot afford to offend most of the so-called single-issue groups. More than anything else, presidential electoral politics is coalitional politics. Interest groups help to get others involved in a campaign. They help mobilize voters on election day. Groups are seldom neutral. They lean to one party or one kind of candidate over others. Groups want access to the political system and they want someone favorable to their interests in the White House.

Candidates find the general election fraught with dynamic tensions. Issue-oriented enthusiasts urge them to "speak out on the vital issues." Party regulars urge them to work closely with the party bosses. Television consultants urge them to devote most of their time to brief television spots and talk-show appearances. Public relations aides urge them to invoke patriotic symbols and quote from prestigious heroic sources. Campaign managers generally argue that debating all the issues and trying to educate the public is the worst possible way to run a campaign. Issues that attract some groups repulse others.

Do elections matter? In a general way they do, although elections rarely give a president or country a specific mandate. The 1996 elections are a good example. Clinton won because of a strong economy and because Dole was a weak candidate who ran a weak campaign. But the voters did not give Clinton a policy mandate. A national election in the United States is not a plebiscite or referendum on a number of specific issues. Because of candidates' policy ambiguity, we often vote without a clear idea of what the candidates will do if elected.

Candidates who win by a large margin usually claim they have won a mandate from the people. Such a claim is often unjustified, however, especially in elections such as those of 1964, 1972, and 1984 where the "landslide" was more a factor of the personal popularity of a president or of the negative perceptions about the defeated than positive perceptions about the victor.

Presidents invariably claim some sort of mandate, believing that if the public and Congress believe the claim they will have an easier time governing. A mandate is based on three features: the *size* of a president's electoral victory; the *type* of election (issue oriented or personal); and the *number* of candidates from the president's party elected to Congress (the president's coattails and the aggregate numbers).

An election can produce one of three types of mandates. A *positive mandate* (e.g., Reagan in 1980) comes as the result of a significant margin of victory in a race in which a significant number of candidates from the president's party were elected to Congress, major issues were contested, and one side won a clear victory. Positive mandates usually mean that the president can accomplish significant elements of his campaign platform. A *negative mandate* (e.g., Nixon in 1968 and Clinton in 1992) is a repudiation of the incumbent president or party and normally does not serve as much of a guide to future action, nor does it grant the victor significant power. A *marginal mandate* (e.g., Reagan in 1984 and Clinton in 1996) normally results from a relatively issueless campaign and confers no measurable margin of power.

The debate over whom to vote for usually centers around which candidate is best equipped to handle the job and the problems at hand rather than around the detailed specifics of how to solve the big issues of the day. Thus, our elections represent mandates to get the job done, but we leave the means up to the judgment of the president. Elections sometimes set limits on what can be done, but they seldom determine the precise content of public policy.[45]

Unlike their parliamentary counterparts, elections in the United States do not confer *power*. They merely grant office, which grants the opportunity to seize power. Presidents must work at translating an electoral victory into political power.

Presidential elections are important in terms of who is to handle the issues and with what broad leeway. But elections seldom are direct policymaking events. The general mood in the nation on issues and the partisan and ideological balance in the recently elected Congress are probably as important, if not much more so, in shaping how a president will act.

The claim that a president has a mandate to govern is further undermined by the fact that voter turnout is so low. Nearly half of voting-age Americans do not vote in presidential elections. During the 1950s and 1960s, voter turnout averaged over 60 percent. In 1992, slightly over 104 million Americans voted in the presidential election; over 85 million failed to vote. Roughly one in three Americans between the ages of 18 and 24 vote; approximately two in three over the age of 45 vote. In 1996, even efforts directed at younger voters such as MTV's "Rock the Vote" failed to yield a significant increase in turnout. Less than 50 percent of eligible voters voted in the Clinton, Dole, Perot presidential election (Table 2.2).

Why? The primary explanations for nonvoting are attitudinal: disinterest, powerlessness, and distrust. Nonvoters ought not be described as a cohesive group. There is about as much diversity among nonvoters as among voters. Moreover, the old view that nonvoters are almost entirely poor, uneducated minorities is not true. Dissatisfaction with the caliber of the candidates, the limited number of candidates, and the difficulty in distinguishing differences among their stands on issues lead many to think that voting is not the effective act it may once have been. All candidates seem pretty much the same, and it is true that voters are offered less choice than in comparable Western democracies. Voters complain that too many politicians "talk in circles" and "say one thing, then do another."

In 1996 the voters essentially had a choice between a character-flawed candidate in Clinton and a charisma-challenged candidate in Dole. Clinton's scandals and his credibility, or lack thereof, turned off many of his would-be supporters. Dole failed to provide a clear reason why he should be elected. He failed to fashion a consistent or coherent message. And he proved to be a weak debater. Most voters also had soured on Perot. All of this contributed to a decline in voter turnout in 1996.

Just as ticket splitting and being an Independent rather than a Democrat or a Republican are becoming fashionable, so also refraining from voting may be becoming an accepted norm. Abstinence occurs in part because people feel that voting for the lesser of two evils is demeaning. Not voting is a way to show they don't approve of either candidate; in effect it is a peaceful protest.

TABLE 2.2
Voting Turnout in U.S. Presidential Elections 1824–1996

Year	% of voting population	Year	% of voting population	Year	% of voting population
1824	9.0	1936	56.0	1964	61.9
1840	33.0	1940	58.9	1968	60.9
1860	32.0	1944	56.0	1972	55.2
1880	37.0	1948	51.1	1976	53.5
1990	35.0	1952	61.6	1980	52.6
1920	44.0	1956	59.3	1984	52.9
1932	52.4	1960	62.8	1988	50.1
				1992	55.2
				1996	49.0

Source: Updated from U.S. Department of Commerce, Bureau of Census, *Statistical Abstract of the United States* (Washington, D.C.: Government Printing Office, 1993), p. 284.

As political scientists Nelson Polsby and Aaron Wildavsky point out, "In two respects, Americans are different from citizens of other democracies. A smaller proportion of Americans will vote in any given election than citizens of other democracies, but Americans collectively vote much more often, and on more matters than anyone else."[46] In the United States, ballots tend to be longer, more complicated, and filled with a greater number of issues than in most other democracies. Given the average citizen's limited interest in politics, many feel ill equipped to make decisions that may include races for the presidency, Senate, the House, state-wide offices, statewide ballot propositions, local offices, and local propositions.

THE ELECTORAL COLLEGE

In spite of what most Americans believe, presidential elections *are not* technically decided by the popular vote. Presidents are elected by the electoral college. Every four years, political observers grumble about our electoral college system—largely because it does not ensure victory to the presidential candidate who wins a majority of the popular votes. But nothing seems to get done. The complexity of the electoral college system discourages the average voter from understanding its deficiencies. The road to electoral college reform is littered with the wrecks of hundreds of previous efforts. Moreover, if the disease is clear, the remedy is not.

Experts and political strategists differ in their estimates of the consequences of electoral reform. Politicians and commentators are divided and deadlocked over whether to retain the electoral college or to amend the Constitution to provide for direct popular election of the president.[47]

Many observers believe it may take the election of another popular vote loser as happened in 1824, 1876, and 1888 to prompt the sustained action needed to amend the U.S. Constitution. Others think merely one more close election, such as that in 1960, 1968, or 1976, may be enough. Whatever change does occur will have important consequences for the health of the political system, for the quality and balance of our political parties, and for the kind of democracy we have. These considerations are not merely housekeeping matters. They speak directly to the integrity of our political system.

The electoral college system for the election of American presidents was much debated at the Constitutional Convention of 1787 and since then it has been about the most widely debated aspect in the Constitution.

THE CASE FOR THE DIRECT ELECTION OF PRESIDENTS

The direct election method means that the person who gets the most votes wins. There is no electoral college. Advocates of the direct popular system contend everyone's vote should count equally, people should vote directly for the candidate, and the candidate who gets the majority of votes should be elected.

With the electoral college method, a president can be elected who has fewer popular votes than his opponent, as was the case in 1824 when John Quincy Adams, with 30.92 percent of the vote, defeated Andrew Jackson, with 41.34 percent of the vote. This happened again in 1876 when Rutherford B. Hayes, with 47.95 percent of the popular vote, won over Samuel J. Tilden, with 50.97 percent of the vote; and in 1888, when Benjamin Harrison, with 47.82 percent, won over Grover Cleveland, with 48.62 percent. This can happen because all of a state's electoral votes are awarded to the winner of the state's popular vote regardless of whether the winning candidate's margin is one vote or three million votes.

Ironically, the major "defect" here, the unit-rule provision, is not part of the Constitution. This winner-take-all formula (unit rule) is merely a state practice, first adopted in the early nineteenth century for partisan purposes and gradually accepted by the rest of the states to ensure maximum electoral weight for their state in the national election. In recent years, Maine and Nebraska have modified the winner-take-all allocation rule. In these states, two electoral votes go to the statewide plurality winner and one vote goes to each plurality victor in each of the state's House of Representative's congressional districts.

But because the rest of the states use the winner-take-all rule, the electoral college benefits large "swing" states at the expense of the middle-sized states. The unit-rule arrangement magnifies the relative power of residents in large states. As political scientist Larry Longley suggests, each of the voters in the ten largest states might, by their vote, "decide not just one popular vote, but how a bloc of 33 to 54 votes are cast—if electors are faithful." Hence, the electoral college has a major impact on candidate strategy. Most recent presidential candidates have spent much of their time in the closing weeks of the campaign in crucial states such as California, Illinois, Michigan, Ohio, Pennsylvania, and New York.[48]

The smaller states are advantaged by the "constant two" electoral votes. Thus, Alaska may get one electoral vote for every 40,000 popular votes, whereas Minnesota may get one for every 200,000 popular votes. In the 1990s, "the seven states with 3 electoral votes each had a ratio of 268,000 or fewer citizens per electoral vote, while every state with 13 or more electoral votes had a ratio of 475,000 or more citizens per electoral vote."[49]

In contrast to the complexities and dangers of the electoral college system, the direct-vote method is appealing in its simplicity. Since it is based on a one-person–one-vote principle, it more clearly makes a president the agent of the people and not of the states. Governors and senators are elected by statewide direct popular voting, and they are supposed to be agents of the state. The president, however, should be president of the people, not president of the states—or at least this is what reformers say.

Most of those who favor a direct popular election of the president base their views on the undemocratic character of the electoral college. They say it doesn't treat voters equally, it discourages turnout, and it discriminates against third parties and independent presidential candidates. They point out that those living in large states and small states get more influence. They point out that Democrats in Kansas or Idaho and Republicans in the District of Columbia can be discouraged from even turning out to vote since votes are cast in bloc and the chances of their party winning are small.

Thus there are major questions of fairness and democracy that constantly get raised in the debate over the electoral college and its alternatives. A direct popular vote would eliminate undesirable and undemocratic biases. Stephen J. Wayne wrote "It would better equalize voting power both among and within states. . . . The large, competitive states would lose some of their electoral clout by the elimination of winner-take-all voting. Party competition within the states and perhaps nationwide would be increased. Candidates would be forced to wage campaigns in all fifty states. No longer could an area of the country be taken for granted. Every vote would count in a direct election."[50]

Critics of the electoral college system invariably also attack the contingency election procedure and would change the contingency procedure to a popular runoff. The current system provides that if no presidential candidate wins an absolute majority of the electoral votes (270 under present arrangements), the members of the House of Representatives choose a president from the top three vote getters. The members vote as a part of their state delegation with each state having just one vote. Thus, Alaska and California, Delaware and New York, Wyoming and Texas get one vote.

If this contingency procedure had to be used in future elections, it would obviously be a big boon for small states. But more important, critics lament, it would be an incredibly undemocratic way to select a president.

Further, the intrigues in the House might prove unseemly, especially in a three-person, three-party race.

Thus, proponents of the direct vote say it is the most forthright alternative and far preferable to the present system. They contend that voters, when they are choosing a president, think of themselves as national citizens, not as residents of a particular state. According to the American Bar Association, the present system is "archaic, undemocratic, complex, ambiguous, indirect and dangerous."[51] The direct popular vote system, proponents claim, is simple, democratic, and clear-cut.

THE CASE FOR RETAINING THE ELECTORAL COLLEGE

Those who favor retaining the electoral college usually argue their case with passion. They emphasize the wisdom of the founders and stress that ours is a federal system, not a unified, centralized, pure hierarchical nationalized system. They remind us that our nation was founded as a republic with a number of checks and balances.

Plainly the founders designed the electoral college in part to modify or check the probability that popular majorities might choose the wrong person for president. Nowadays, the principal effect of the electoral college is to give added weight to the large, highly populated, and typically industrial states, offsetting in part the greater clout of the smaller and more rural states in Congress.

Advocates for retaining the electoral college contend we should not lightly dismiss a system that has served us so well for so long.

Defenders stress that, despite some imperfections, the system works. We have not had a popular-vote loser elected by the college in well over a hundred years. The chance of this happening, they say, is not as dangerous as the likely consequences of a move to a direct vote.

Supporters of the electoral college argue that eliminating this fixture would:

- weaken the party system and encourage splinter parties, triggering numerous contingency elections
- undermine the federal system
- lead to interminable recounts and challenges and encourage electoral fraud
- necessitate national control of every aspect of the electoral process
- give undue weight to numbers, thereby reducing the influence of minorities and of the small states
- encourage candidates for president who represent narrow geographical, ideological, and ethnic bases of support
- encourage simplistic media-oriented campaigns and bring about drastic changes in the strategy and tactics used in campaigns for the presidency

An additional factor has motivated groups such as the Americans for Democratic Action and the National Association for the Advancement of Colored People (NAACP) to oppose the direct vote and defend the electoral college. One of the few ways American minorities can have an impact on the electoral process is by being the deciding factor in determining which candidate wins a given state and receives its electoral votes. In this way, urban and rural interests and blacks, Latinos, and other minorities can compete for some attention and some share of public policy. If we had direct election of the president, defenders of the electoral college claim, the necessity to take into account the needs and desires of minorities would no longer be as pressing. Candidates could campaign for the American middle and ignore various underrepresented groups. America's minorities could easily suffer. With the direct vote, the weight given to small blocks of voters would be far less than it is today.

CONSEQUENCES FOR THE TWO-PARTY SYSTEM

Proponents of the electoral college system say it minimizes the impact of minor parties and, because of the unit-rule provision (used by all states except Maine and Nebraska), encourages a politics of moderation. Under the present system losers at party nominating conventions generally abide by their party's choice. With a direct vote, these same losers could be tempted to go after the presidency anyway, hoping to force a runoff election. John Sears, campaign manager for Ronald Reagan's failed presidential bid in 1976, says that if the direct-vote method had been in operation, he would have counseled Reagan to bypass the Republican convention altogether.

Critics of the direct vote suggest that the major parties are, on the national level, only loosely assembled aggregates of state party organizations. Whatever internal discipline they possess comes primarily from their ability to make their nominations stick. They do this primarily because winner-take-all discourages disgruntled losers from launching a campaign on their own. If winner-take-all is removed, say the opponents of direct election, the major parties will lose one of their most potent weapons for enforcing their nominating decisions.

Historian Arthur Schlesinger, Jr., feared that tiny parties or single-cause candidates would be able to magnify their strength through the direct-vote scheme.

> Anti-abortion parties, Black Power parties, anti-busing parties, anti-gun control parties, pro-homosexual-rights parties—for that matter, Communist or Fascist parties—have a dim future in the Electoral College. In direct elections they could drain away enough votes, cumulative from state to state, to prevent the formation of a national majority—and to give themselves strong bargaining positions in a case of a run-off.[52]

Under the present system, the votes for these marginal parties or single-issue candidates have little or no impact, although they sometimes can help one of the major parties and hurt the other. For example, in 1976, independent candidate Eugene McCarthy may have cost Jimmy Carter as many as four states. But under the direct-vote system, each vote cast for a minority party would count. Carried over from state to state, this vote might add up the 19 percent, as the 1992 Perot vote did, or more. This would, it is argued, increase the incentives for these kinds of parties to send a message, register their strength, flex their muscles—and, it is claimed, cause the proliferation of splinter or third parties.

Proponents of the direct vote just as strongly say their plan will not undermine the two-party system. They point out that we have a direct vote in the states and that the two-party system is safely intact in the states. They also assert the two-party system is shaped and sustained not by the electoral college and the winner-take-all provision but by the election of almost all public officials in the United States by single-member districts. Citing the writings of French social scientist Maurice Duverger, they note that almost every government in the world that elects its officials from single-member districts and by plurality vote has only two major parties, whereas countries that use multimember districts and proportional representation have a multitude of parties.

They point to George Wallace's strategy (even though it failed) in 1968 under the present system. Wallace was encouraged by the electoral college arrangements to try to carry enough states in a three-way race to enable him to force the two major candidates into a deadlock and bargain with them in the House vote. In effect, the electoral college could sometimes reward regional third parties (like Wallace's American Independent Party) and punish parties with a national constituency. The electoral college system did not discourage independent candidate H. Ross Perot from waging an effective campaign in 1992, in which he captured nearly a fifth of the popular vote (but no electoral votes).

The direct-vote plan stipulates that a candidate must obtain at least 40 percent of the popular vote to win. To prevent that and cause a runoff, a third party or several smaller parties would have to win more than 20 percent of the vote and the two major parties would have to split the rest almost evenly. Says journalist Tom Wicker, "That's no more incentive, and probably less, to a minor party than its chance, under the present system, to prevent an electoral majority and throw a presidential election into the House."[53]

Finally, under the direct-vote system, organizers and supporters of third parties would be able to take votes from the major-party candidate closest to them in policy convictions. Thus, their party could become a spoiler party and thereby ensure the victory of the least preferred party.

Some contend that with the direct vote we would eliminate the requirement that pluralities be created state by state. This could easily undermine the remaining basis of party competition. The likely result is that more voters would move into the primary of the majority party. Fewer would move into the party of the minority. The minority party would become more extreme until it eventually disappeared.

CONSEQUENCES FOR DEMOCRATIC LEGITIMACY

The direct-vote method could easily produce a series of 41-percent presidents, thereby affecting the legitimacy of the winner. Lincoln was the only president with a vote that small, but he wasn't on the ballot in several states. In an era when confidence and trust in the national government have eroded, the direct vote would almost ensure that we would have minority presidents—persons who won with less than 50 percent of the vote—most of the time. We have had sixteen minority presidents, yet the present electoral college system is a two-stage process in which the popular votes are converted into electoral votes. In every election this has the effect of magnifying the vote margin of the winner, so much so that only once in more than a hundred years has a president received less than 55 percent of the electoral college vote (Wilson received 52 percent).

No electoral system is neutral. Trade-offs are very much in evidence in this continuing debate. Electoral college supporters clearly want to discourage third parties and protect a party system that pits one moderate-centrist party against another moderate-centrist party. Moreover, electoral college sympathizers strongly prefer the idea of cross-sectional concurrent majorities. Direct popular vote advocates clearly want all votes to count equally and the candidate with the largest number of popular votes to win. Both schools are concerned with the legitimacy, or public acceptance, of the election process and outcome. As political scientist William Keech has noted: "A decision about which system of electing the president to prefer depends on values, on priorities among those values, and on estimates of the likely consequence of change."[54]

There are liabilities or likely adverse effects with both the direct-vote and the electoral college systems. Opponents of the direct vote may have overstated their case. The dire consequences they foresee are probably exaggerated, and their view about the role the electoral college plays in maintaining and ensuring the vitality of federalism doubtless gets overstated. Still, they pose enough uncertainty about the possible undesirable side effects of a national direct vote that even the most ardent and democratic populist should pause and consider yet other alternatives, other possibilities.

THE NATIONAL BONUS PLAN COMPROMISE AND OTHER ALTERNATIVES TO THE ELECTORAL COLLEGE

In the late 1970s, a task force created by the New York-based Twentieth Century Fund (a nonpartisan, not-for-profit foundation) proposed a compromise method of presidential selection that attempted to deal with some of the major problems inherent in both the electoral college and direct election plans.[55] The compromise plan retains the existing 538 state-based electoral votes but adds a national pool of 102 electoral votes that would be awarded on a winner-take-all basis to the candidate who wins at least 40 percent of the popular vote nationwide. There would thus be a combined total of 640 state and national electoral votes, and the candidate with a majority would be the winner. As a consequence, the existing federal bonus in the current electoral college system (of two electoral votes for each state plus the District of Columbia) would be balanced by this national bonus given to the nationwide popular winner.

This National Bonus Plan, a reform alternative to the existing system, would virtually eliminate the major flaw in the electoral college arrangements. That is, the new system would make it unlikely for the popular-vote winner to lose the election, as happened in 1888 and could have happened in 1960 and 1992.

Because a constitutional amendment is required to establish the National Bonus Plan, a few other changes for simplicity could also be incorporated. Under the National Bonus Plan, there is no need for the office of elector or for the electoral college. The practice of having designated electors would be eliminated and all electoral votes, those now assigned and those proposed under the National Bonus Plan, would be allocated automatically on a winner-take-all basis to the popular-vote winner in each state and in the nation as a whole. Thus, the "faithless elector" problem would be eliminated.

In the unlikely event that no candidate receives a majority of the total electoral vote under the National Bonus Plan, a runoff would be held between the two candidates receiving the most popular votes. This runoff would be held within thirty days of the first national election, and the candidate who won a majority of electoral votes would be elected president. A reasonable alternative would provide for a direct-popular-vote runoff.

These innovations should bring about an improvement in the fairness, and an enhanced public acceptance, of the electoral process without being a drastic or sweeping restructuring of the system. The plan retains some familiar features such as the federal principle, yet it eliminates or minimizes most of the problems that plague the existing presidential election process.

Compared with the existing system, this set of reforms would:

- go a long way toward ensuring the election of the candidate with the largest number of popular votes
- reduce the possibility of a deadlock and make the contingency election procedure more representative
- eliminate the so-called faithless elector
- enhance voter equality
- encourage greater voter turnout

Against direct election, it would:

- avoid a proliferation of candidates and help maintain the two-party system
- preserve the federal or cross-sectional character of the presidential election process
- lessen the likelihood of minority presidents
- lessen the likelihood of runoff or second elections
- lessen the likelihood of regional or sectional candidates emerging as major candidates

Thus, this proposed system would remedy most of the problems of the electoral college system and avoid many of the potential problems and risks that might be encouraged with a direct-election process.

Congressman Jonathan Bingham, a New York Democrat, introduced the National Bonus Plan as a constitutional amendment in the Ninety-sixth Congress and it received several editorial endorsements. However, it won little popular support and was soon forgotten.[56]

The National Bonus Plan may bring to a common position those who advocate direct election of the president and those who support the electoral college. Direct-vote advocates, who believe reform should go all the way to a one-person–one-vote system, may be harder to win over than defenders of the electoral college. Yet one leading direct-vote advocate called the National Bonus Plan an innovative proposal that would break the long-standing logjam on electoral college reform. Though he favors a simple direct-election amendment, he concludes, "Between the existing system, with all its perils, and the National Bonus Plan, I find the National Bonus Plan infinitely preferable."[57]

Sooner or later there is going to be another presidential election in which one candidate wins the popular vote but loses the presidency. Or some election will once again have no majority winner of electoral votes and cause the election to be thrown into the House of Representatives for resolution. At that time the furor and heated emotion of the moment may cause a rush to change the system in a radical or overly simplistic way. The National Bonus Plan should at such a time be dusted off and given consideration as a reasonable alternative.

There are, to be sure, obvious drawbacks to this National Bonus Plan. First, it is complicated and is a bit of a Rube Goldberg device. Smaller states would have their vote diluted a bit. Southern states, in general, have a lower turnout and consequently in a nationwide race would lose some voting power to states such as Minnesota, which has high turnout. At least one critic worries that the National Bonus Plan could be deficient in that it might allow a highly sectional base of voting support to win the 102 bonus votes and thereby win the presidential election. The National Bonus Plan is also, in some respects, a direct-vote scheme that is only partially in disguise. Yet, its defenders note, it would help to preserve centrist and two-party politics in the United States, and that might be worth all the trouble.

But the National Bonus Plan is merely one of several alternatives that should be considered. Students of the presidency and the Constitution should examine other possibilities as well, ranging from modest reforms such as eliminating electors, to allocating votes according to some form of proportional representation or approval voting method.[58]

Careful students of presidential elections should begin by trying to develop as rich an understanding as possible of both the existing process, generally known as the electoral college system, and its chief rival, the direct, one-person–one-vote plan.

What is needed is the same level of imagination, focus, and long-term appreciation for posterity that the best of our constitutional framers displayed in Philadelphia in the late eighteenth century. In some ways, the constitutional convention of 1787 is still in session. The debates continue. The delegates today are the legal scholars, political writers, judges, legislators, and informed citizens who care enough to reinvent, modify, or reconfirm our political institutions and processes.

CONCLUSION

Our presidential selection process is neither tidy nor easy to understand. It has evolved in varying and often unpredictable ways since the framers met in Philadelphia. It seeks to achieve a variety of often contending and conflicting purposes. Yet it is one of the grand compromises so often found in the structure of the American political system that seek to paper over regional, political, and even ideological differences. The old tensions of how strong a central government we really want and of how strong and powerful a central leader we are willing to tolerate are never far behind the scenes. Then, too, questions of just how much democracy we really want and how much we actually trust the judgment of the average citizen are often involved in our attitudes about the presidential election system. Finally, the whole selection process is part of the continuing American quest to preserve a politics of compromise, coalitions, moderation, and pragmatism. The politics produced by such a quest is not always pretty, is not always wholly democratic, and is full of puzzles and paradoxes.

CHAPTER 3

Evaluating Presidential Performance

Great presidents possess, or are possessed by, a vision of an ideal America. Their passion is to make sure the ship of state sails on the right course. If that course is indeed right, it is because they have an instinct for the dynamics of history. . . . Great presidents [also] have a deep connection with the needs, anxieties, dreams of the people.

Arthur M. Schlesinger, Jr., "The Ultimate Approval Rating,"
The New York Times Magazine, December 15, 1996

Who were my heroes among past presidents? Harry Truman. Without a doubt I believe he was the best president in my lifetime. He was honest, bold and set a moral standard. Then I'd say Woodrow Wilson, because he was a basic definer of moral leadership. I'd add Teddy Roosevelt in the environmental field. Then I guess everyone's choice would have to include Lincoln. Also Thomas Jefferson.

Jimmy Carter, seminar discussion, *U.S. Air Force Academy,*
August 26, 1985

Lincoln never ceased to be a common man: that was his source of strength. But he was a common man with genius, a genius for things American, for insight into the common thought, for mastery of the fundamental things of politics . . . for judging men and assessing arguments.

Woodrow Wilson, *Mere Literature and Other Essays* (1896)

A nation reveals a lot about itself by the leaders it remembers and honors, by those it rates as great and by those it forgets. American historians and biographers agree that Lincoln, Washington, and Franklin D. Roosevelt were our great presidents, followed in the distance by Jefferson, Wilson, Jackson, and possibly Theodore Roosevelt. Presidents Polk and Truman also win favorable approval.[1]

The average American is ahistorical. Americans like presidents whom they know or at least about whom they have first-hand stories. Thus the recent Gallup poll popular favorites are John F. Kennedy, Ronald Reagan, Harry Truman, FDR, and to a lesser extent Dwight Eisenhower.[2]

Both experts and citizens either forget or want to forget Buchanan, Pierce, Harding, Fillmore, Hoover, and Andrew Johnson, among others.

Americans were tough on King George III, and we continue to be rough in how we treat our own governing elites. We love our country but we have never loved our government. Presidents, as symbols of government, are both highly scrutinized and regularly blamed for the ills of government. They have to prove themselves or else pay an exacting penalty.

The most glaring paradox in how we evaluate presidents is the confused standards by which we judge leaders. While we are "results oriented" we are also deeply concerned about means. We demand that presidents succeed, and criticize them if they do not. But they must not demand too much of us, push the system too far, call for too many sacrifices, or trample on our rights and liberties.[3] Like all people everywhere, we want to have our (political) cake, and eat it too.

We want a lot of leadership, but we are notoriously lousy followers.

What follows is a discussion of how what we expect of presidents today differs in many ways from what the framers wanted in the 1780s. What we want from presidents has changed over time. Moreover, as discussed in Chapter 1, we not only hold clashing expectations about our presidents but we have redefined their job in such a way that it is nearly impossible for anyone to excel in it.

WHAT DO WE EXPECT OF OUR PRESIDENTS?

Alexander Hamilton wrote in 1788 that the office of president would never fall to the lot of anyone not in an eminent degree endowed with the requisite qualifications. Indeed he predicted the office would be regularly filled by individuals noted for ability and virtue. This was too optimistic a prediction; our presidents have varied dramatically in skill and character. Moreover, what we expect of a president has varied over the course of our nation's history. The framers of the Constitution were not of one mind, and the final constitutional provisions were compromises, often even guesses. They left much to be worked out.

One expectation, however, has remained constant over the years. We want prudent and intellectually honest leaders who exercise their powers to the fullest when emergencies arise or when the country or the Constitution is threatened. We want presidents who will place the country and the Constitution ahead of their own personal or partisan interests. At the same time, the framers and citizens worried that a president capable of exercising robust executive power in a crisis situation could also become a demagogue and the worst kind of tyrant. Hence, checks, both formal and informal, would always be needed to safeguard liberty. An appreciation of this paradox has given rise to an ambivalence toward presidents and presidential power.

As America expanded presidential powers and responsibilities, and as we moved away from Whig notions of limited presidential leadership, this ambivalence only grew more intense. We yearn for effective leaders to become president and we want them to have the powers and resources necessary to do the job. Yet events throughout the twentieth century again and again raised concern about the abuse of power and the possible irresponsible actions of an even moderately empowered president. We have come to understand that presidential power can be used to achieve both noble and ignoble ends. "What seems like a paradox at first now acquires logic."[4] Limited tenure, as now required by the Twenty-second Amendment, and additional statutory checks and accountability procedures such as inspectors general, independent counsels, and ethics codes are some of the devices we now use to try to keep strong power from becoming irresponsible power.

The growth of presidential power came in fits and starts over our first 140 years. Since Franklin D. Roosevelt, the office and its power have steadily grown. It did not grow because of presidential fiat or reckless abandonment of earlier constitutional principles. It occurred because the times changed, popular expectations changed, and Congress passed laws and delegated increased responsibilities to presidents, especially during crisis periods. As the United States emerged as a world economic and military power, presidential power grew.

WHAT DID THE FRAMERS EXPECT OF PRESIDENTS?

The framers of the Constitution were torn between their fears of executive tyranny and the need for executive power. With King George III and royal governors in mind, they feared the possibility of an arbitrary and ruthless leader. However, even the most vigorous champions of a strong executive insisted that the president would not become a monarch. Alexander Hamilton, in *Federalist* 69, compared the powers of a president with those of a king, and argued persuasively that presidents would possess far less power than the English monarch.

The framers were deliberately vague about the precise character of presidential powers, yet they provided the office with a potential for growth. Over the short run, they wished a president to be no more than an equal partner in a triumvirate, to be restrained by the judicial and especially by the legislative branch. They took care, in drafting the Constitution, to construct a governmental system that did not depend on a strong, popular leader. Even to admit the need for strong leadership was to leave open the possibility for the exercise of discretion and thus power. Unchecked power was thought dangerous. According to many historians, the framers, in a burst of intellectual cleverness, devised a constitutional system that made great political and presidential leadership all but impossible, except in times of crisis. Yet, Alexander Hamilton believed prosperity would be realized not from the absence of power, but from the presence of firm decisive executive power and leadership.

The framers hoped for a wise and virtuous statesman to be president, a person who would work closely with Congress and respond to the "sense of the community." Implicit, if not explicit, in their early expectations was the notion that the president, as well as members of Congress, would be preoccupied with what ought to be done rather than with what shortsighted temporary majorities might desire. They thus sought to insulate the president from the pressures of popular whim, and likewise to protect the other branches from the dangers inherent in a president armed with the support of the masses.

The reason for providing distance between the people and the presidency is clear. It was to allow ample scope for leadership and statesmanship. In a representative democracy, or republic, public opinion would not always be identical with the public interest. There is more than a hint in Hamilton's conception of national leadership that sometimes a president would have to have the power "to do what the laws would do if the laws could foresee what should be done, as well as the power to make exceptions in the execution of the laws, exceptions governed by a judgment as to whether it would be good to apply them or not."[5]

Hamilton's prediction came true more quickly than he had imagined. President George Washington set a few precedents for unilateral executive action not exactly envisioned by the framers. He saw to it that he alone would lead the executive branch of government. He issued the Neutrality Proclamation of 1793 without the consent of Congress. He withheld information from Congress he thought should not be disclosed for reasons of national security. Washington seldom assumed these leadership responsibilities over the objection of the other two branches of government or the people. Congress willingly conceded to Washington most of the executive powers he exercised, especially those in foreign policy matters. By making the president commander-in-chief and allowing him to appoint ambassadors and to negotiate treaties, the framers provided important foreign policy and symbolic roles for the newly invented presidency.

Once Washington established these early practices, however, executive assertiveness was somewhat limited. The monarchial fears of the anti-federalist opponents of the Constitution were not realized. With only a few exceptions, Adams, Jefferson, Madison, and Monroe regularly deferred to the Congress. In essence these early presidents did so because they believed the voice of the people and their duly elected representatives in Congress must be heeded. The framers, and the American voters of the day, did not expect regular and explicit domestic and economic policymaking leadership from presidents, only administrative efficiency. Washington, in taking executive action yet always acknowledging the power and importance of Congress, set the prudent example.

What was expected of our early presidents was an executive role with neither tyrannical nor hereditary powers, restrained by representatives of the people yet able to protect the people, at least on occasion, from their own representatives. Under George Washington, Americans in the 1790s got what they wanted and more. Although Washington was effective more for who he was and how he did things than for what he did, widespread acceptance of his decisions suggest he assumed only necessary executive powers.

WHAT WAS EXPECTED IN THE NINETEENTH CENTURY?

Even with the unanticipated evolution of political parties, the early presidents did not have to conduct presidential campaigns as we know them today. Due to the natural restrictions of travel and the absence of more modern media tools, they remained distant from the average citizen. To be sure, presidents traveled about the country for occasional ceremonial visits. The early presidents also held weekly social gatherings and were accessible even to casual visitors in the nation's capitol. Many presidents deemed it a requirement of the job to take an occasional "bath in public opinion," to paraphrase Lincoln.

Presidents in this period, however, were not expected to provide popular leadership. Policy suggestions, should not be made at the grass-roots level, but at the representative level. Representatives would be better able to make important decisions than the average citizen. Being physically removed from the workings of government, the general public had fewer expectations of their representative, and of the president. Moreover, the impact of the national government on the average citizen was decidedly less than it is today. Not until later in the nineteenth century did the notion take hold of a president as a direct representative of the American people. Even then, the general public asked less and expected less than we do today.

The typical nineteenth-century president was expected to be a "constitutional executive," a dignified presider over the administrative branch who would not tread on the responsibilities of the other branches. Although the nineteenth century eventually witnessed remarkable industrial growth based on individual initiative, most of that growth operated independently from any governmental restrictions. Freedom *from* government seemed to permeate all aspects of economic and social life.

The doctrine of laissez faire was widely accepted. Government might help, yet only in a limited way. Public expectations of the president were not geared toward greatness, but toward efficiency in those limited responsibilities assigned to the presidency.

Other factors also made the public expect less of presidents in the nineteenth century than today. Presidents before McKinley (1897–1901) were not expected to make the United States a world power. Thus, when we look back and judge presidents, we rate them largely according to domestic affairs, forgetting they had fewer responsibilities than twentieth-century presidents. Presidents in the nineteenth century were also not expected to contend with, or preside over, a large and complex federal bureaucracy or such a large, culturally diverse, and urbanized population. The evolution of increasingly complex problems, a world steadily developing into bipolar spheres of influence, and America's constant growth in power during the twentieth century, led to increased expectations of modern presidents. To say that nineteenth-century presidents had less to do does not mean crises and emergencies did not arise. Assuredly the Civil War was the greatest crisis the nation and our constitutional system ever faced. Rather it means the public seldom viewed presidents the way we do today.

With the exception of Lincoln's crisis leadership, the presidency was rarely a dominant and often not even a particularly visible leadership institution in America. For much of the first half of the century, government was dominated by men in Congress such as Daniel Webster, Henry Clay, and John Calhoun. Andrew Jackson, forceful chief executive that he was, did not approve of a forceful national government prior to taking office. Abraham Lincoln, before he was elected president, counseled against a strong presidency. And save in matters of war or security, President Lincoln regularly deferred to Congress and his cabinet.[6]

Not until Theodore Roosevelt did the general public consciously expect a president to provide affirmative leadership, in minimizing the detrimental side effects of the industrial revolution. Populist manifestoes of the late nineteenth century began to change attitudes, as people turned to popular movements, popular political figures, and the central government as possible remedies for their economic problems. The crusade to elect William Jennings Bryan was illustrative of this new development. Attitudes toward presidents and presidential leadership began to change.

Teddy Roosevelt used the "bully pulpit" of the presidency to set the nation's programmatic agenda. And in doing so over time he also expanded expectations for the office.[7] Woodrow Wilson echoed and extended Roosevelt's conception. Wilson wrote that the presidency should respond to and help enact the progressive sentiments of the people. Since the country had expanded in so many ways, now the presidency must expand. The people, Wilson suggested, should look to their presidents to promote the conscience of the nation.[8]

Were Roosevelt and Wilson responsible for changing the expectations of the presidency, or were they merely in office when the public began to expect and demand more? Probably more the latter than the former, but doubtless both factors were at work.

PUBLIC ATTITUDES TOWARD PRESIDENTS IN RECENT TIMES

Public attitudes toward the presidency are subject to predictable cycles. After strong presidents, who are often crisis or war presidents, the public yearns for a lessened presidential role, a return to normalcy. After a weak leader, and especially after a series of weak leaders, we yearn for strength. Presidents live in the shadows of their immediate predecessors, and each president pays for the sins of a predecessor. After the presidencies of Theodore Roosevelt and Woodrow Wilson, the Harding-Coolidge-Hoover administrations assumed a more passive posture. Coolidge once said nine out of ten problems brought to his office could be safely ignored and did not require his leadership attention. After Franklin Roosevelt served there was a backlash; the public supported the Twenty-second Amendment in an apparent effort to punish the institution (as well as reaffirm the two-term tradition). After the Kennedy-Johnson-Nixon administrations there was another backlash to presidential power, as Ford and Carter provided a return to normalcy and rectitude in the White House. However, Ford and Carter paid for the political excesses of Nixon by being severely constrained by a more energized Congress; later Bush and Clinton paid for the economic excesses of Reagan by being constrained in policy terms.

Yet seldom does a president actually relinquish any powers associated with the office. The more passive ones may not use the powers in a vigorous way, yet even the three Republican presidents of the 1920s (Harding, Coolidge, Hoover) attempted to protect the powers of the office they inherited by "consciously trying to steer a middle course between the extreme activism of Wilson and the passivity that had destroyed Taft."[9]

President Franklin D. Roosevelt set a precedent for executive leadership that remained popular as the national government became more and more involved in everyday life. Roosevelt established a pattern that most chief executives would inevitably follow: proposing legislation, lobbying to get proposals enacted, rallying public opinion in support of his measures, creatively using the newly invented electronic media, and inserting himself into on-going international diplomatic negotiations.

In the beginning of the Republic there was at best grudging acceptance that a president would intervene in the affairs of Congress. Now it is taken for granted that presidents regularly initiate and seek to win support for their measures. The Roosevelt performance irretrievably altered people's views of presidents and the presidency. General Dwight Eisenhower added the reputation of a national hero to the luster of the office. Then, partly due to the youthful enthusiasm and glamor of the Kennedy administration, people and scholars were captivated by the magic of the office. The mystique of the "textbook presidency" prevailed. After Lyndon Johnson won a landslide reelection and enjoyed a two-year honeymoon with Congress, he was able to pass vital civil rights legislation and he got the Great Society programs off to a rousing start. The power of the modern presidency appeared virtually unlimited. Public and scholarly celebration of the presidency reached a peak.

Then came Vietnam, increased secrecy and deception, Watergate scandals, obstruction of justice, and an increasingly frustrated American public. For a period, this dissatisfaction was reflected by a profound anti-Washington sentiment. Public approval of presidents declined and presidential expectations were briefly lowered.

Presidents Johnson, Nixon, Ford, and Carter were criticized and found wanting. Yet after a brief infatuation with the idea of turning to Congress for national leadership (notably in the 1973–80 period), Americans revived their demand for decisive and assertive presidential leadership. Ronald Reagan won election in part because he played upon this yearning. "Reagan's apparent self-assurance, good humour, decisiveness and faith in the eternal verities seemingly struck a responsive chord among a demoralized public," according to two public opinion specialists. "His presentation of self, whether natural or stage-managed, communicated a sense of pride in the nation and its past."[10] Reagan's likability, his talent for speaking plainly, his ability to convey his love of the country, and his ability to give the people he was talking to, whether an individual or the public at large, the impression of liking them made him a rare political phenomenon. Reagan called for and promised a more powerful presidential performance. He charmed and often entertained the nation. Reagan and his hand-picked successor, George Bush, also were able to celebrate the end of the Cold War, yet the national debt and annual budget deficits soared and Bush was denied reelection in 1992.

Americans in the post-Vietnam and post-Watergate years may often dislike individual presidents, yet they still express confidence in the power of the presidency. Presidents are expected to be symbols of reassurance and to possess extraordinary nonpolitical personal qualities long associated with the legendary hero presidents such as Washington, Lincoln, the Roosevelts, and more recently John F. Kennedy.

Are Americans realistic? Rather than becoming more realistic about the presidency, perhaps we have become only more skeptical about the individuals we expect to perform presidential responsibilities.

QUALITIES AMERICANS LOOK FOR IN THEIR PRESIDENTS

We appear to have a tough, unwritten code of conduct for our nation's chief executive. We demand much. Part of this comes from our inflated image of what "the great" presidents did. Our selective memories about the past glories and victories under our favorite presidents result in our holding incumbents to unusually high standards. A president is expected to be an agenda setter—"greater than anyone else, but no better than anyone else," writes John Steinbeck. "We are related to the President in a close and almost familial sense; we inspect his every move and mood with suspicion. . . . We subject him and his family to close and constant scrutiny and denounce them for things that we ourselves do everyday."[11]

According to polls, how we evaluate presidents varies with the times and depends partially on who has recently been president and on who is now in office. Polls in the late 1950s and early 1960s showed the general public wanted strong presidential leadership. They also wanted checks on presidents. People recognized presidents had to play an increasingly central role as a foreign policy leader. People had considerable faith in the dedication of presidents to the public interest.[12] In the mid-1970s, that faith in the notion that "the president knows best" sharply diminished, only to increase again in the mid-1980s.

Polls in the post–Cold War era are more ambiguous, perhaps reflecting a confusion about the role of the national government. In 1992 voters rejected George Bush. In 1994 voters rejected the Democratic leadership in Congress and elected more than seventy Republicans to the House, enabling them to take control of the House of Representatives. In 1996, in large part because of a sound economy, the public voted to divide power, keeping Clinton in the White House but ensuring he was well checked by a clear Republican majority in both chambers of Congress.

Experts usually rate experience, vision, and intellectual capacity over honesty. But this is not the case with the average person. *Honesty* and *credibility* are top qualities the public wants in a president. The average voter wants a leader whose words he or she can both understand and believe and whose intentions are trusted. When George Bush enjoined the public to "Read my lips, no new taxes," and later was compelled to break that vow, the public was unforgiving. Likewise, when candidate Clinton promised a middle-class tax cut but broke that vow, he lost political capital as well as credibility.

Surveys show the public rates honesty as the most important quality a president should have, followed, in descending order of importance, by compassion, intelligence, toughness/decisiveness, and decision-making ability.[13] Experience and knowledge of economic issues are also cited frequently (see Tables 3.1 and 3.2).

HOW AMERICANS EVALUATE INCUMBENT PRESIDENTS

Do presidents shape public opinion or are they slaves to public opinion? This debate is as old as it is unanswerable. Long before the advent of polls, President Lincoln said public sentiment is everything. With public support, nothing can fail; without it, nothing can succeed. Lincoln, who was fortunate to live through the most trying of times without Gallup and

TABLE 3.1
Assessing Candidates: Criteria and Skills Sought

Q. What is the single most important standard by which you judge a presidential candidate?

Honesty	41%
Political beliefs same as own	37
Moral character	15
Party affiliation	2
Personal appearance and presence	1
Don't know/No answer	3

Q. What is the single most important skill you look for in evaluating a candidate's credentials for the presidency?

Experience in national-level politics	32%
Knowledge of economic issues	30
Experience in foreign affairs	20
Experience in business or other non-political fields	10
Other	2
Don't know/No answer	5

$N = 1001$.
Source: "The American Public's Experience with Recent Presidential Elections," a national survey of public awareness and personal opinion (New York: The Hearst Corporation, 1988). Interviews were done in late October 1987. Also see *Times Mirror Center for the People & the Press*, "Voter Anxiety," *Los Angeles Times*, November 14, 1995, p. 22; and *The Roper Center*, "Americans Rate Their Society and Chart Its Values," *The Public Perspective* 8(2) (February/March 1997).

Harris polls reporting his considerable public disfavor, would warn us today that in evaluating presidents we should take care not to be too profound, for the sources of positive or negative evaluations are often superficial. He would doubtless also say the influence between a people and their president is reciprocal: presidents must be both leaders and followers of public opinion.

With the coming of professional polling organizations in the 1930s, we began to accumulate extensive data about whether people approved or disapproved of presidents. The Gallup Poll organization surveyed the general public on this question occasionally in the late 1930s and regularly under Harry Truman. Several patterns are evident as one examines the ups and downs of presidential approval over a fifty-year period.

- Presidents are more popular in their first year than they are later.
- Presidential approval ratings often rise during the "honeymoon" early months in office.
- Over time, however, approval ratings usually drop, with a number of zigs and zags caused by specific events, conditions, or crises.
- Presidents typically lose popularity during mid-term elections when the "guns" of the opposition party are aimed directly at the White House.
- Domestic crises typically cause approval evaluations to drop.

- International crises, regardless of what a president does, typically cause us to rally around our president in the short run.
- Republican presidents are typically less approved at the beginning of their presidencies than Democrats who served just before them. (Eisenhower, Nixon, Reagan)
- Americans give an initial high approval evaluation to vice presidents who are thrust suddenly into the presidential job at a time of death or emergency. (Truman, LBJ, Ford)
- Presidents often enjoy a rebound late in the first term—especially when they begin to be evaluated in comparison with likely opponents.
- If a president is reelected, the pattern often repeats itself, although approval ratings drop more quickly and tend to be lower than during the first term.
- Post-Watergate presidents have started out with lower approval ratings.

While these patterns are persistent, they are not inevitable. It is premature to be definitive when we are examining just ten presidencies.

One of the more confusing aspects of presidential accountability is the way the American people find it convenient to blame presidents for a whole range of problems, regardless of whether the problems have been subject to presidential influence and solution. Presidents are the focus of attention and receive blame, and sometimes praise, for events over which they have little or no control. We generally withhold our applause when a president's work is good, yet we seldom fail to hiss blunders. No matter what presidents do, their popularity generally declines. Reagan's Chief-of-Staff Howard Baker once commented, "the opportunities to get in trouble in this place are absolutely limitless." When news is good, the president's popularity goes down or stays about the same; when news is terrible, popularity merely goes down faster and farther. The unfortunate fact is that the majority of the public is politically inattentive and will notice only mistakes, while failing to appreciate successes. The decline in approval of the president is partly a function of the inability and unlikelihood of a president to live up to the buildup he receives during the presidential honeymoon (or perhaps the hopes and promises he raised during his campaign). Heightened expectations are invariably followed by disappointment. Sometimes the disappointment warrants despair and retribution; often unfairly people turn on a president almost as if he were the sole cause of everything that is wrong in their lives.

One of the most difficult paradoxes for a president to deal with stems from the public's desire to have contradictory things of government. The public demands budget cuts yet not service cuts (for them at least); we want tax cuts yet also demand that entitlements be fully funded. In a recent USA Today/CNN/Gallup poll, 82 percent of the sample reported that it was a "major concern" that the Republican majority in Congress "reduce the budget." At the same time, 69 percent responded that it was a "major concern" that the Republicans provide a "tax cut for most Americans."[14]

Most presidents concede that those who use presidential power must understand that by using it they may dissipate it. No president has had as much power in his last year in office as he had in his first. Several presidential resources wane as the years roll by. The presidential honeymoon comes to an end. Prominent politicians who are rivals for the office begin, after a short grace period, to attack a president's positions, partisan followings crystallize, and the ranks of those who disapprove of the president begin to swell. Presidential promises, at least some of them, go unachieved. Presidential achievements often fail to get acknowledged, or at least remembered. Factions in the president's own party inevitably develop. Press criticism increases. And persons whose pressing claims go unheeded become disaffected.

A vicious cycle can develop. As popularity wanes a president loses valuable public support, and he may eventually find himself in an ever-deepening hole of credibility. Unpopular chief executives are far less able to influence public opinion. Such a president may often avoid divisive issues in favor of "not making waves." Instead he could engage in short-term "quick fixes" to maintain his popularity or stage media events and photo opportunities in hopes of appearing to be effective or successful. And the more presidents strive to achieve favorable evaluations and thus protect their public support, the more they engage in short-term transitory policies—policies that may not be good for the country in the long run. We also "have the paradox that governments following public opinion polls," writes Samuel Kernell, "begin to look more and more incompetent. As they look incompetent, confidence in government begins to disintegrate."[15]

The frequency and duration of these dynamic ebbs and flows of approval between the president and the public are not well understood. Virtually all presidents lose popularity during congressional elections when the opposition party takes the offensive against the incumbent administration, and most presidents' popularity declines considerably in the public-opinion polls over time (Table 3.2). Eisenhower, however, was generally immune to this cyclic factor, perhaps due to his "don't rock the boat" demeanor in office. Legislative activists like Truman, Johnson, and Clinton suffered most in the polls, whereas Eisenhower and Nixon, until Watergate, enjoyed more stable support. Is more expected of Democrats? Johnson enjoyed a comfortable margin of American trust for at least two and a half years, then suddenly it was irretrievably gone. Reagan's popularity declined in late 1982 due in large part to the worst recession in post–World War II history. Yet a year later his approval ratings climbed. And he went on to win an impressive reelection victory in 1984, sweeping forty-nine states. His ratings climbed further in 1985 and early 1986 before the Iran-Contra affair and other setbacks brought increased disapproval.

TABLE 3.2
Average Yearly Presidential Approval Rating, 1953–1996

Year	Approval	Year	Approval	Year	Approval	Year	Approval
1953	68%	1964	75%	1975	44%	1986	61%
1954	66	1965	66	1976	49	1987	48
1955	71	1966	50	1977	63	1988	57
1956	73	1967	44	1978	45	1989	64
1957	64	1968	43	1979	38	1990	68
1958	54	1969	63	1980	42	1991	71
1959	64	1970	58	1981	58	1992	40
1960	61	1971	51	1982	44	1993	49
1961	76	1972	58	1983	44	1994	46
1962	71	1973	43	1984	56	1995	48
1963	65	1974	36	1985	61	1996	49

Source: The Gallup Organization, Princeton, New Jersey.

George Bush rose high in the polls during and after the war to liberate Kuwait in 1991, yet he suffered the greatest loss of public approval any president has ever suffered over the next year and a half. And Bill Clinton, elected with only 43 percent of the popular vote, had difficulty winning majority approval—even his reelection in 1996 was with just 49 percent.

Legislative activists suffered great declines in public approval, in part because they were outspoken on domestic issues. (Reagan was often outspoken as well, yet he always portrayed himself effectively as reasonable and willing to compromise.) Perhaps the expectations generated by these administrations were more difficult to fulfill, and the greater heterogeneity of the Democratic party made opposition to any presidential action more likely. Each time presidents are forced to act on such controversial domestic issues as racial integration, health care reforms, or the separation of church and state they risk losing the support of intense minorities. How does a president please both conservative southerners and blacks, Baptists and Catholics, farmers and consumers? The temptation to become the people's source of inspiration for every worthy policy can overcommit the political resources, as well as the institutional prestige, of the presidency. Presidents learn, sometimes after the fact, that they must follow public opinion as well as lead it.

The public's rhythmic "issue-attention cycle" tends to restrict the president's leeway in seeking solutions to the nation's domestic problems. Economist Anthony Downs persuasively argues that "American public attention rarely remains sharply focused upon any domestic issue for very long, even if it involves a continuing problem of crucial importance to society. . . . Each of these problems suddenly leaps into prominence, remains there for a short time and then, though still largely unresolved, gradually fades from the center of public attention."[16]

Another problem occurs when a president is forced to act on a controversial issue; he is likely to create intense and often unforgiving opponents among his former supporters. President Clinton's efforts to allow gays to serve in the military galvanized opposition from many members of Congress in his own party and somewhat undermined his desire to be seen as a "New Democrat." Even without taking action, a president can come under attack from those who voted for him but find themselves neglected in the distribution of federal patronage or contracts or otherwise disappointed by their inability to influence future presidential policies. Because modern press and television coverage gives so much emphasis to the presidency, it serves to quicken and intensify these reactions. Precisely because a president can gain immediate publicity, he is expected to communicate his views quickly and solve problems without delay. Precisely because he is supposed to be a shaper of public opinion, he is expected to inspire the country to great causes.

Leading the public is believed to be one of the key sources of power for a president. But the ability to move the nation and generate public support takes time and effort; it is not automatically conferred upon assuming office, nor is it always translated into political clout. Presidents spend considerable time and effort trying to build popular support for both themselves and their programs. Their efforts usually come to little.[17]

A president with popular support (and other skills to complement popularity) can exert pressure on Congress and is more likely to get his program passed.[18] As a result, presidents are keenly aware of fluctuations in popularity, and routinely engage in efforts aimed at dramatizing themselves in the hope of increasing their popularity.

Is popularity a source of power? Many scholars believe that popularity is indeed a convertible source of political power; that presidents can convert popularity into congressional votes, better treatment by the media, and less criticism by political opponents. While hard evidence to support these claims is scarce, one must remember that in politics, *perception* often counts as much as and sometimes more than *reality*. If a president can convince Congress that he is popular and that to defy him is politically dangerous, the battle is half won. In his first two years in office this is precisely what Ronald Reagan was able to do. Democrats in Congress were afraid Reagan's popularity might really have been as wide and as deep as Reagan claimed, and thus many in Congress were more inclined to vote for his legislative proposals. This perception of power strengthened Reagan's bargaining position.

Yet we must not make too much of Reagan's or any other president's ability to move public opinion. After all, on one of the issues nearest and dearest to Reagan's heart, aid for the rebels in Nicaragua (the Contras), Reagan failed, despite Herculean efforts, to sway the public or Congress to support his goals. In the end, Reagan was forced to act unilaterally, but he was never able, in spite of his much deserved sobriquet "the great communicator," to convince the public they should accept his proposals.

Can presidents convert popularity into power? Political scientist George Edwards is skeptical. After a statistical study of the relationship between the two, Edwards concluded that while popularity may help, there is no certainty that popular presidents will get their legislative programs through Congress.[19]

In *Presidential Power,* Richard Neustadt suggests public support can be a helpful source of potential power. He suggests too that its absence can weaken a president: "The weaker his apparent popular support, the more his cause in Congress may depend on negatives at his disposal, like the veto, or 'impounding.' He may not be left helpless, but his options are reduced, his opportunities diminished, his freedom for maneuver checked in the degree that Washington conceives him to be unimpressive to the public."[20]

Public support may empower presidents but it is not an easy thing to attain. And that is the challenge for presidents, to gain public support *and* convert it into political power.

But how can presidents *lead* public opinion? Franklin Roosevelt said "All our great Presidents were *leaders* of thought at times when certain historic ideas in the life of the nation had to be clarified." His cousin Theodore Roosevelt observed, "People used to say of me that I . . . divined what the people were going to think. I did not 'divine' . . . I simply made up my mind what they ought to think, and then did my best to get them to think it."[21]

Great presidents take risks. Those who merely want to be loved sometimes sacrifice conviction for wanting to be loved. Harding was popular but also a failure. JFK was popular but often cautious. FDR once said, "Judge me by the enemies I've made." Historian Arthur M. Schlesinger, Jr. suggests that effective presidential leadership is provided by those who follow their convictions and, at least on important issues, ignore middle-of-the-road pragmatism:

> We hear much these days about the virtues of the middle of the road. But none of the top nine (highly rated presidents) can be described as a middle-roader. Middle-roading may be fine for campaigning, but it is a sure road to mediocrity in governing. The middle of the road is not the vital center: it is the dead center. . . . The Greats and Near Greats all took risks in pursuit of their ideals. They all provoked intense controversy. They all, except Washington, divided the nation before reuniting it on a new level of national understanding.[22]

Between 1953 and 1965 (excluding only 1958) the average yearly presidential Gallup poll approval rating hovered at 60 percent or better (Table 3.2). But starting in 1966, roughly the time when the most recent era of "failed presidents" begins, the level of popular support declines dramatically. Presidents who attained a 50 percent support level were unusual. In the past thirty years, the average yearly popularity of presidents has dropped roughly 10 percent. No wonder these presidents have a hard time leading.[23]

If popularity equals (to an extent) power, and if efforts to manipulate levels of popular support are limited, presidents are caught between a rock and a hard place. They must constantly be concerned (obsessed) with their popularity, and yet their ability to generate support is often out of their control. Gone are the days when Harry S. Truman could publicly show his disdain for approval ratings with comments like "I wonder how far Moses would have gone if he'd taken a poll in Egypt? What would Jesus Christ have preached if he'd taken a poll in Israel? . . . It isn't polls or public opinion of the moment that counts. It is right and wrong and leadership . . . that makes epochs in the history of the world."[24]

Polls and popularity ratings "have altered the time frame of democratic government, from quadrennial election to monthly review."[25] They force the president to dance to a different and much faster tune. With one eye always focused on popularity, presidents may be less likely to make the tough choices, the hard decisions.[26] Flattering the public has a greater short-term benefit than speaking a harsh truth. Truman would know—his public approval ratings in his last year barely averaged 30 percent. Now, however, he is rated as well above average and by many as at least a near-great president.

The relationship between presidents and the public they are to serve and lead is a complex one. To govern, a president must have a fairly high level of popular support. But the avenues open to a president for attaining that support are limited. Public demands and expectations are high and often contradictory. The public wants presidents to deliver good news, not hard reality.

WHY DOES POPULARITY DECLINE?

As suggested, presidents who have served during the past two generations have generally lost popularity the longer they remained in office. Eisenhower is an exception. Different administrations have encountered different problems, advocated different programs, and presented different images to the public. Presidents during these years have employed a number of distinctive leadership styles. Inevitably the public becomes disenchanted and their earlier more optimistic expectations diminish.

Today, when 55 percent or even less of eligible citizens vote, a president can and often is elected by only slightly more than a quarter of the adult population. Moreover, many of these people frequently, and rationally, vote retrospectively rather than prospectively. That is, many people vote against an incumbent because of the incumbent's record and failings rather than for the prospective and only dimly understood programs of the challenger. This was clearly the case in 1976, 1980, and again in 1992. The new president "won" because we were voting out the incumbent president.

Presidents have usually had the support of a majority of Americans just after their inauguration. A rallying around the newly elected president customarily takes place. Many people, including many who did not vote or who did not vote for the victor, give new presidents the benefit of the doubt. Most Americans want to have positive feelings about the presidency, and the inauguration is a great American political and ceremonial event. In the election campaign leading up to the inauguration, voters are told to expect more from the president than they are currently receiving. Promises are made: "I'll get the nation moving again"; "I will never lie to you"; "I'll balance the budget"; "No new taxes." In many cases expectations naturally become unrealistic.

The heroic view of what might be accomplished is promoted by journalists who write or talk about the possibilities of the first one hundred days and in doing so sometimes attribute to the president the entire responsibility for the nation's economy. The problem is compounded by Americans who want to believe their presidents can make a major difference. Polls regularly provide persuasive evidence that people do believe a president can make a difference, that Franklin Roosevelt did make things happen, that America "would have been much different if John F. Kennedy had not been assassinated," that presidents can reduce unemployment or inflation, increase governmental efficiency, pursue an effective foreign policy, and strengthen the national defense.

To win the presidency, presidents have to make promises. Once in office, they find themselves battling with the Congress and colliding with the Supreme Court on occasion, unable to exert as much influence over the federal bureaucracy as they had presumed and constrained by public opinion, the press, interest groups, and other forces. A president's power today to make and implement policies is limited by the decentralized character of politics and government in America, which often results in gridlock. America's growing interdependence with the rest of the world and the economies of other nations, as well as the relative decline of U.S. power, also serves to limit the flexibility of the office.

Thoughtful people realize presidents are indeed constrained and should be constrained. But no matter, they still hold presidents responsible. Presidents are a convenient, highly visible scapegoat for the nation's problems. We oversimplify issues, we personalize them, and television is constantly featuring the president on the evening news. When things go wrong, people notice. Yet when things go right, people are usually complacent. Reality may be far more complex, yet it is easier for busy people to ascribe blame to a specific person.

In recent years scholars have tried to determine why presidential popularity declines. While findings from this research are inconclusive, the more compelling suggestions for why presidents lose public approval are the following:

- Expectations are too high at the outset.
- There are negative perceptions of the economic direction of the nation.

- Major negative or divisive events, such as Vietnam, Watergate, the Iranian hostage episode, or Whitewater, occur and drag on as an unresolved condition.
- The president's economic policy performance is viewed as weak, indecisive, or unwise.
- There are continuous bad news and negative media coverage associated with the government's performance in economic and foreign policy matters.

Studies find the American people evaluate presidents on the basis of their perception of presidential job performance. Presidents are held accountable for the quality of social, economic, and international policy outcomes during their terms. Yet the public's perceptions of presidential performance are usually inexact or hazy. One scholar concludes, "the public evaluates the president more on the basis of how it thinks the government is performing on economic policy than how it thinks the economy is performing."[27] Solving economic problems ranks high on the American public's roster of priorities, and the relative weight individual citizens give to economic progress or lack thereof significantly affects their evaluations of a modern president. Yet their evaluation may not be based on their own situation as much as on their general judgment of how the nation is doing. "When the economy falters, support for the president erodes not so much because citizens blame the president for their private hardships, however vivid, immediate, and otherwise important they might be—but because citizens hold the president accountable for the deterioration of national economic conditions. In evaluating the president, citizens seem to pay more attention to the nation's economic predicament," [than to their own].[28]

One study of the fluctuations of presidential popularity acknowledges some support for a cycle theory in which a brief honeymoon period is followed by a decline over the next three years followed by an upswing of approval in the fourth year. This fourth-year upswing occurs because the public is comparing the incumbent with an all-too-human alternative (i.e., the other party's presidential candidate) or because of a reluctance to make a break from the familiar to the unknown. Yet the economy's impact on presidential popularity is an additional key variable explaining public approval and disapproval of presidents. Unemployment, especially among white-collar workers and married men, is the single most important factor in an incumbent's approval rating. A cycle phenomenon and a citizen's rational response to economic developments are both at work.[29]

While nearly all scholars agree economic realities such as inflation and disposable income usually do have an effect on how people judge presidential performance, the overall picture of economic variables shaping presidential popularity remains muddled.[30] Other kinds of domestic, international, or political events have a way of arising to override people's normal concern with economic issues. "It may well be true that in the absence of foreign or domestic crises for extended periods of time, people evaluate presidential performance largely on the basis of economic conditions. But "normal times," it appears, are not the norm." Certainly, some people hold presidents responsible for the economy. "Perhaps everybody does at one time or another. But in general it appears that current events—both international and domestic—and the day-to-day policy decisions that the president must make take precedence in people's minds."[31] Thus sometimes the economy takes a back seat, or so this argument goes.

This verdict is persuasive for many if not all cases. Lyndon Johnson lost public approval more because of civil rights disturbances and the divisiveness of his Vietnam policies than because of economic developments; the economy was actually quite good at the time of his major popularity losses. Nixon lost his popularity and his claim to office not because of economic factors, but because of all the things associated with Watergate. Ford experienced one of the biggest single Gallup poll declines in polling history, more than twenty points, because of his pardon of former President Nixon. And Reagan's public approval took a twenty-point nose dive after the Iran-Contra affair disclosures. On the other hand, Carter, Reagan, and Bush usually seemed to be judged according to the general public's perceptions of inflation and the so-called "misery index" of leading economic indicators. In spite of a robust economy, Bill Clinton had difficulty reaching the 50-percent approval level, though his 1996 reelection must in large part be attributed to good economic times, or the general peace and prosperity of the period.

Another reason presidents lose popularity is because we have faulty memories. We forget past presidents also had difficulty controlling events and solving problems quickly. We also forget that all leaders, even our heroes, made mistakes. We often judge our presidents according to a model of the perfect, ideal, heroic president. Interestingly, in their fourth years many presidents gain back some of their public support. Truman, Eisenhower, Nixon, and Reagan are illustrative. This is partly so, as suggested earlier, because in their fourth year presidents get compared with the real-life alternative—to Dewey, Stevenson, McGovern, and Mondale, respectively, not to the model of the perfect presidency.

When asked, people often acknowledge they hold unfair and excessive expectations about what presidents can accomplish. When people are asked, "Why do you think presidents almost always lose popularity the longer they stay in office?," they suggest the blame is cast for these reasons:

- Presidents can't please everybody.
- People only see their good points at first.
- Presidents are scapegoats for our problems.
- The job is too much for one individual.
- They make too many promises they can't keep.

- They're only human—they make mistakes too.
- Some presidents can't cope with new problems.
- Familiarity and overexposure breeds either contempt or boredom—and sometimes both.
- Presidents often have to make unpopular decisions.
- Presidents are not as powerful as people often think.
- People don't always look at the overall record of a president.
- People are fickle.

In sum, the American people do hold presidents accountable and responsible. They do judge them on their record, on whether they follow through on their pledges, and on their overall job performance. Our support for presidents is influenced by how long they have been in office, but for the most part questions of peace, prosperity, and integrity are more important. The public is relatively quick to reward but quicker to punish, depending on a president's performance and actions in office. Hence, although presidents have limited opportunities to shape or lead public opinion, they are forced by upcoming elections, by on-going polls, and by their desire to be responsive to act on the desires of the American public.

HOW AMERICANS EVALUATE AND RANK PRESIDENTS

"Americans pick a President every four years," said Adlai Stevenson, "and for the next four years we pick him apart." Why are we so tough on them?

Mid-term elections are rarely American's finest hour. The guns of the out-of-office party pound away at the president. Then the president fires back. This is, of course, part of the tribal ritual in which the "outs" trash the "ins" and the "ins" retaliate by castigating the obstructionist opposition, the mean-spirited media, and the selfish special interests who are all depicted, of course, as thwarting "progress"—progress, at least, as defined by those in the White House.

But this president bashing of ours is about something more than ins and outs. We beat up our presidents because they invariably disappoint us. And they disappoint us precisely because we still hold exaggerated expectations of what presidents might be able to achieve. While it is true that the mistakes and misdeeds of Vietnam, Watergate, Iran-Contra, and Whitewater have made us more skeptical about presidents, we nevertheless continue to believe that the vitality of our constitutional democracy depends in large measure on creative presidential leadership. The unwritten presidential job description, the one we carry around in our heads, calls for a president to be a visionary problem solver, a unifying force for the nation, a healer and sage. Thus, every four years we search anew for a fresh superstar who is blessed with the judgment of Washington, the brilliance of Jefferson, the genius of Lincoln, the political savvy of FDR, and the youthful grace of JFK. Expectations always rise. Yet an exaggerated sense of the possibilities of the perfect, or what can be called our notion of "the textbook president," inevitably blind us to the limits of what a president can accomplish.

It is instructive, both for future presidents and for our understanding of the presidency, to know more about what the people judge to be effective presidential leadership.

One difference in the evaluation between the public and experts is the perspective of time. The public is plainly influenced by what historians have termed *presentism*. The people are more likely to remember presidents who were good ones in their own lifetimes, rather than past presidents who were perhaps great in the judgment of history. People who remember the youth and excitement of John Kennedy, especially in comparison to his successors who seem ineffective or dishonest, rank Kennedy as greater, much greater than Thomas Jefferson. (See Tables 3.3 and 3.4.) Also, two of the presidents most highly regarded by the public—FDR and Reagan—were two of the biggest "deficit presidents" in history.

Popularity while in office is not a useful gauge for judging reputation. After all, Harry Truman received some of the lowest popularity ratings in history, yet his reputation today is very high. As political scientist James Ceasar reminds us,

> It is worth emphasizing how little approval ratings have to do with any lasting judgment of presidential performance. A President's legacy derives from his accomplishments or failures, and no President will be long remembered for having an average approval rating of more than 60 percent, nor quickly forgotten for having an average lower than 45 percent. As an instrument of presidential power, a high approval rating has some value as a reminder to others of the potential "cost" they might have to pay in opposing a popular President. Yet it is important to remember that an approval rating is a lag, rather than a lead, indicator. What determines the score will be the public's assessment of conditions, performance, and persona. An astute President should accordingly be prepared in most cases to sacrifice his standing today, if by doing so he can affect positively the future assessment of these factors.[32]

It is helpful to look at the Gallup poll surveys asking the public to choose past presidents and what they remembered about them. Again the top choice was Kennedy, because he was "a strong leader," "concerned for the people," and "had the confidence of the people." Number two, Franklin Roosevelt, provides an exception to the rule. He was chosen for what were viewed as good policies and programs, his ability to handle economic problems, and his image as an effective leader. The public remembered Truman for his forcefulness, his forthrightness and outspokenness, and his leadership. The public admired Eisenhower because he had the confidence of the people, he was a leader, and he seemed forceful. "Greatness," in the public eye, is based upon certain intangible qualities, those chiefly being leadership and the way the president made the people feel. Enthusiasm gives the nation a lift. Inspiration or bold and contagious self-confidence can cover up a lot of ineffective or inadequate policy. In choosing their most admired or least admired president, what criteria are used by the general public?

What Americans expect of a president depends of course on varying factors. Not least of these is who is president, who was recently in office, and what the times demand. If recent presidents appear to have been weak, the public will likely call for more decisiveness. If the recent presidents or president seem to have been too dominant in relation to Congress, the public demands honesty or a cooperative president.

In short, the American public likes assertive, activist, and honest presidents. We set extraordinarily high personal standards for our presidents, and most of us, in addition, demand them to solve major problems. The American public does not always base its evaluation of presidents on policies, issues, or achievements. The public remembers presidents who seemed to be great leaders, and just as important, ones who made them proud of their country. They then use these historic examples as paragons to judge candidates and incumbents. Our evaluations of presidents are based partly on the times they served. In this sense the public is sometimes guilty of a lack of perspective. Experts, on the other hand, are more often concerned with results, the content of a president's program, and the president's success in carrying it out.

Both the expert and the average citizen alike are influenced by the cycles of American politics. People cannot quite emancipate themselves from their own age and country. To be sure, the historian and biographer are expected to transcend the present, but even they never entirely succeed.

TABLE 3.3
Ranking of American Presidents in the Gallup Polls, 1975, 1985, and 1991 (percent)

President	Year		
	1975	1985	1991
John F. Kennedy	52	56	39
Abraham Lincoln	49	48	40
Franklin D. Roosevelt	45	41	29
Harry S. Truman	37	26	17
George Washington	25	25	21
Ronald Reagan	—	21	19
George Bush	—	—	18
Dwight D. Eisenhower	24	16	18
Richard M. Nixon	5	11	10
Jimmy Carter	—	9	10
Thomas Jefferson	8	7	6
Theodore Roosevelt	9	7	8
Lyndon B. Johnson	9	5	3
Woodrow Wilson	5	1	3
Herbert Hoover	0	1	—

Source: Compiled by the authors from Gallup poll data.

TABLE 3.4
Presidential Favorites

Question: "Looking back on the past American Presidents, living or dead, which ONE of them would you want to have running this country today?" (July 1996)

John F. Kennedy	28%
Ronald Reagan	13
Franklin D. Roosevelt	8
Abraham Lincoln	8
Harry S. Truman	8
George Bush	5
Dwight D. Eisenhower	4

$N = 979$.
Source: From The New York Times/CBS News Poll, July 11–13, 1996. Copyright © 1996 by the New York Times Co. Reprinted by permission.

HOW EXPERTS EVALUATE PRESIDENTS

There is no commonly accepted standard on what constitutes presidential greatness. In general experts judge presidents on the basis of four factors: the scope of the problems they faced; their efforts (actions) and intentions (vision) in dealing with these problems; what they were able to accomplish; and what the long-term results of their actions were. Also, we are critical in judging presidents who were corrupt, whose administrations were rife with corruption, or who were wholly lacking in character.

This last point raises an interesting dilemma. While we are clear that public corruption, that is, abuse of public trust, criminality, improper use of government for personal gain, leads to negative ratings, what of private corruption? Does a president's (or candidate's) private behavior, his personal character, give clues to performance in the White House? How important is what we call character?

The word *character* comes from the Greek *charakter,* meaning the mark of a coin or seal. Euripides defined character as "a stamp of good repute on a person." Some of our highly regarded presidents were men of dubious personal character in their private relations. Recently press attention has focused on extramarital relations presidents and candidates may have had, the implication being that such affairs should disqualify a candidate from the presidency. Yet if that is our standard of judgment, several of our most popular presidents—Jefferson, FDR, Eisenhower, Kennedy, Clinton, and others—would be disqualified from office.

While the president represents the nation and serves as a symbol of who we are, and while logic tells us that character is important, in terms of presidential performance there is no correlation between "high" personal moral character and performance in office. Some of our presidents with checkered backgrounds performed well, and others of the highest private character were political failures.[33]

The process of rating presidents is fraught with potential biases. The actions of presidents look different from different historical vantage points. There are dangers in trying to assess the effectiveness and achievements of a presidency too early, say within twenty-five years. Sometimes a spate of new biographies about a former president will cast a favorable light upon a presidency, or conversely, may condemn a previously respected administration.

The charge is made, too, that most of the experts polled in these surveys are university professors or writers who tend to be more liberal, progressive, or Democratic in their political orientation. While this is true, one thorough study found the ranking by conservatives and liberals "remarkably similar." Serious disagreement in the ranks of the top ten most favored presidents apparently involved only the relative placement of Franklin Roosevelt.[34] Still the charge cannot be entirely ignored. Experts on the presidency are generally biased in favor of activist Democratic presidents and those who succeeded as wartime presidents.

Presidents who served during times of crisis are often ranked higher. Crisis, of course, does not ensure greatness. Madison, Pierce, Buchanan, Andrew Johnson, and Hoover all faced crises yet did not respond well. On the other hand, Lincoln, Wilson, FDR, and Truman all rank as great or decidedly above average because they responded to crises effectively. Although severe crises do not necessarily bring forth great leadership, the so-called "greats" served in periods of military, social, and economic upheaval, and to the extent the tests were met justifiably enhanced their reputations. The old question of whether great times make for great leadership or great leaders make great times is never satisfactorily settled.

Activist, combative presidents who contested rivals and rival institutions and expanded the normal responsibilities of the presidency and left the office larger than they found it are invariably rated as great or near-great presidents. Could it be that historians and political scientists, especially those who get polled in these presidency evaluations, are biased in favor of a vigorous, vital presidency? Presidents like John Quincy Adams, Taft, and Coolidge suffer in this context because they are often compared, and unfavorably so, with their contemporaries such as Jackson, Theodore Roosevelt, Wilson, and FDR, all of whom infused enormous personal energy into the presidency. Experts, along with the general public, seem to have a bias in favor of presidents with energy and presidents who, because they are unsatisfied with the status quo, enlarge the presidential role.

Another bias that renders rankings problematic is the difficulty in judging some presidents solely on their performance in office. Their achievements over a lifetime are often a complicating factor. Thus Washington, Jefferson, Madison, and Eisenhower were distinguished Americans quite apart from their presidencies. Grant was a great general yet a weak president. Madison was a superb constitutional architect, yet barely an average president. Hoover had been an outstanding success in business and a splendid war relief administrator and cabinet member, but was far less successful as a president. It is difficult to separate the presidential years from the overall career.

Whether a president dies in office from natural causes or is assassinated may also affect later rankings. While Harding and McKinley provide less evidence for this supposition, the public surely views Lincoln, FDR, and Kennedy differently, and one suspects much more favorably, because of their deaths in office.

An additional problem arises. Can we objectively compare presidents from different eras? Just as presidents cannot be experts on all areas of policy, historians cannot be expected to be experts on all presidents. Yet, isn't this what is necessary to make fair, objective judgments? Seldom are presidents faced with the same situation. Thus, we honor Jefferson and Polk for adding vast areas to the territory of the United States, yet twentieth-century presidents cannot duplicate that feat. Sometimes it is not possible to distinguish vision and foresight until years later, as with Andrew Johnson's purchase of Alaska in 1867.

WHAT IS PRESIDENTIAL GREATNESS?

Any attempt to define "presidential greatness" is necessarily subjective. We all have biases. We all live in a fixed historical era. We almost always approach the task with an eye on contemporary problems and with a partisan bias of some kind. Still, it is worthwhile to attempt such a definition.

Experts judge presidents as effective or outstanding when they acted greatly in challenging times. Did the president have the courage to fight for what was right? Did they bring out the best in the American people? Did they display common sense, decisiveness, and good judgment in picking priorities, associates, and fights? The lesser presidents often shy away from rather than join a conflict. They fear that decisiveness or polarization will lead to blood in the streets, even though history often teaches that courageous and early confrontation of divisive issues often avoids even worse violence later.

Also, moral courage stands out. Were they willing and able to stand up and be counted, as Lincoln and FDR were, when the nation and the nation's allies were challenged by the forces of hate and oppression? The great presidents were persons of principle and commitment, prepared to sacrifice popularity to what they knew was right.

The maxim that what is popular is always what is right seems, on cursory examination, to be untrue. As it concerns presidential action, after more reflection, it becomes even more untrue. An almost universally accepted aspect of great leadership is the willingness and ability to take risks, to act boldly when situations necessitate vision. Presidents are asked to educate and shape public opinion—even as they consult it. The question, to be sure, is more difficult than it first seemed, because determinations of what is popular and what is right are hard to make. Still, we can look back and find examples of presidential courage, of bucking the popular trends, and conclude satisfactorily that these actions were correct.

Thus George Washington faced intense hostility over the Jay Treaty, yet saw it through. Adams was prepared to break up his party to avoid war with France, and did both. Lincoln defied public opinion by removing General McClellan of the Army of the Potomac, and Truman displayed similar courage nearly a century later in removing General Douglas MacArthur. Though these actions were unpopular at the time, history vindicated them.

Experts look for accomplishments and whether the president provided leadership for the common cause. Historian Henry Steele Commager in 1977 formulated this list of five common denominators for those he termed great presidents.

- All believed the president should be both a symbol and leader, all identified themselves with the whole people, all had programs they wished to see adopted, and all worked actively to get them through Congress.
- All ranged themselves on the side of the people, of an enlarged scope for government, and of progress and reform.
- Few were good administrators, nor was the art of the politician a contribution to greatness.
- All possessed wisdom, sagacity, and intelligence.
- All were men of principle, prepared to sacrifice popularity for what they thought was right.[35]

Political scientist Clinton Rossiter once suggested that presidents who wanted to be judged as "truly eminent" by historians must stretch the Constitution. Other students of the presidency have observed that, much as they may admire the checks and balances, presidents must do what needs to be done. Sometimes this is called the doctrine of necessity. Thus it is said of Lincoln that he violated the niceties of the Constitution yet saved the union, and in doing so, he saw to it that the Constitution endured. So also in various ways Jefferson, FDR, and others either stretched or ignored their explicit oaths to follow the Constitution to pursue what they thought was right.

Those presidents generally considered "great" were also often the ones who gave Congress the hardest time. They were the ones who exercised many of the powers that were thought to be solely congressional. Some of them even seemed to enjoy "kicking Congress" (and sometimes the Supreme Court) around.

While these vigorous presidents may win praise from both the expert and public, they may also make life exceedingly difficult for those who come after them. Thus, Lincoln "made the office so strong that, like Jefferson and Jackson, he weakened it for his successors in the inevitable morning after."[36] Richard Nixon and some of Reagan's National Security Council aides tried to ignore Congress and, on occasions, the Constitution as well. But they were condemned and punished rather than praised. Just how far can the Constitution be stretched?

What did the great presidents do? George Washington, more than any other individual, converted the paper Constitution into an enduring document. With commanding prestige and extraordinary character he set the precedents that balanced self-government and leadership, constitutionalism and statesmanship. Washington's success made possible the success of the Republic. Jefferson, a genius in his generation, was a skilled organizer and a resourceful chief executive and party leader. He made his share of mistakes yet he adapted the office to countless new realities. His expansion of territory with the Louisiana Purchase, an achievement breathtaking in its consequences, assures him special status. Lincoln saved the union and will be remembered as the foremost symbol of democracy, union, and tenacious leadership in our hour of ultimate crisis. FDR saw the nation through its worst economic crisis and rallied the nation and the world to defeat nazism and Hitler. In a time of dictatorships he managed a democratic response to the depression.

These accomplishments are admirable. Still, in this line of reasoning, they raise as many questions as are answered. If presidents alone were responsible for all those notable and enduring accomplishments, it is little wonder that our expectations are so high. Plainly, a variety of people and institutions contributed to the achievements.

History seldom adequately honors the economic and movement leaders and reform activists who often provide as much or more of an era's leadership than does the president of the time.[37] Rarely do we take opportunity levels, resources, situation, or context into account when we judge presidents. Yet the best of the good ones will surround themselves with talented independent-minded aides, advisers, and administrators. Rarely do we judge presidents—on their own terms—on how successful they were in achieving their goals. Great leadership depends on its surroundings, on teams of leaders, and on people who demand leadership. Doubtless luck—sometimes "dumb luck"—also plays a part in greatness. Timing and being at the right place at the right time also help. Many of "the great" presidents also made great mistakes. The great presidents typically look better in death than they did while they were among the living, a proposition all of them would affirm.

We learn perhaps as much by those we judge as failures. Harding is ranked near or at the bottom by the experts because his administration was so corrupt. Loyalty to his friends was fatal. Others rank near the bottom because they lacked integrity, a program, or political skill. While Nixon had notable achievements, particularly his China policy, he is viewed as one of the worst because he lied to the American public, plotted a cover-up of illegal and unconstitutional activities, resigned in disgrace, and, in the eyes of many, thoroughly disgraced the integrity and dignity of the presidency. Some at the bottom of the lists were good and decent individuals but were mediocre leaders, overshadowed by other people, overtaken by events, for example, Buchanan, Pierce, Grant, Fillmore, Andrew Johnson, Tyler, and Taylor.

Surveys typically ask scholars to pay special attention to presidential performance in ranking presidents. Thus the 1996 *New York Times Magazine* poll asked historians to judge performance in office. "The standard was not lifetime achievement but performance in the White House. The scholars were to decide for themselves how presidential performance was to be judged." Historian Arthur M. Schlesinger, Jr., added, "It was supposed that historians would know greatness—or failure—when they saw it. . . ."[38] And so they ranked presidents as great, near great, average, below average, and failure. (See Table 3.5.)

A *Chicago Tribune* survey specifically suggested leadership categories and asked scholars and biographers to award points on the basis of these qualities, skills, and achievements. Presidents in that survey were compared on the basis of leadership qualities, accomplishments and crisis management, political skills, appointments, character, and integrity. These are sensible if still general categories. (See Tables 3.6 and 3.7.) Future surveys should use these and similar probing categories. Nonetheless, its findings were generally the same as more informal surveys.[39]

Interpreting and evaluating the Reagan presidency raises additional problems. Reagan enjoyed unusually high personal popularity at the end of his first term and in the early part of his second term. His landslide reelection had a lot to do with his personal popularity, as did his uncanny ability to portray himself as a supremely self-confident, strong, and "effective" president. His early success in winning many of his important measures in Congress added to this aura of competence. Yet, with adversity and scandal, Reagan also experienced somewhat greater disapproval.

Contrary to conventional wisdom, Reagan's first-term public approval ratings were not consistently high. People approved of *him* more than they approved of his initiatives or his performance. Reagan's policies were considerably more controversial than Reagan himself. While he proved remarkably adept at deflecting personal criticism, communicating with the

TABLE 3.5
New York Times Magazine Ratings of Presidential
Performance—1996

Great:	Lincoln, FDR, Washington
Near Great:	Jefferson, Jackson, Wilson, Theodore Roosevelt, Truman, Polk
Average (High)	John Adams, Eisenhower, Cleveland, McKinley, Kennedy, LBJ, Monroe
Average (Low)	Madison, J. Q. Adams, Van Buren, Hayes, Arthur, Harrison, Taft, Ford, Carter, Reagan, Bush
Below Average	Tyler, Taylor, Fillmore, Coolidge
Failure	Pierce, Buchanan, A. Johnson, Grant, Harding, Hoover, Nixon

Source: Arthur M. Schlesinger, Jr., "The Ultimate Approval Rating," *The New York Times Magazine*, December 15, 1996, pp. 48–49. Copyright © 1996 by the New York Times Co. Reprinted by permission.

TABLE 3.6
President Ranking by "Experts" (Murray/Blessing Poll)

Great	Near Great	Above Average
Abraham Lincoln	Theodore Roosevelt	John Adams
Franklin D. Roosevelt	Woodrow Wilson	Lyndon B. Johnson
George Washington	Andrew Jackson	Dwight D. Eisenhower
Thomas Jefferson	Harry S. Truman	James K. Polk
		John F. Kennedy
		James Madison
		James Monroe
		John Quincy Adams
		Grover Cleveland

Average	Below Average	Failure
William McKinley	Zachary Taylor	Andrew Johnson
William Taft	Ronald Reagan	James Buchanan
Martin Van Buren	John Tyler	Richard Nixon
Herbert Hoover	Millard Fillmore	Ulysses S. Grant
Rutherford B. Hayes	Calvin Coolidge	Warren G. Harding
Chester A. Arthur	Franklin Pierce	
Gerald Ford		
Jimmy Carter		
Benjamin Harrison		

Source: Robert K. Murray and Tim Blessing, *Greatness of the White House* (University Park: The Pennsylvania State University Press, 1994), p. 16. Copyright 1994 by the Pennsylvania State University. Reproduced by permission of the Publisher.

public, and maintaining reasonably congenial relations with Congress, his policies sometimes divided the country. His tax cuts, for example, produced the largest deficits in the nation's history. His aid to the Contra rebels and his Iran-Contra scandal disappointed many people.

Political scientists and historians rate Reagan less highly than the average citizen does. Whereas the public judges him largely on his rhetoric and means, at least in the short run, the experts pay more attention to policy outcomes and results. Experts also differentiate popularity from greatness. They will ask whether Reagan was good for the country. They weigh the substantive ends of his administration in their evaluations. In the long run Reagan will probably be praised more for his political skill in dealing with Congress and his public relations skills than for what he stood for.

The experts rate George Bush's presidential performance as decidedly average.[40] And there is little hope that he will be judged as great or near great in the future. Strangely, however, the same American people who voted him out of office in 1992 and barely accorded him a one-third positive approval rating in late 1992, four years later viewed his presidential performance far more favorably. This again raises questions about how moody or changeable or even fickle the public's judgment can be.

Our fascination with presidential effectiveness and failure reveals much about ourselves, about qualities we admire and those we dislike. It

TABLE 3.7
The Best and the Worst Presidents, According to the Experts (58 Presidential Historians and Political Scientists)

The Ten Best	The Ten Worst
1. Abraham Lincoln	1. Warren Harding
2. George Washington	2. James Buchanan
3. Franklin D. Roosevelt	3. Franklin Pierce
4. Thomas Jefferson	4. Ulysses Grant
5. Theodore Roosevelt	5. Andrew Johnson
6. Woodrow Wilson	6. Millard Fillmore
7. Harry S. Truman	7. Richard Nixon
8. Andrew Jackson	8. John Tyler
9. Dwight Eisenhower	9. Calvin Coolidge
10. James Polk	10. Herbert Hoover

Source: Steve Neal, "Putting Presidents in Their Places," *Chicago Sun-Times,* November 19, 1995, pp. 30–31.

tells us, too, about our dreams and national character. We yearn for heroism in literature, film, and our own lives. And at times, we look for certain heroic qualities, Mount Rushmore qualities, in our presidents. We delight in presidents who have triumphed when they faced ultimate tests. We are diminished by those who reveal the darker side of human nature and public life.

Leadership at the presidential level is full of risks. Sometimes risk-taking does not pay off, as with Johnson's Vietnam policies. FDR risked his political life and reputation by arranging secret naval deals with the British, and his policies were handsomely rewarded. Nothing may be more dangerous than to be constantly guided by instant and temperamental national poll results. For while the public often affords a president popularity on the basis of means, it is ends that ultimately determine the common good. The means may require sacrifice and alienation. "Pandering to the polls" can be a highly undesirable practice in the White House. The presidency is not and should not be a popularity contest. In one sense,

> Dr. Gallup renders us a disservice by his repeated polls on popular standings. Excessive acclaim, while personally gratifying, may be a bad sign as far as eminence is concerned. . . . Presidents who achieved reform, like Wilson and Franklin Roosevelt, were bound to make foes among the dislodged. President Eisenhower was so popular as to raise the suspicion that he had not disturbed enough the entrenched interests.[41]

One instance of a president appearing to pander to the polls and public sentiment came from President Ford. At a press conference he was asked whether he favored a stiff tax on a gallon of gasoline as a form of price rationing to dampen demand. Columnist George Will recalls his answer and condemns Ford's reasoning: "No, today I saw a poll that showed that 81% of the American people do not want to pay more for a gallon of gasoline. . . . "Therefore," said Ford, "I am on solid ground in opposing it." The problem is that all ground seems solid when your ear is to it, and as Churchill said, "It is hard to look up to someone in that position."[42]

A democratic Republic puts its faith in the people, faith that they will not merely elect presidents who will be responsive to their desires, but also do what is right. Americans want to be heard, yet they also want leaders who will independently exercise their own judgment. We honor courageous leaders who refuse to be intimidated by contrary public opinion data. Although presidents, to be sure, must take care not to be so self-assured they become insensitive to criticism and counsel, they also need a certain inner sense of what has to be done. In the end, the best of presidents are likely to be those who can accurately interpret or anticipate the sentiments of the nation and rally the people and other political leaders to do what must be done, and what the public will later learn to respect. These presidents will commonly be strong political leaders with a vision, if not a detailed program, of where they think the nation should go. We will ask of them: Did they give it their best? Did they preserve, protect, and enrich the liberty, the rights, and the economic opportunities of all the people? Did they so engage with the public to educate and build support for what was right and to encourage the many alternative sources of leadership in a nation with such ample talent? And, fair or not, we will ask: Did they help to make us proud of ourselves and our country?

THE POPULARITY DILEMMA

Presidents serve a variety of functions for the American public. Political scientist Fred Greenstein summarizes the psychological connection between president and people noting that the president serves as a symbol of national unity, stability, and the American way of life; an outlet for effect, for feeling good about America; a cognitive aid, simplifying complexity into a single symbol; and a means of vicarious participation in the political world.[43]

Because the president serves as both head of government (the nation's chief politician) and head of state (the symbolic representation of the nation), the president is simultaneously the chief divider of the nation and its chief unifier. Ronald Reagan was a masterful head of state. In the aftermath of the Challenger disaster Reagan took on the role of high priest and national healer for a nation devastated by tragedy. This role allowed Reagan, the nation's top politician, to rise above politics and become symbol for the nation as a whole.

Invested in the office of the presidency is high respect, but also high expectations (added to by presidential campaign promises). Yet the president's powers are not commensurate with the responsibilities or the public's expectations. Often this leads to frustration, and also to a decline in popularity. Likewise, if the public continues to demand that the government deliver on contradictory expectations (e.g., lower taxes *and* more government services), presidents are put in no-win situations. Given this paradox, what's a president to do?

Political scientist Theodore Lowi claimed that presidents resolve this dilemma by resorting to rule by political manipulation. Impression management replaces policy achievement. Appearances are everything.[44] Political scientists Paul Brace and Barbara Hinckley maintained that excessive reliance on appearances has led to the development of a "public-relations presidency . . . concerned primarily with maintaining and increasing public support."[45]

What then is the proper function of a leader in a democratic system? Should the leader follow the public's wishes? Or attempt to educate the public? Should the president attempt to *act,* to move the government in the direction he feels is best? Should a president find out what the people want and merely attempt to give it to them? Or should presidents speak truth, however unpleasant, to power?

Leading in a democratic society unveils a web of paradoxes and contradictions that can never be wholly resolved. And while the founders feared a president who might fan the embers of popular passions, it is clear that in our time the president has become the embodiment of the nation's government and is looked to as symbol-in-chief.

During his presidency, Jimmy Carter often tried to educate the people on the limits of America's power, thereby treating the public with the respect due an educated citizenry in a democracy. But Carter was unable to convince people that they had to settle for less, make sacrifices, and accept limits. This failure opened the door for Ronald Reagan who mocked limits, scoffed at sacrifice. He promised easier solutions to complex problems. He flattered the public and told them to buy and spend. They did. And at the end of the eight years of the Reagan presidency, the United States had gone from being the world's largest creditor nation to the world's largest debtor nation. Carter spoke of a tough reality, Reagan offered an easy path with less sacrifice. Carter lost, Reagan won.

THE DOUBLE-EDGED SWORD OF THE MEDIA

The lens through which we observe the presidency is the media. Some argue that television has created an "electronic throne" from which the president gains access to the people. Others argue that the media are a constant thorn in the side of presidents, always tearing them down. The paradox here is that to a degree, both views are correct.

Television has both *enlarged* and *shrunk* the presidency. It has enlarged the presidency by focusing a disproportionately large amount of attention on the presidency at the expense of Congress (if there is a balance of power, it rarely shows on television). It has shrunk the presidency by an overexposure that has led to the trivialization and/or intrusion of the press into the most personal and private of presidential activities. Before the 1992 election, Bill and Hillary Clinton were asked by Mike Wallace of "60 Minutes," "Are you prepared to say you never had extramarital affairs?" And after the election a young woman, with TV cameras running, asked Clinton, "Mr. President, the world's dying to know, is it boxers or briefs?"

The media are a two-edged sword that cuts both ways. Some presidents use the media. Others are used by it. Kennedy and Reagan, handsome, poised, witty, self-assured, were able to use the media to their advantage. Nixon, Ford, Carter, and Bush, stiff, nervous, awkward, fared less well. Clinton often, but not always, has been able to win good media coverage, largely due to his superior interpersonal skills.

Generally, presidents get somewhat favorable press coverage.[46] But in the aftermath of Vietnam and Watergate, reportage has become much more critical. In an age of cynicism, the media have seen their role as more adversarial and, along with the proliferation of cable television, more personal and intrusive.[47]

Before Vietnam and Watergate, the press was more deferential to presidents. Certain private peccadillos, while known, were deemed unfit to report. Even so obvious a handicap as Franklin Roosevelt's paralysis was not mentioned, nor were pictures plastered on the front pages showing a handicapped president struggling to walk. Today, TV programs such as "Hard Copy" would flood the public with images of a "weak and struggling" FDR, and it is unclear that such a person could be elected.

We generally want to know about personal foibles, in detail, of our celebrities and public officials. With the increased number of media outlets comes greater competition for a limited viewing audience. Thus, sensationalism increases, responsible reporting is pushed to the side, and the personal and private become fodder for investigative reporting and the White House press corps.[48]

As political scientist Larry Sabato noted:

> It has become a spectacle without equal in modern American politics: the news media, print and broadcast, go after a wounded politician like sharks in a feeding frenzy. The wound may have been self-inflicted, and the politician may richly deserve his or her fate, but the journalists now take center stage in the process, creating the news as much as reporting it, changing both the shape of election-year politics and the contours of government.[49]

Even the more responsible media outlets have been affected. Both print and electronic journalists have had to respond to market demands, with print imitating the *USA Today* "light news" format, and television adopting the fast cut, short, quick shots of MTV. Style wins out over substance. The average "sound bite," the time devoted to candidate statements, has shrunk from forty-two seconds in 1968 to seven seconds in the 1990s.[50]

Today, presidents must do what they can to manage the news. They attempt to elicit positive stories and favorable coverage, and can leave little to chance. They must have a clear strategy for dealing with the media. The strategy attempts to generate stories that add to the mystique of the president's *image* (strength, knowledge, political skill), *personality* (intelligence, family involvement, reassuring style), and *leadership* (in command, decisive, in control of events).

It was hard enough for presidents to manage the news when there were only three major TV networks and a handful of newspaper reporters to worry about. Today, everywhere presidents turn, cameras and microphones are thrust in their face.

News coverage has altered presidential politics. The impact of the Cable News Network (CNN), with twenty-four hours of coverage, instant information, and live on-the-spot reporting from every corner of the earth, forces presidents to face some issues they otherwise might wish to (perhaps need to) avoid. Because it was on TV, in the living rooms of the public, presidents often feel pressured to act in remote areas on issues not vital to the nation's national interest. Issues may not be able to ripen, presidents may feel compelled to make quick decisions. Speed may seem more important than deliberation.

How would the presence of CNN have altered the Cuban missile crisis? If Kennedy had been forced to show his hand earlier, when more aggressive military options seemed likely, the crisis could have ended in war. As it was, Kennedy could deliberate away from the glare of television, the issue could ripen, wiser heads could prevail. Is this less likely in an age of instant communications?

As television images show hunger, oppression, famine, and war around the world, the president may face great pressure to intervene for humanitarian reasons. Disturbing pictures can create a crisis atmosphere in which a president must act lest his leadership gets called into question. Inaction may be wise, but it may also appear as weakness.

When the public sees pictures of a captured U.S. soldier dragged through the streets of Mogadishu, Somalia, the cries for action can be deafening. But would action be wise? Once hidden from view, now every issue becomes closer, more proximate, larger than it may be. And the pressure to act, to solve the problem, may be irresistible.

Cable TV and talk radio have altered the political landscape. CNN and C-SPAN have increased coverage of political events and issues, and more specialized cable stations such as the Christian Broadcasting Network have given special interest groups their own sources of information and influence.

The impact of talk radio has been astonishing. Talk radio in the 1990s became a popular venue for people to vent their spleen. Called "hate radio" by some, this raw form of media democracy has fanned the embers of America's fears. Talk radio has taken up the role of mediating institution that the parties have forfeited, and has given the general public a

sense of connection to politics. Generally conservative and male oriented,[51] these programs offer an outlet and a vehicle for the antipolitics,
politics of the 1990s.

Politically, the president is the nation's center of attention. But rarely
can a president reach out and touch "real people." The media serve as the
president's conduit for reaching the public. Through the media, especially
television, the public sees presidents, hears about their ideas, gets interpretations of how well or poorly they are doing their job. In this way,
the media are a prime intermediary between a president and the public.
Presidents cannot succeed without the media, yet the media often are a
president's toughest critics.

The media's fascination with presidents, their families, even their pets,
has led to increased attention lavished on the president who serves as the
nation's "Celebrity-in-Chief." The gravitational pull of the presidency gives
the president a readymade bully pulpit, yet it also has a down side. In an
era of "in your face" investigative reporting, everything and anything a
president does is fodder for tabloid journalists. The micro coverage of the
president sometimes trivializes the office, the person, and politics itself.

A study of periodical literature shows that Franklin D. Roosevelt averaged 109.5 stories per year about his presidency, John Kennedy averaged
about 200, and Jimmy Carter averaged 407 per year. Ronald Reagan averaged over 500 stories per year.[52] By contrast, network news coverage of
Congress declined by roughly 50 percent between 1975 and 1985.[53]

Since the media serve as the president's conduit to the public, and
because so much is riding on good press coverage, no president can afford to ignore, or even leave to chance, relations with the media. Presidents must devise and implement strategies for attaining good press, and
efforts at media manipulations are quite common.

Because presidents are limited in what they can accomplish, and
given that the public expects presidents to deliver the goods, sometimes,
when faced with an inability to achieve policy results, presidents will resort to political grandstanding to give the impression that they are doing
something. Few White House operators used image management better
than the Reagan team.[54]

In a perverse example of the tail wagging the dog, policy is often used
for the public relations advantage it might deliver. And presidents can use
selected pictures and images to erase political reality. At the time President Reagan proposed a cut in funds for assisting the handicapped, the
president went to a Special Olympics event and the pictures on the evening news showed a caring, concerned Ronald Reagan while the reality
was that the president wanted to slash funding. And when polls revealed
that the public disapproved of Reagan's civil rights policies, the president
paid a public visit to a black family that had a cross burned on its front
lawn. Reagan appeared to be concerned and caring. If his policies revealed
an attempted reversal in civil rights enforcement, then his pictures had to

supplant reality. If carefully managed, a picture can sometimes be worth at least a thousand words.

This obsession with image has forced presidents to campaign 365 days a year. As journalist Hedrick Smith noted, "In the image game, the essence is not words, but pictures. The Reagan imagemakers followed the rule framed by Bob Haldeman, the advertising man who was Nixon's chief of staff, the governing principle for politics in the television era: The visual wins over the verbal; the eye predominates over the ear; sight beats sound. As one Reagan official laughingly said to me, 'What are you going to believe, the facts or your eyes?' "[55]

The strategy of presidential power in the media age is called "going public." Samuel Kernell saw an unending political campaign where presidents go public in an effort to increase political clout within the Washington community.[56]

In this, the United States has come a long way from what the creators of the presidency envisioned. The inventors of the presidency feared a president who was too closely linked to the public because this might undermine deliberation and reason, subverting the routines of representative government. Over time, such concerns gave way to the unstoppable logic of the electronic media and the need to use whatever levers a president could find to increase power.

Today, presidents go public because they believe they have few other options. In a political landscape cluttered with one-term presidents, an incumbent must use whatever resources are available, and governing by appearance, where governing by accomplishment fails, is something to which presidents sometimes resort.

Upon entering office, presidents feel compelled to announce their intention of having an "open presidency." Such efforts are always short-lived. Presidents soon become concerned with the release of bad news, or leaked information, and begin to engage in news (or information) management and "spin control."

Presidents can wield a carrot or a stick in dealing with the press. Some presidents revert to a "blame the messenger" strategy and attack the media. While press harassment can have an effect, most presidents mix strategic approaches ranging from co-optation to condemnation.[57]

It is important to link the media strategy to the president's political agenda. In this, presidents are advised to follow the Haldeman strategy. Richard Nixon's chief-of-staff, Bob Haldeman insisted that before something went on the president's agenda or schedule, the staff member who proposed it had to tell (1) what the expected headline would be; (2) what the photo accompanying the headline would be; and (3) how the first paragraph of the story would read. The Reagan administration followed a variant of this model. Key aides met early each morning to decide what the story of the day would be. They also developed long-term strategies for shaping the agenda.

By thinking strategically with a media focus presidents can command media attention and shape coverage of the president's agenda. Presidents often choreograph every step, "wallpaper" stories with attractive visuals, and when things sour, send the spin doctors out to do damage control.

Former Vice President Walter Mondale may be partially correct when he says that the electronic media are largely to blame for turning the presidency into the nation's "fire hydrant,"[58] but the media are also a source of presidential power. There are times when the media are the lapdogs for the president,[59] and at other times, the media check presidential power. If a president does not actively use the media, he may become its victim.

Are the media biased? If so, are they biased in favor of the left or right of the political spectrum? During the twelve years of Republican presidents from 1980 to 1992, conservatives insisted that the press was hostile first to Ronald Reagan and then to George Bush. But was this accurate? The media do a strange mating dance where presidents are concerned, and one may be inclined to level the charge of bias when your preferred president is being criticized in the press. Indeed, George Bush did get rough treatment from the media in 1992. If that is the point at which one measures bias, it might appear that indeed the press was out to get Bush. But if you choose a different point of time to measure bias, say, the first year of the Clinton presidency, a different picture emerges. Clinton was pummeled in the media during his first several months in office.[60] And this was merely the beginning for Clinton.

Rather than look on the media as primarily leftish, it may be more helpful to see the media as following rather than leading public opinion. When a president is up, they tend to be somewhat gentle (Bush in 1991, Reagan in 1984, Nixon in 1972); when a president is faltering, they become more critical (Nixon in 1974, Bush in 1992, Clinton in 1993).[61]

Presidents dislike the press because it catches their inconsistencies, mistakes, and failure to deliver on promises. But, to govern, presidents must work with and through the media, and presidents must think and act strategically where the media are concerned. Good press makes governing a bit easier; bad press makes governing difficult. Of course, the best way to get a good press is to do a good job. But this is never guaranteed.

CONCLUSION

We use varying and often unfair standards with which to judge presidents. We expect a lot and leave little room for error. The opinions of experts differ from what the general public believes. Members of the working media have their own views and are shaped by distinctive incentives that influence their treatments of presidents. Our expectation of presidential performance is based in large measure on the Mt. Rushmore presidents, and on our memories of their good days, yet few can live up to such high standards.

Paradoxically, we want someone who can summon us to live up to our shared ideals of a generous, benevolent democracy, yet we also want a president who will pay close attention to our immediate needs such as jobs, safety, cheap oil, lower taxes, peace, and prosperity.

CHAPTER   4

Presidential Power and Leadership

Taken by and large, the history of the presidency is a history of aggrandizement, but the story is a highly discontinuous one. Of the . . . individuals who have filled the office not more than one in three has contributed to the development of its powers; under other incumbents things have either stood still or gone backward. That is to say, what the presidency is at any particular moment depends in important measure on who is President. . . .

Edward S. Corwin, *The President: Office and Powers*
(New York University Press, 1957), pp. 29–30

Though the powers of the office have sometimes been grossly abused, though the presidency has become almost impossible to manage, and though the caliber of the people who have served as chief executive has declined erratically but persistently from the day George Washington left office, the presidency has been responsible for less harm and more good, in the nation and in the world, than perhaps any other secular institution in history.

Forrest McDonald, *The American Presidency: An Intellectual
History* (University Press of Kansas, 1994), p. 481

Anyone can lead where people already want to go; true leaders take them where only their better selves are willing to tread. That's where the leader's own values come in. They must want to do something with their power, not just for the powerful.

Jonathan Alter, *Newsweek*, December 3, 1990

"Power" is an often used yet little understood concept. The problem is particularly vexing when we wrestle with the concept of presidential power. What do we mean by power? How much power does a president have? How much power should the president wield in our system? What are the appropriate limits to be placed on presidents?

The very term *democratic leadership* is paradoxical. How can a democracy—government of, by, and for the people—countenance one person, or even a team of people, leading? It sounds so antidemocratic.[1] The words *leader* and *leadership* imply that someone provides command or direction. The word *democracy* implies widespread participation and rule by the people. The two seem irreconcilable.

While there is certainly a tension between democracy and leadership, we believe it can be a creative tension that enhances and supports both democracy and leadership. Yet, all too often this potentially creative tension degenerates into forms of behavior that can undermine both democracy and leadership.

Before we can examine what democratic leadership might look like, we will discuss presidential power and leadership. For while democratic leadership may be our goal, the road to achieving it may be more difficult than we think.

THE PRESIDENT AND THE SYSTEM

The presidency is part of a system. This system consists of a constellation of political participants such as the Congress, the courts, public, media, interest groups, and bureaucrats, all of whom share a claim to power. The president is a part of this system, often a most important part, yet a president is not in control of the system.[2]

Ours is a constitutional system of shared, separated, overlapping, dispersed and fragmented powers. No one influential figure controls the levers of power; they are spread out across the system. This lessens the likelihood of tyranny. But it also makes it difficult to achieve necessary change. The system is slow and cumbersome, designed to make leadership deliberate and usually difficult as well.

Leadership is difficult precisely because the framers of our Constitution wanted it, except in emergencies, to be so. The men who invented the presidency did not wish to create a *ruler;* their experience with kings of England soured them to this prospect. Instead, they hoped to create the conditions where *leadership* might from time to time flourish. A ruler commands; a leader influences; a ruler wields power; a leader persuades. Presidents have *limited and shared powers*. The paradox here is that our expectations are high, we demand much of our presidents, but the resources at the disposal of the president are limited and the system in which a president operates can easily frustrate efforts at presidential leadership. We have created a "leadership aversion system"[3] and demanded that presidents take control and lead. Is it any wonder that we are so often disappointed in presidents? In effect, we set them up to fail and then turn on them when they do.

AMERICA'S ANTILEADERSHIP CULTURE

Several hurdles stand in the way of strong presidential leadership. A system organized around a separation of powers as opposed to a fusion of power (Great Britain fuses legislative and executive power, for example) structurally inhibits strong leadership. America's *political culture* also undermines strong leadership.

Ours is in many ways an antiauthority and therefore an antileadership culture. Political scientist Samuel Huntington refers to "the American Creed," which consists of a commitment to individualism, egalitarianism, democracy, and freedom.[4] Admirable goals all, yet not especially supportive of strong leadership or strong deferential followership. Political scientist Clinton Rossiter captured this paradox when he wrote, "We have always been a nation obsessed with liberty. Liberty over authority, freedom over responsibility, rights over duties—these are our historic preferences. . . . Not the good man, but the free man has been the measure of all things in this "sweet land of liberty"; not national glory, but individual liberty has been the object of political authority and the test of its worth."[5]

How does this American creed affect our attitudes toward leadership? While it creates a fundamentally antileadership disposition, the American propensity for individualism does generate a type of hero worship that feeds into the heroic model of the presidency. We thus honor Washington the patriot, Jefferson and Jackson the great democrats, Lincoln and FDR the saviors. But such hero worship masks, in some ways, the stronger cultural pull that is antileadership and antiauthority.

In general, the antiauthority strain of the American creed has personified our culture. Max Lerner has written that "American thinkers have been at their best in their anti-authoritarianism."[6] From Jefferson and Madison to Calhoun and Thoreau, some of America's most profound and influential thinkers reflected deep strains of individualism and antiauthoritarianism. The implications of this for leadership were not lost on that thoughtful French observer of the American scene, Alexis de Tocqueville, who wrote,

> When it comes to the influence of one man's mind over another's, that is necessarily very restricted in a country where the citizens have all become more or less similar . . . and since they do not recognize any signs of incontestable greatness or superiority in any of their fellows, are continually brought back to their own judgment as the most apparent and accessible test of truth. So it is not only confidence in any particular man which is destroyed. There is a general distaste for accepting any man's word as proof of anything.[7]

It is difficult, except in a crisis, for presidents to ask the people to make major sacrifices for the common good, to move beyond self interest. When Jimmy Carter asked citizens to turn their thermostats down to 68 degrees Fahrenheit in an oil crisis, he was rebuked. When President Clinton called on the American people to provide military peace forces in Somalia, Haiti, and Bosnia, Americans were hesitant if not hostile to such urgings.

The United States lacks a defined "public philosophy" of the role and purpose of government in a post-Cold War era. Americans have always been somewhat ambivalent about power and government, and suspicious of authority. The lack of a public philosophy often reflects a lack of consensus as to what we as a people should do, where we as a nation need to move. As sociologist Michel Crozier has written, "the land of consensus turned into the land of disorder and tumult."[8] In the absence of a national consensus, presidents have a difficult time moving our separated system.

Our sense of a shared community, or of a social contract, has generally remained murky. We are not "one nation indivisible" as much as a collection of individuals. We rarely pull together for the common good. In part, we are a "nation," a loose-knit collection of individuals, and in part we are a confederation of tribes.[9]

Yet effective leadership requires a supporting consensus behind governmental authority and presidential initiatives. Disharmony and conflict have often characterized the American political system. A divided nation, a polarized nation, cannot be a well-governed nation. "Where did we go wrong?," leadership expert Warren Bennis asked, "America has always been at war with itself. We have always dreamt of community and democracy but always practiced individualism and capitalism. We have celebrated innocence but sought power. We are the world's leading sentimentalists, and it's a very short step from sentimentality to cynicism."[10]

While Americans demonstrate patriotic fervor, they do not often demonstrate a deep sense of community. From a strong commitment to community, one can elicit a consensus upon which to build leadership. When the public lacks a strong sense of community, leaders have a difficult time pulling the nation together (except in a crisis) to respond collectively, politically, as one nation.

THE MOODS AND CYCLES OF AMERICAN POLITICS

Leadership does not occur in a vacuum. There are different "seasons" of leadership, times when presidents are afforded more or less room to exercise power.[11]

These cycles take many forms: the business cycle of economic growth followed by recession; the political pendulum of liberalism followed by a conservative period; strong presidents followed by weak ones; the mood swing of public confidence in government followed by a retreat into private interests and cynicism; the foreign policy shifts from isolationism to international involvement. These cycles are, for the most part, beyond the control of presidents, but they do have an impact on presidential power.[12]

There are times when, and conditions under which, presidents are afforded considerable leverage and power. There are also times when presidents are kept on a short leash.

Thus, in the aftermath of the war in Vietnam and of Watergate, the public turned against the government and presidential power and the United States entered a period when we questioned everything a president did. Thus Presidents Ford and Carter were restricted in their opportunities to exercise power. The public was more cynical and suspicious, the press was more skeptical, the Congress was willing to reassert its authority, and if that weren't enough, the economy was sluggish. Even if Ford and Carter had been highly skilled, gifted politicians, their level of political opportunity was so low there was little way they could have been successful.

FDR, by contrast, a highly skilled politician, also benefited from a level of opportunity that was unusually high. He came to office at a time when the public demanded strong leadership, when the Congress was willing to accede power to the president. Thus the mixture of high skill and high opportunity allowed FDR to pursue both power and success.

Political time is also important.[13] During periods of great change and significant social disruption, certain types of leadership will be more necessary, while during periods of normalcy, a different type of leadership may be required. Likewise, a different sort of leadership is required during periods of crisis. Clearly there is not one leadership style for all seasons. Effective leaders are able to adjust to these different seasons, they "style-flex."

Several cycles are especially relevant to presidential leadership. One is the *succession cycle*. Political scientist Valerie Bunce noted that major policy shifts are most likely when new leaders come to power.[14] Thus, leadership change is connected to policy change. Another cycle is the *cycle of decreasing popularity*. Over time, presidents tend to lose popular support (as stated in the last chapter), which makes it more difficult for them to put together political coalitions as time progresses. Another presidential cycle is the *cycle of growing effectiveness*.[15] The president's learning curve is at its lowest at the beginning of a term and rises as time goes by. Yet the president's power is usually at its zenith early in the terms, when knowledge is lowest.

The cycle that seems most relevant to presidential leadership deals with the long-term ebb and flow of American politics. Historians Arthur M. Schlesinger and Arthur M. Schlesinger, Jr., suggest that this cycle is like a pendulum swinging back and forth. The United States alternates between periods of "conservatism versus innovation" (Emerson), "diffusion versus centralization" (Henry Adams). Arthur M. Schlesinger, Jr. describes these mood swings as "a continuing shift in national involvement, between *public purpose* and *private interest.*"[16]

The roots of this repeating cycle doubtless lie deep in human nature and the U.S. structure of government, and the pattern follows roughly a thirty-year alternation between the pursuit of public purpose and of private interest. In the twentieth century there were a few periods of high governmental activism in support of public purpose: Theodore Roosevelt and Woodrow Wilson from 1901–9 and 1913–17, Franklin D. Roosevelt from 1933–35, and John F. Kennedy and Lyndon Johnson from 1963–66. There have also been "high tides of conservative restoration—the 1920s, the 1950s, the 1980s."[17]

We must view political change as a connection between leader (office holder) and follower (citizen) at different points in time. This allows us to broaden our horizons to the complex and interconnected nature of political change in a democracy. Leaders do not always lead; followers do not always follow. Different political climates require different types of political leadership and a different set of political skills.

THE VAGARIES OF PRESIDENTIAL POWER

All presidents on assuming office receive similar formal constitutional powers. Sometimes, however, the amount of power exercised exceeds the legal limits. The founding fathers arranged for an office broadly defined as well as vaguely outlined. The power granted, they felt, should be a broadly discretionary residual power available when other governmental branches failed to meet their responsibilities or failed to respond to the urgencies of the day. Power was available, but it would be subject to various checks and balances. The president is open to challenges and veto points from a number of other political institutions.[18]

As Edward S. Corwin and others have pointed out, there is a plasticity in our fundamental conception of the presidency and its powers.[19] The exact dimensions of executive power at any given moment are largely the consequence of the incumbent's character and energy combined with the overarching needs of the day, the challenges to system survival and regeneration. Some presidents have been power maximizers. Jackson, Lincoln, and the Roosevelts are illustrative. Certain of them became shrewd party leaders. Some saw themselves as direct agents of the American people, as the people's choice with mandates to carry out in exchange for the grant of powers. Still others employed the "take care that the laws be faithfully executed" clause of the Constitution to broaden the notion of executive power well beyond the boundaries envisioned by most of the framers of the Constitution. Plainly, an office underdefined on paper has become enlarged with the accumulation of traditions and with the cumulative legacy of some often brilliant achievements.

Tocqueville rightly observed that if executive power was weaker in the United States than in European countries (in the early 1800s), the reason was more in circumstances than in laws. He added that "it is generally in its relations with foreign powers that the executive power of a nation has the chance to display skill and strength."[20] If the United States had remained a small nation, isolated and removed from the other nations of the world, the role of the executive would doubtless have remained weak, certainly much weaker than the presidency we have come to know in the late twentieth century. Just the opposite has been the case, of course. Had the presidential office been incapable of expanding, the nation may well not have survived. Presidential powers and presidential leadership have been essential for both progress and stability.

One of the more persisting of presidential paradoxes from the standpoint of presidents is the realization that the office carries much less power, however power is defined, than the candidates had thought when they ran for the office. Lyndon Johnson did his best to pyramid available power resources to the office soon after he found himself there. But he never stopped complaining that his responsibilities always exceeded his powers. Thus it is not surprising that although LBJ did not have much advice for his successor, he did have these words of caution:

> Before you get to the presidency you think you can do anything. You think you're the most powerful leader since God. But when you get in that tall chair, as you're gonna find out, Mr. President, you can't count on people. You'll find your hands tied and people cussin' you. The office is kinda like the little country boy found the hoochie-koochie show at the carnival, once he'd paid his dime and got inside the tent: "It ain't exactly as it was advertised."[21]

Political scientist Richard Neustadt's celebrated *Presidential Power,*[22] first published in 1960, studied this same problem. His work is a veritable manual of personal power: how to get it, how to keep it, and how to use it. James Reston called it the nearest thing America has to Machiavelli's *The Prince.* Presidents Kennedy and Johnson were influenced by Neustadt's thinking, as were other presidents. Kennedy studied it and employed Neustadt as a consultant on staffing and varied policy questions. Johnson almost naturally seemed to conduct himself according to the Neustadt power principles.

Presidential Power originally was a call for presidents to exercise their power wisely and fully. In subsequent years Neustadt updated and revised his theses in the light of scholarly commentary and subsequent administrations. His book remains one of the influential works in presidential scholarship. How do his views hold up as we enter the twenty-first century? After Vietnam, Watergate, Iran-Contra, and Whitewater, does Neustadt's position still inform? Have the intervening decades validated his basic premises?

CONCEPTIONS OF PRESIDENTIAL POWER

Presidential Power was hailed as a pioneering contribution to our understanding of the operational realities of presidential leadership. It was widely read in 1960 and it remains a classic, especially in seminars on the American presidency.

Neustadt's *Presidential Power* broke away from the traditional emphasis on leadership traits, the compartmentalized listings of functional tasks, and the then-dominant tendency to study the presidency in legal or constitutional terms. Instead, Neustadt used organizational and administrative behavior as frames of reference for the study of what presidents must do if they want to influence events and why and how presidents often lose the ability to influence. His stress was on the *shared* powers of the office rather than the separation of powers. He emphasized the reciprocal character of influence, the constant personal calculations and trade-offs that motivate people to cooperate or not to cooperate with presidential initiatives.

The message of this book is that the presidency is neither as powerful as many people think nor should its strengthening be feared as many others believe. Neustadt called on future presidents to acquire as much power as they could, for he noted that the formal institutional powers were fragile, even puny, compared with the responsibilities.

Neustadt says presidential power is the power to *persuade,* and the power to persuade comes through *bargaining.* Bargaining, in turn, comes primarily through getting others to feel it is in their own self-interest to cooperate. Presidents are depicted as being constantly challenged by threats to their power and constantly needing to enhance their reputation as a shrewd bargainer. Tenacity and proper timing are also essential.

Neustadt viewed Franklin D. Roosevelt as the many-splendored prototype. Roosevelt had that rare combination of self-confidence, ambition, political experience, sense of purpose, and Neustadt's elusive term *sensitivity to power,* that was necessary to harness formal authority with effective personal performance and thus make the presidency work. "Roosevelt had a love affair with power in that place. It was an early romance and it lasted all his life."[23] "[He] saw the job of being President as being FDR."[24] "Roosevelt always knew what power was and always wanted it."[25] In short, Roosevelt had the will to power, the driving ambition and uncommon sense of knowing how to deal with people, and these combined almost always to leave him in the driver's seat.

Neustadt was especially impressed with the way Roosevelt juggled assignments, kept people guessing, put men of clashing temperaments, outlooks, and ideas in charge of his major projects. FDR loved dividing authority and keeping his organizations temporary and overlapping. To some people this might have been the art of manipulation, but to Neustadt it was the essence of leadership. Sixteen years later, Roosevelt remained the Neustadt model, particularly his robust, playful temperament: "Roosevelt's sense of fun combined with Roosevelt's sort of confidence remain for me what they were . . . a target at which to aim."[26]

There was widespread agreement among political scientists in the early and middle 1960s that Neustadt's analytical treatment of presidential power was a perceptive description of how to strengthen the hand of a president one liked. Neustadt's propositions made sense to most people. Here is a list of several of them, slightly rephrased:

1. To be a leader, a president must have a will for power. If he lacks a consuming hunger for power and a penchant for shrewdly handling people, then he is not suited for the office.

2. The skill of a president at winning others over to his support is the necessary energizing factor to get the institutions of the national government into action. He and only he supplies the agenda for action for the Washington political community.

3. A president cannot be an introvert, or above the battle, or above politics, or simply work from within the confines of his own ideas; a sensitivity to the thoughts and feelings of others and an ability to create solutions that compromise contesting points of view are what distinguishes effective leadership from nonleadership.

4. The members of Congress act the way they think they have to in order to get reelected and to "look good." A President's job is to get members of Congress and other influential members of the government community to think his requests are in their own best interests.

5. A president has to ride events and crises to gain attention. Most Americans grow attentive only as they grow directly concerned with what may happen in their lives.

6. A president should never rely on others to determine his power stakes. He should be his own intelligence officer, his own expert on crucial power relationships. Moreover, he should be concerned with details, gossip, and the intricacies of human sensitivities. A president who delegates the job of being chief politician to others is a president who will not exercise much influence.

7. The lesson from cases is that when it comes to power, nobody is expert but a president, and if he too acts as a layman, it will be fatal.

8. Popularity and public prestige produce favorable credit for a president among the professional Washington community. But public disapproval heightens and encourages resistance from the Washington community (defined as members of Congress, bureaucrats, the press, governors, diplomats, etc.).

9. Presidential power is not easy to come by and even the most skillful of presidents will have to be flexible, always sensitive to the need for multiple channels of information, always frugal in using power re-

sources to get their way, and always employing the art of persuasion in a bargaining situation, thereby avoiding at almost all costs the direct issuing of a confrontational command.

Thus, *Presidential Power* held out hope that a shrewd and artfully manipulative leader could and would be a powerful Hamiltonian engine of change. An aggressive, ambitious politician, determined to get his way and ever distrustful of the motives of others, seemed to be the remedy for the post-Eisenhower years. Neustadt concluded that the foremost problem of the presidency in the late 1950s was how the presidency could regain control over the drifting Washington policy apparatus. Forceful leadership was needed, and only the president could fill the leadership vacuum.

Neustadt's greatest contribution to the understanding of the presidency is his notion that the power of a president rests ultimately on his ability to persuade others. Because a president shares authority with other institutions, he cannot merely command. Rather he must constantly bargain and trade in an effort to gain his ends. Thus, Neustadt argued, don't study the institution of the presidency or its formal powers as ends in themselves; seek instead to understand how these are used to enhance the personal power of individual presidents, and thus their success as leaders and policymakers.

This was wise advice.

PRESIDENTIAL POWER RECONSIDERED

In the 1970s and 1990s, Neustadt published reflections on the post-Eisenhower presidencies. In spite of the many abuses of presidential power in the Johnson and Nixon years, Neustadt contended his earlier conclusions were still valid. Although he offered a few amendments to his earlier views, he implied generally that, concerning presidential power, the more things change, the more they stay the same:

> Since April 1960, when *Presidential Power* first came out, there have been events aplenty, not alone Vietnam and Watergate, but assassinations, riots, inflation, recession, even a nuclear confrontation, to say nothing of changes in our world relationships, political *and* economic. Still, these do not appear to have altered very much the general character of presidential power. Nor do they disclose to me a likely shift of central role from President to Congress or the courts, or elsewhere, of the sort that Wilson found emergent in a turn from Congress to the Presidency.[27]

He now found that Eisenhower's concern with his own and his office's prestige, and hence credibility, was a more impressive contribution to the presidency than had been previously credited to him. It had been difficult for Neustadt as a young, liberal, Democratic professor to accept Eisenhower's low-profile leadership. An older Neustadt, comparing Eisenhower with

Johnson and Nixon rather than with FDR, writes that Ike's shunning of controversy and his conscious enhancing of the presidency's legitimacy were eminently sensible in that they added to Ike's potential to govern. Although he does not develop this point, it may be that an older Neustadt now realized, as most people did after the 1960s and 1970s, that presidents should be measured not only by what they do but also by what they do not do. Eisenhower's refusal to be drawn into Indochina would, in retrospect, win praise.

Neustadt appears to criticize Johnson and Nixon for attempting to amass too much power. Maximizing power, he suggests now, cannot be a president's sole criterion for choices. There had been little in his 1960 treatise to anticipate this. In terms of power as a source of clues to policy, we are told that for both Johnson and Nixon, the short-run tangibles seem to have overshadowed long-run risks. Their sensitivity to their power stakes simply didn't work for them. Johnson chose to gamble that the maximum duration of the Vietnam War would be the two-and-a-half-year figure some of his aides passed along to him. Nixon, similarly, was woefully inexpert in distinguishing the crucial from the irritating. Nixon also was inexpert in distinguishing the implications over time from the effects of the moment. Wrote Neustadt in 1976: the 1960 discussion of power stakes is still valid but "not nearly as helpful as I once had hoped."[28]

Neustadt admitted also that he underestimated the negative influence of palace-guard loyalty on a president's power stakes. He lamented the fact that several recent presidents had allowed a swelling of the presidency. Thus, Nixon's White House aides put the whole elaborate center of their system in the White House, attracting vast amounts of second-string activity, and at the same time tried to spare Nixon the details. Thinking they were doing him a service, they were in practice helping to do him in. Roosevelt still provides the model, wrote Neustadt. "For Roosevelt, the President was not the Presidency, both ought to be staffed, the President should weigh advice from both. He sought advice as well from everybody else that he could get his hands on. . . . Roosevelt never thought that staffs had a monopoly on judgment or on information either."[29]

Thus, Neustadt affirmed most of his earlier views. Presidents still share most of their authority with others and are no more free today to rule by sheer command. "Persuasion in a sense akin to bargaining remains for major purposes the order of his day."[30] Despite Watergate and Vietnam and the strengthening of Congress, the presidency's advantages as a leadership institution have not suffered. Congress, cabinet, party leaders, the establishment, and others need the president rather more than the president needs them. Indeed, wrote Neustadt, today they are less colleagues than customers, to whom he has to sell more than consult.

But the gap between presidential responsibility and capability is an ever-widening one. Neustadt bemoaned the post-Watergate outpouring of legislation that made a bad president weak because it enfeebled the good presidents as well. He dismissed the idea we could succeed for long in lowering our expectations of the office. He has accepted the inevitability of high expectations.

UNRESOLVED QUESTIONS

What of the problems and confusion raised by *Presidential Power?* What of the book's worth as a guide to understanding presidential leadership?

Perhaps the most frequent complaint about the 1960 study is that it seemed too preoccupied with the will to power, the acquisition of power, and stockpiling the power, divorced from any discussion of the purposes to which power should be put. Such an emphasis on means without any clear discussion of ends left the impression that the art of leadership is tantamount to the art of manipulation. Neustadt was faulted too for his failure to emphasize the role that a "sense of direction" plays in presidential leadership, and how a president would call on or consider ideological values in his power exchanges.

One critic went so far as to suggest that Neustadt "baptizes" political ambition just as Dale Carnegie and self-help manuals baptize greed. Does power tend to purify, and absolute power to purify absolutely? Many readers would have liked a more thoughtful discussion of the ends of presidential power, of the ethical boundaries. What are the higher claims on a president and how does the creative president join together the ethic of responsibility and the ethic of ultimate ends?

Put another way, Neustadt's methodology "does not allow political science to distinguish between the use and the abuse of presidential power."[31] Neustadt made little claim beyond this, although he did suggest that a president's sensitivity to the means of power has a good deal to do with "viable" public policy. To some people, this seems as if in Neustadt's terms, public purpose *is* linked directly to presidential power.

In any event, Neustadt did not apologize for his focus on the means of leadership. Although certain critics did, and with some reason continue to contend that the analysis and discussion of power devoid of content is a shallow, even an empty, exercise, in Neustadt's defense it can be said that the ends were, and perhaps still remain, implicitly embedded in the liberal vision of the 1960s.

A second criticism of Neustadt's book was that it is too worshiful of presidential power. It seems to say: Find the right president and teach him what power is all about and progress will be realized. It portrays presidents as potential saviors. If only we had the Second Coming of Franklin Roosevelt, all would be well! It comes close to suggesting we need a charismatic larger-than-life figure on whom to lean to make the system work. Further, the book speaks almost exclusively about the Washington political community and makes little or no reference to the people's role or to social movements.

Two problems arise from this emphasis on FDR and on presidents as the answer to our needs. First, Neustadt failed to take into account the degree to which presidents are almost invariably stabilizers or protectors of the status quo rather than agents of redistribution or progressive change. Neustadt gave little attention to the way the prevailing American elite values, and market capitalism,[32] often limit a president's freedom. Bruce Miroff wrestles with some alternative interpretations in his *Pragmatic Illusions: The Presidential Politics of John F. Kennedy.*[33] One gets the impression from reading Neustadt that he thought a president can roam at will, providing he is shrewd enough to be able to persuade others that their interests are the same as his. In fact, however, all of our presidents have had to prove their political orthodoxy and their acceptability to a wide array of established powers, especially corporate leaders, entrenched interest-group leaders, and so on. Thus, Neustadt raised hopes that the presidency will be an instrument for the progressive transformation of American politics. He apparently saw no other alternative.[34]

Just how much and how often can we turn to the White House and hope that a benevolent and bright president will provide truly inspired leadership? Sometimes we must turn there, yet a reading of history suggests that breakthroughs and leadership often come from the bottom (or at least the middle) up. Civil rights workers, consumer organizers, women's rights activists, environmental protectionists, tax-revolt champions, and antiwar protestors are illustrative of the catalysts that more often than not bring about policy change in the United States. John W. Gardner, himself a citizen-activist, cautions that "crusading citizens' groups may not always be wise, witness the Prohibitionists. Or they may be wise but unsuccessful in persuading their fellow citizens. For every citizens' group that changes the course of history, there are thousands . . . that never create a ripple."[35]

In Neustadt's defense it can be said that although his treatise reads like a hymn of praise for FDR, the author really did not go so far as to say: Defer completely to your president and trust him. He appears instead, at least most of the time, to say that without a good engineer the train just won't go.

There is also an elitist cast to Neustadt's formulation. There is an implicit, if not explicit, fear of the masses juxtaposed with a robust faith in the great leader. This was, of course, a prevailing view among most academics at the time. Roosevelt may have been a grand manipulator and sometimes deceptive as well, but wasn't this increasingly necessary? Leaders need a free hand. It was as if Roosevelt on occasion deceived people, including the American people, much as the physician who lies to the patient "for the patient's own good." Walter Lippmann, George Kennan, Hans Morganthau, and others had previously and often eloquently argued a similar case. What was needed was an activist helmsman to define the national interest, to subdue or avoid the passions of the electorate and the parochialisms of the special-interest groups and Congress. Look to your presidents for leadership, for in Lippmann's oft-quoted words, "The unhappy truth is that the prevailing public opinion has been destructively wrong at the critical junctures."

Neustadt was influenced by this tradition just as this tradition would be reinforced by him. The message is plain: the whole system revolves around an activist, persuasive president who knows how to avoid the pitfalls and the sand traps of the Washington obstacle course. Neustadt believes this still. Presidential power is as contingent, as uncertain, as tenuous as ever. The press, the attentive public, or a mindless Congress can too easily do enormous harm to a president's ambitions. Do not, Neustadt tells us, worry too much about an imperial presidency. Do not weaken the powers of the presidency. In Watergate and Vietnam we witnessed the abuse of power, not an excess of power. The absence or weakness of power can also corrupt.

Ultimately Neustadt is a realist. For those who cherish more open and more participatory forms of governmental arrangements and for those who like to view Congress as the people's branch, the seat of republican government, and the agency of popular action, his message is unpleasant—at times even chilling. He knows that so much depends on clever leadership in the White House. He knows, too, how to help those presidents whom we want to succeed, although ironically his analysis and prescriptions are just as available to those presidents we wish to constrain. His optimism about the presidency may be a bit too comforting. Thus he dwells in his discussions of Nixon and Johnson on their deficiencies as men and seldom on the possibility that the presidency itself or our political arrangements could have contributed to each man's downfall. We need more attention to the persistence of the conditions that encouraged the imperial presidency.

Taken together, Neustadt's 1960 book, and its updates, are among the best analytical treatments we have of the modern American presidency. Those who find fault with them do so primarily because they wish he had significantly broadened his scope. Undoubtedly, too, they rest uncomfortably with his realistic, rather than idealistic, frame of reference. He did not dream about how to fashion a more democratic presidency. He has not outlined the "just society" our presidents should help us to achieve. Rather, he has treated, as best he could, the realities of Washington politics and brilliantly described for us how presidents might just possibly be able to make something of their increasingly no-win situation.

Did Neustadt's call for strong presidential leadership betray a liberal bias? For most of the post–World War II era, the heroic/strong presidency that Neustadt promoted was considered the model for the political left. Neustadt and others advocated a powerful activist presidency in pursuit of more government intervention in problem solving. From the 1950s through the 1970s, conservatives advocated more limited government, less presidential power, and a smaller role for the federal government.

But as testimony to the seductive lure of the heroic presidency model, by the 1980s even conservatives had been converted to presidential power. The heroic presidency could, conservatives found, be used to pursue the conservative agenda. Emboldened by Ronald Reagan's rhetorical skills and the promise of a "Reagan Revolution," conservatives saw the presidency as a vehicle for political power. Conservative centralizers sprang up all over.[36] Neustadt's strong presidency view became a philosophy for multiple political leanings.

POWER-MAXIMIZING STRATEGIES

Any essay dealing with presidential power must deal with presidential weakness as well as strength. The founders left the president vulnerable to several potential veto points, most notably the Congress. The framers did not want executive impotence—had that been their goal they could have stuck with the Articles of Confederation. But neither did they want executive dominance—had that been their goal they could have adopted the British parliamentary model with a fusion of executive and legislative power. They chose instead an executive office of limited powers that operated in a separation-of-powers model which fragmented power. Of course, such a system had consequences for the president's ability to lead.

Left to its own devices, the presidency has limited powers and is vulnerable to the will or whim of Congress and others. Neustadt rightly focuses our attention on the need for presidents to be conscious of their power positions.

As any student of the presidency knows, rarely can a president act alone. A president is usually dependent on the compliance or cooperation of others to achieve his or her goals. The presidency gives the occupant of the White House little more than the opportunity to govern.

In this Neustadt points us in the right direction. Presidents cannot merely command, they must persuade. But they must do much more. For presidents to be effective, and govern within the bounds of democratic leadership, they must also develop a sophisticated "strategic sense." They must know what to accomplish and how to accomplish it. And, their means and goals must aim to further the will of the people, empower citizens, and strengthen democratic accountability. Success is not measured merely by getting one's way but also by the ends to which policies are directed.

By strategic sense, we mean the ability to devise an overall plan that is designed to integrate means and ends, that shows what to do and how to do it, that links smaller tactical steps to a broad guiding vision. It is the president's game plan.

Surprisingly, few presidents take the time to think strategically. They tend to think short term, in an "I'll cross that bridge when I come to it," manner. They are more reactive than proactive. The world has a way of forcing presidents to respond rather than create. Someone once said that presidents cannot see beyond the next election. Today the time frame of presidential activity has shortened. Due to the proliferation of opinion polls, instant television analysis, and talk radio punditry, presidents feel compelled to worry how something will look on CNN in the next hour or two. This is unfortunate.

To overcome the roadblocks built into the system a president must develop a strategic sense of governing. Some elements of presidential leadership remain fairly constant over time—the Constitution, the separation of powers. Others are variable—the skill level of a president, the context or situation at hand, the nature of a president's mandate, the president's party strength in Congress. The president must develop a strategy designed to maximize political capital.[37]

Presidential leadership refers to more than mere officeholding. Leadership is a complex phenomenon involving *influence*—the ability to move others in desired directions. Successful leaders are those who can take full advantage of their opportunities, resources, and skills. Institutional structures, the immediate situation, the season of power, the political culture, the regime type, and the dynamics of followership define the opportunities for the exercise of leadership. The resources at a leader's disposal include constitutional and statutory powers, intermediaries, media, level of popularity, nature of the congressional majority, and policy arena. We often mistake resources for power. A successful leader converts resources to power. This does not happen automatically. To convert resources to power requires skill.

The leader's style, political acumen, experience, political strategy, management skills, vision, ability to mobilize political support, character traits, and personal attributes provide a behavioral repertoire, a set of skills. Opportunities, resources, and skills interact to determine the potential for success or failure in attempts to lead and influence.

Presidents who can play to the optimum their cards of opportunity, resources, and skill have a chance of succeeding.

But few presidents are FDR, and few face the opportunities for leadership that FDR enjoyed. Power is dispersed, it "exists only as a potential. Leadership is the means by which the president can exploit that potential. This is no easy task."[38]

The president has vast responsibilities, with high public expectations and limited power resources. But as an institution, the presidency has proven to be elastic; it stretches to accommodate skilled leaders in situations of high opportunity, but contracts to hem in less skilled leaders. Noting the variable nature of the power potential in the presidency, Robert Dahl has written, "The Presidency is like a family dwelling that each new generation alters and enlarges. Confronted by some new need, a President adds on a new room, a new wing; what began as a modest dwelling has become a mansion; every President may not use every room, but the rooms are available in case of need."[39]

The variable nature of presidential power has left presidency scholars a bit schizophrenic regarding the proper scope and limits of presidential power. When presidents performed credibly (1930s to early 1960s), they called for an enlarged institution. As presidents faltered (1960s to 1980s), they called for restrictions on presidential power. This variability is reactive, it is based on the subjective judgments derived from the short-term political performance of presidents. Such evaluations miss the key point: the *normal* state of the presidency is one of constrained power. To be successful, a president must overcome nature, or at least the nature of power built into the system. Thus, it is better to look at the requirements for leadership generically, at the preconditions necessary for the exercise of political leadership, than merely to react to the current temporary occupant of the Oval Office. In doing this we can see which threads run through the *presidency* as well as see how well a *president* performed.

The founders wanted a government of energy, but a specific type of energy, energy that resulted from consensus, cooperation, and collaboration. What model or type of leadership is required to move such a system? A model based on consensus, not fait accompli; influence, not command; agreement, not independence; cooperation, not unilateralism. Such a model of power requires presidents to think strategically.[40]

PRESIDENTIAL LEADERSHIP

The words *leadership* and *power* are often used interchangeably. This is misguided. Leadership suggests influence; power is command. Leaders inspire and persuade; power wielders order compliance. Leaders induce followership; power holders force compliance. Officeholders have some power by virtue of occupying an office. Leaders, on the other hand, must earn followership. The officeholder uses the powers granted by virtue of position. The leader tries to reshape the political environment, "he seeks to change the constellation of political forces about him in a direction closer to his own conception of the political good."[41]

Occasionally, a president can act on his own authority, "independent" of others. President Bush's 1989 decision to overthrow the government of Manuel Noriega in Panama is an example of this form of claimed power. But such unilateral acts are the exception, not the rule. In most cases, presidents share power. Therefore, the informal "powers" usefully discussed by Neustadt become important to presidents who wish to promote political change.

All presidents have some power, some ability to command. Yet such power is short-term and limited. It ceases to exist the moment the president leaves office. Its effects can often be undone by a new president.

A president's formal power, for so long the focal point of presidential studies,[42] includes constitutional authority, statutes, delegated powers, and those areas where others follow a president's command. But with the publication of Neustadt's *Presidential Power,* the ability to persuade took center stage in presidential studies. Rather than take an "either/or" approach to the debate over formal versus informal powers, one should see these two potential sources of strength as complementary, as presidential options in the pursuit of their goals. While presidents derive some of their power from constitutional sources, they derive other parts from political and personal sources. Effective presidents use *all* the resources available to them. They assess the situation and determine where their best chance for success rests, and are flexible enough to make the adjustments necessary to optimize their power. True leadership occurs when presidents are able to exploit the multifaceted nature of opportunities to both command and influence.

For better or worse, only presidential leadership can regularly overcome the natural lethargy built into the American system and give focus and direction to government. Congress can, on occasion, take the lead on select policy, as was the case when the Congress overcame presidential opposition from Ronald Reagan and imposed economic sanctions on the white minority government of South Africa. The Gingrich-led Congress in 1995 provides a few examples. But such cases are infrequent. Congress simply is not institutionally well designed to provide consistent national leadership over extended periods of time. In the second half of the twentieth century, the citizens have most often looked to the White House for leadership and direction. If the president does not lead, gridlock usually results. In this sense, John F. Kennedy's view that "The presidential office is the vortex into which all the elements of national decision are irresistibly drawn,"[43] is persuasive. Presidential leadership usually remains the key for moving the machinery of government.

THE BUILDING BLOCKS OF PRESIDENTIAL LEADERSHIP

Vision

The most important "power" a president can have is to present to the public a clear and compelling *vision*. An attractive, meaningful, positive vision that is rooted in building blocs of the past, addresses needs and hopes of the present, and portrays an image of a possible future opens more doors to presidential leadership than all the skills and resources combined. A powerful vision can transform a political system, recreate the regime of power, and chart a course for change. Vision energizes and empowers, inspires and moves people and organizations. A president with a compelling vision can be a powerful president.

Few presidents use to the fullest what Theodore Roosevelt referred to as "the bully pulpit" to develop a public philosophy for governing. Rather than attempting to educate and lead the public, most presidents serve as managers or clerks of public business. But if presidents wish to craft significant change, and not merely preside, they must use the bully pulpit to promote a moral and political vision in support of change.

A visionary leader gives direction to an organization or nation, gives purpose to action. Visionary leadership charts a course for action.

Visions are empowering. They are derived from the core values of a community, flow from the past, are about the future. Visions inspire, give meaning and direction to a community, are roadmaps. Visions are about achieving excellence.

"Leaders," wrote Warren Bennis and Burt Nanus,

> define what has previously remained implicit or unsaid; then they invent images, metaphors, and models that provide a focus for new attention. By so doing they consolidate or challenge prevailing wisdom. In short, an essential factor in leadership is the capacity to influence and organize meaning for members of the organization. . . . Communication creates meaning for people. Or should. It's the only way any group, small or large, can become aligned behind the overarching goals of an organization.[44]

Visionary leaders are remembered and continue to have an impact long after they have left office. Thus, long after FDR, John Kennedy, Martin Luther King, Jr., and Ronald Reagan were gone from public life, the power of their ideas and the impact of their words remain powerful forces in the political arena.

Ronald Reagan was able to mobilize and inspire his followers because he was skilled at presenting his vision to the public. In contrast, George Bush, who admitted he had trouble with "that vision thing," was often an uninspiring officeholder.

No other politician in America is better positioned to present a vision to the public than a president. Already the focus of much media and public attention, presidents can become "highlighters" of important issues to be addressed as a part of the president's agenda.

Skill

For successful political leadership, skill is important, but skill is never enough. Even the most skilled of individuals face formidable roadblocks. Skill helps determine the extent to which a president takes advantage of or is buried by circumstances, but circumstances or the political environment set the parameters of what is possible regarding leadership. President Reagan referred to the "window of opportunity," his way of talking about how open or closed circumstances were for exercising presidential leadership. Skilled presidents facing a closed window (e.g., the opposition party controlling Congress during a period of economic troubles where the president's popularity is low) will be limited in what they can accomplish. Presidents of limited skill, when the window of opportunity is open, will have much greater political leverage, even though their skill base is smaller. It is thus entirely possible for a president with limited skill to be more successful than a president with great skill. If one is dealt a weak hand, there is only so much skill can do. But if one is dealt four aces, what Machiavelli called "fortuna" (fortune), not skill, determines outcome.

This is not to say that skill is unimportant. But in the constellation of factors that contribute to success or failure, skill is but one, and probably not the most important element.

Before an election one often hears political cynics whine, "It doesn't matter who gets elected, they all end up doing the same thing anyway." Social science lends some support to this cynical view. After all, social scientists widely believe the institution, role, expectations, time, and other factors play a significant role in determining behavior. While the cynic has a point, it should not be taken too far. Individuals *are* constrained, but individuals do matter.[45]

One way of looking at the skill/opportunity dilemma is to focus on what is referred to as a president's "political capital." Is a president's political capital like a bank account where a one-time deposit is made at the beginning of his term and he invests carefully and draws on that account prudently, lest he run out of resources? Or can a president add to his bank account from time to time, renewing his resources?

President Bush seemed to think that his capital was a fixed sum, and in spending it, he dissipated it. Bill Clinton, on the other hand, did not view his political capital as a fixed asset. He regularly spent, then tried to replenish his assets as best he could. As President Clinton noted,

> . . . even a President without a majority mandate coming in, if the President has a disciplined, aggressive agenda that is clearly in the interest of the majority of the American people—I think you can create new political capital all the time, because you have access to the people through the communications network. If you have energy and sort of an inner determination that keeps you at the task, I think you can re-create political capital continuously throughout the Presidency. I have always believed that.[46]

Do leaders make a difference? Yes, sometimes. As political scientist Erwin C. Hargrove writes, "Surely the issue is not, Do individuals make a difference? but under what conditions do they make a difference? The relative importance of leaders varies across institutions and across time and place . . . The task of scholarship is to integrate the study of individuals with the social and institutional forces that move them and that they, in turn, may influence."[47]

At what point do individuals matter? When is skill an important variable in presidential success? While it may sound obvious, it is nonetheless important to note that "effectiveness in achieving goals was enhanced if skill and task were congruent."[48] To that we would add a third element: opportunity level. Skill, matched to task when the window of political opportunity is open, usually leads to goal attainment. Plainly the effectiveness of a leader is greater or lesser depending on the historical situation and the opportunities available to them.[49]

The impact of individual presidents who made a difference in specific policy areas is unmistakable: Lyndon Johnson and civil rights and the war on poverty, Richard Nixon and China, Jimmy Carter and human rights, Ronald Reagan and tax cuts, George Bush and the Gulf War, Bill Clinton and the Brady bill and NAFTA (North American Free Trade Agreement). These presidents made choices, moved in new directions, made a difference. Yes they were limited, yet they were able to overcome the natural lethargy of the system and succeed in selected policy arenas. That in most areas and on most issues presidential behavior appears more similar than dissimilar should not obscure the fact that on some issues at some times, some presidents can make a difference.

The problem is that determining what role skill played in these events is a slippery task. Could another president have opened doors to China? Perhaps, yet maybe not. Could any president emphasize human rights in his foreign policy? Probably. So where are we left? When does skill matter? How can we recognize or measure political skill? Regrettably, it is probably too elusive a concept to measure precisely. Suffice it to say that high levels of skill, task congruence, and high opportunity usually lead to presidential success, and low levels of the aforementioned usually lead to failure.[50]

To beg the question: If skill is of some importance, it is useful to ask, what skills are most useful to a president? Political experience is often cited as a requirement for effective leadership, and while this sounds like common sense, the correlation between experience and achievement is not strong.[51] After all, some of our most experienced national leaders were, in many respects, failures (Hoover, LBJ, Nixon, and Bush come to mind). But overall, more experience is better than less experience, although by no means a guarantee of successful performance. This suggests that other factors determine to a great degree the success or failure of a president.

Ronald Reagan and Bill Clinton serve as excellent examples of amateurs, in Washington, D.C. terms, who sometimes behaved amateurishly in the White House. Reagan in Beirut, and Clinton in Somalia, placed American military personnel in harm's way without sufficient support or protection and against the recommendations of more experienced advisers. In both cases, Americans were attacked and some killed (over 220 marines in Beirut and a dozen in Somalia) before changes in policy led to increased safety precautions. In both cases, leaders with little or no foreign policy experience made blunders that could have been, and probably should have been, avoided. More experienced hands might have known better.

Political skills are always in short supply, and a president with a high level of political acumen can sometimes gain the upper hand. By political skill we mean a good sense of timing (knowing when an issue is "ripe," when to move, when to hold back), task competence, a power sense, situational skills (crisis versus routine decision skills), policy skills (to develop sound workable programs), and political savvy (you need to love playing politics, as FDR did, not hate politics, as Nixon did). Of FDR's love of and for politics, James MacGregor Burns has written, "If other leaders bent under the burdens of power, Roosevelt shouldered his with zest and gaiety. He loved being President; he almost always gave the impression of being on top of his job. Cheerfully, exuberantly, he swung through the varied presidential tasks . . . The variegated facets of the presidential job called for a multitude of different roles, and Roosevelt moved from part to part with ease and confidence."[52]

Presidents also need people skills. They must know how to persuade, bargain, cajole, and co-opt. They must be masters of self-presentation. They must be able to motivate and inspire, to gain trust and influence. Charisma helps. In short, occupants of the White House must master the art of "presidential schmoozing."

Personality skills are also a significant part of the arsenal of presidential requirements.[53] All presidents have a strong drive for power. Some hold it within healthy bounds; others, like Nixon, destroy themselves. To be effective, presidents should be self-confident, secure, flexible, and open. Presidents who are consumed by self-doubt, insecurity, and rigidity (Barber's "negative" characters) are often dangerous and more apt to abuse power.

Presidents also need self-knowledge. We all have our weaknesses, but we do not all let our weaknesses consume us. Good presidents recognize their strengths *and* weaknesses, and attempt to deal constructively with weakness, come to grips with failings, offset the negative. For example, presidents who are inexperienced in foreign policy may need to compensate by surrounding themselves with very experienced foreign policy insiders.

Managerial skills are also important if a president is to succeed.[54] Upon taking office presidents tend to see only the personal as important; the historical and institutional are often downplayed.

And a president must have personal skills. By this we mean he must be disciplined, intelligent, have stamina, show sound judgment, and act with maturity. Good presidents are creative, empathetic, and expressive. They should be relentlessly optimistic. They must also have a sense of humor, and learn to control their temper. President Reagan's self-effacing sense of humor served him well as president, it disarmed opponents and won over much of the public.

Political Timing

The "when" of politics also matters greatly: when major legislation is introduced, when the public is ready to accept change, when Congress can be pressured to act, when to lead and when to follow, when to push and when to pause. A sense of political timing, part of the overall "power sense" all great leaders have, helps a president know when to move, when to retreat, when to push, and when to compromise. The transition and the honeymoon period are especially important periods.

Getting a good start ("hitting the ground running") is a key element of political success, and during the transition, the eleven-week period between the November election and the January inauguration, some of the most important work of an administration is done. Here the groundwork is laid for much that will follow and a tone is set that shapes the way others see the new administration. During the transition, the president makes key decisions on who the top advisers will be, who will fill important cabinet positions, how the staff will be structured, what decision style will be employed, whether to pursue a partisan strategy or try to woo the opposition, how to mobilize the public, and what issues the administration will push during the first year. These "who, what, where, why, and when" of politics set the stage for not only how the administration will operate but how the public, Congress, and media will view and respond to the new administration.

"Power is not automatically transferred, but must be seized. Only the *authority* of the presidency is transferred on January 20; the *power* of the presidency—in terms of effective control of the policy agenda—must be consciously developed."[55] To seize power, presidents must adopt a strategic approach to the transition, one that leaves little to chance and that deals self-consciously with the use of power.

Presidential scholars agree that a president's first one hundred days and first six to ten months are crucial periods for setting the agenda. It is important to win at least a few key legislative victories as soon as possible.

Strong twentieth-century presidents such as Woodrow Wilson, Franklin Roosevelt, Lyndon Johnson, and Ronald Reagan began with clear goals and pushed Congress to approve bold new programs. Indeed, each new president is in effect invited in the first year to share a vision and a national agenda with attentive Washington, national, and international audiences.

When Reagan took office, for example, his administration self-consciously chose to focus on a select few big-ticket items: tax cuts and increased defense spending. Nearly all else, including important foreign policy matters, were put on the back burner. This allowed the Reagan administration to concentrate its energies and also conveyed an image that this president could succeed. He appeared to be, and for a time was, influential.

Scholars often point to the Reagan honeymoon as an example of how to get the most out of early opportunities. Reagan had an attractive personality, a compelling television presence, a vision that he ably shared, and a clear focus on only a few key ideas. During his transition, his key advisers put together, with little involvement from the president-elect, a narrow, focused agenda and initiated a public relations and legislative blitzkrieg that ended in the president gaining much of his early legislative agenda. This in turn added to the aura of power surrounding Reagan, and made it easier for him to win subsequent contests because the Congress "feared" him.[56]

In contrast, George Bush virtually relinquished the advantages offered by the honeymoon period. He got off to a slow start and appeared more concerned with managing a response to events as they happened than with shaping events by his actions. His early weeks were marked by caution and political orthodoxy.

Bush signaled early on that there was not going to be a Bush Revolution, a New Deal, or a Great Society. Pundits searching for a label suggested it might be the "Beige" or "Bland Deal." Conservative columnist George Will noted the absence of any large purpose or specificity and labeled Bush the "pastel president." A *Wall Street Journal* evaluation of Bush's first one hundred days found him blending into the landscape, reacting to events, and allowing Congress or others to set the national agenda. Bush rarely gave the American people a sense of where he wanted to lead. He was generally unclear on what he wanted to do with the presidential powers he had won in the 1988 presidential election.[57]

A president's strategic sense must also take into account other key elements of presidential power maximization: popularity and public opinion, relations with and the number of the president's party members in Congress, media relations, management skills, control of the agenda, and coalition and consensus building.

MORE POWER TO THE PRESIDENCY: AN ENDURING THEME

What about the prevailing attitudes toward the presidency today? Are they following the Neustadt line or are new models available? Even as the presidency was being soundly criticized for its abuses of power in the late 1960s, early 1970s, and the mid-1980s, it was simultaneously portrayed by many people as alarmingly weakened by Vietnam, Watergate, and Iran-Contra. The ranks of the defenders of presidential government may have been temporarily thinned in the mid-1970s, but as of the mid-1990s Neustadt's view, only slightly modified, is alive and well.[58]

Americans may have lost confidence in many of their leaders, yet they have not lost hope in the efficacy of strong purposive leadership. Back in the 1970s, all the revelations about the crimes of Watergate and the dramatic resignation of a president lulled people into believing perhaps that "the system worked," that checks checked and balances balanced. Perhaps the very cataloging of the misuses of presidential powers solved, or seemed to solve, the problem. By 1980 people were asking, "Whatever happened to the imperial presidency?" Since the end of the Watergate era, the public has been more comfortable with efforts at strong presidential leadership. They welcomed Ronald Reagan's efforts at moving the system, and have generally been more responsive to calls for presidential power. Now, observers are more likely to complain of an imperial judiciary, an imperial bureaucracy, an imperial Congress, or an altogether too powerful press and special interests.

In the wake of the wounded or imperiled presidency of the Watergate era, could Congress furnish the leadership necessary to govern the country? Most scholars and writers said no. Even in the wake of the Gingrich-led Republican House of the mid-1990s, most commentators questioned the ability of Congress to present coherent national policies. The conventional answer was that "we will need a presidency of substantial power" if we are to get on top of our domestic problems and maintain our leadership position in foreign affairs.

Defenders of a powerful presidency, such as Samuel Huntington and others, wonder how a government could conduct a coherent foreign policy if legislative ascendancy really meant the development of a Congress into a second U.S. government. Could the United States afford to have two foreign policies? A nation cannot retain for long a leadership role in the world unless it is both clear and decisive. In the absence of those cherished Hamiltonian virtues, chaos would reign. They feared, too, that in establishing its foreign policy decisions, congressional government would operate almost entirely on the basis of domestic politics, a purview that limits its competence in the field of foreign affairs.

Political scientist Huntington, writing in 1975, urged readers to recognize the legitimacy and the necessity "of hierarchy, coercion, discipline, secrecy, and deception—all of which are, in some measure, inescapable attributes of the process of government." "When the President is unable to exercise authority . . . ," wrote Huntington, "no one else has been able to supply comparable purpose and initiative. To the extent that the United States has been governed on a national basis, it has been governed by the President." [59]

The same verdict is heard from those who yearn for strong creative leadership in domestic or economic matters. Thus, Arthur Schlesinger, Jr., even as he condemned the imperial presidency, said that "history has shown the presidency to be the most effective instrumentality of government for justice and progress." [60]

Time and again, people caution against overreacting to Watergate and Iran-Contra. Do not be ahistorical, they argue. Quoting Harold Laski, they say, "Great power makes great leadership possible."

As almost always during the twentieth century, advocates of a strong presidency continue to lament that presidential powers are not stronger. They lament as well, along with Neustadt, that recent presidents have not been effective seekers and wielders of influence. The presidency, they still contend, is America's strongest weapon against such banes of progress as sectionalism, selfish or overly concentrated corporate power, and totalitarianism abroad.

Americans still long for dynamic, reassuring, and strong leadership.

Watergate and Iran-Contra notwithstanding, we still celebrate the gutsy, aggressive presidents—even if many of them did violate the legal and constitutional niceties of our separation-of-powers ideal. We remember fondly Kennedy's command of press conferences and Lyndon Johnson's mastery of Congress. The day of the strong president is here to stay. People want it that way.

The central challenge, then, is not to reduce the president's power to lead, to govern, and to persuade but to ensure that the means to lead, to govern, and to persuade are not corrupted. Such is the neo- or modified-Neustadt conception as we end the twentieth century.

How you stand on the question of how strong the presidency should be depends usually on what policies you favor and how these policies are advanced or hampered by the president or by Congress. It matters too, of course, whether you like the person who is in the White House. If you approve of a president and most of that president's policies, the tendency is to believe that the president should be, to paraphrase Woodrow Wilson, free to be as big a leader as he or she can be.

CONCLUSION

Bold, effective presidential leadership (true leadership as opposed to mere officeholding) is uncommon. In those rare moments of leadership, presidents are able to animate citizens and mobilize government, develop a vision and establish an agenda, move Congress and push the bureaucracy. Such presidents recast the arena of the politically possible. It has not happened often. The forces arrayed against a president usually have the upper hand. The power of lethargy is considerable.

The preconditions for effective presidential leadership are rarely in syncopation: skill, the right timing, a consensus, a governing coalition, high popularity, vision, and a clear mandate. Presidents, to be effective leaders in a democratic system, must bring Hamiltonian energy to a Madisonian system for Jeffersonian ends. Few presidents can achieve this.

To succeed, presidents must be masters of the light (education, vision, mobilization) and the heat (power, bargaining). Given the incredible array of skills and circumstances necessary for presidents to succeed, no wonder many if not most of them fail to live up to our expectations.

Some of the difficulty rests with individual presidents who lack valid programmatic ideas as well as sufficient political skill. Part of the challenge rests with the structural design of the American constitutional system, which separates and fragments power.

In sum, the American presidency, public and international misperceptions to the contrary, is in many ways a limited office with limited powers.

A tension will always exist between the need for leadership and the demands of democracy. The English historian, James Bryce, reminded us that "Perhaps no form of government needs great leaders so much as democracy."[61] But what kind of leadership? The strong, powerful direction of a heroic leader, or the gentle guiding hand of a teacher? One old saw warns us that "strong leaders make a weak people," but can the people come together and accomplish their goals with weak leadership?[62] Proponents of robust democracy realize that leadership has rarely fit comfortably with democracy in America. "The claim of leaders to political precedence violates the equality of democratic citizens," writes Bruce Miroff. "The most committed democrats have been suspicious of the very idea of leadership. When Thomas Paine railed against the 'slavish custom of following leaders,' he expressed a democrat's deepest anxiety."[63]

This tension has not prevented Americans from looking to strong leaders to guide the republic, especially during an emergency. As historian Arthur M. Schlesinger, Jr., noted,

> The American democracy has readily resorted in practice to the very leadership it had disclaimed in theory. An adequate democratic theory must recognize that democracy is not self-executing: that leadership is not the enemy of self-government but the means of making it work; that followers have their own stern obligation, which is to keep leaders within rigorous constitutional bounds; and that Caesarism is more often produced by the failure of feeble governments than by the success of energetic ones.[64]

There are ways to bring leadership and democracy together in a creative tension that calls for a role for the leader while also promoting participation and practices among the citizenry. Just as Abraham Lincoln gave us a succinct definition of democracy as "government of the people, by the people, and for the people," so too did one of America's other Mt. Rushmore leaders give us an appropriate definition of democratic leadership. Thomas Jefferson believed that the primary duties of a leader in a democracy were "to inform the minds of the people, and to follow their will."[65]

Informing the minds of the people speaks to the role of leader as educator. In a democracy, the leader has a responsibility to educate, enlighten, and inform the people. He or she must identify the problems and mobilize the people to act. By informing or educating the citizenry, the leader also engages in a dialogue, the ultimate goal of which is to involve leader and citizen in the process of developing a vision, grounded in the values of the nation, that will animate future action.

The leader's task in a democracy is to look ahead, foresee problems, focus the public's attention on the work that must be done, provide alternative courses of action, chart a path for the future, and move the nation in support of ideas. Former President Jimmy Carter nicely adds that "a government's ultimate goals are to preserve security and to insure justice: to treat people fairly, to guarantee their rights, to alleviate suffering and to try to resolve disputes peacefully."[66] No small order.

The second element in Jefferson's definition is also important. After educating and involving the people, the leader must ultimately follow the will of the people. There is, of course, a distinction between the whim of the people (temporary and changing) and the will of the people (deeply held truths that speak to the nation's highest aspirations). A president's initial responsibility is to inform, educate, and persuade the public to embrace and work for a vision that resonates with the deeper truths and higher purposes of this nation. Then, the leader must serve the people, pursue their best interests, and ultimately follow their direction. This form of leadership is demanding, time consuming, and fraught with pitfalls. Yet this is the type of leadership that strengthens both citizens and constitutional democracy.

CHAPTER 5

The Presidential Job
Description in a
Separated System

And people talk about the powers of a President, all the powers that a Chief Executive has, and what he can do. Let me tell you something—from experience! The President may have a great many powers given to him in the Constitution and may have certain powers under certain laws which are given to him by the Congress of the United States; but the principal power that the President has is to bring people in and try to persuade them to do what they ought to do without persuasion. That's what the powers of the President amount to.

Harry S. Truman, *Public Papers of the Presidents of the United States* (U.S. Government Printing Office, 1949), p. 247

I find that when things go badly, it becomes our business. When the stock market goes down, letters are addressed to the White House. When it goes up, we get comparatively few letters of appreciation. But when you have high unemployment, it is because the President hasn't gotten the country moving again.

John F. Kennedy, quoted in H. W. Chase and A. H. Lerman, eds., *Kennedy and the Press* (Crowell, 1965), pp. 426–27

You have heard about the man tarred and feathered and ridden out of town on a rail? A man in the crowd asked him how he liked it, and his reply was that if it wasn't for the honor of the thing, he would much rather walk.

Abraham Lincoln, reply to a friend who asked how it felt to be president, c. 1862

Anyone searching for the paradoxes of the American presidency need look no further than the president's informal as well as formal job description: high expectations yet limited power; many demands, few resources. We expect presidents to accomplish great things, yet tie their hands. Of course, the job of the president has changed and enlarged dramatically since 1789. While the constitutional provisions relating to presidential power remain virtually unchanged, the responsibilities and expectations placed on the office have increased significantly.

Historians still debate the motives of the framers of the U.S. Constitution. Were they great democrats or protectors of the propertied interests? Did they attempt to empower or enfeeble government? Did they retreat from full democracy because they feared the people or because they were responsible leaders who saw a representative government as most likely to protect liberty and achieve stability?[1]

Whether great advocates of or traitors to robust democracy, it is clear the framers harbored concerns about both mass-based democracy and executive tyranny. These concerns are evident when one examines the inventing of the American presidency.

To understand what is expected of presidents today, we need to examine the invention of the presidency as well as its evolution over time. Only then can we fully appreciate the many roles and demands placed on the presidency.

THE PRESIDENCY AS DEFINED AND DEBATED IN 1787

The presidential job description as outlined in the Constitution was a medley of compromises. The framers wanted a presidency strong enough to do what was asked of it and yet not one that would use governmental authority for selfish ends or contrary to the general welfare. In almost every instance presidential powers were shared powers. Perhaps only the pardon power was a truly imperial grant of power that allowed presidents alone to monopolize an area of policymaking.

Despite the administrative, diplomatic, commander-in-chief, and veto powers granted to them, presidents found they had to act within a set of strong constitutional, political, and social restraints. They had to be sensitive to the dominant elites, the cultural "rules of the game," and, of course, the threat of being impeached or turned out of office at the next election.

The writers of the U.S. Constitution created, by design, what some call an "antileadership" system of government. Their goal was less to provide for an efficient system of government than one that would not jeopardize liberty. Freedom was their goal; governmental power their concern. The individuals who toiled through that hot summer of 1787 in Philadelphia thus created an executive institution, a presidency, that had limited powers.

Primarily, the framers wanted to counteract two fears: the fear of the mob (democracy, or mobocracy), and the fear of the monarchy (centralized, tyrannical executive power). The menacing image of England's King George III, against whom the colonists rebelled and whom Thomas Paine called "the Royal Brute of Britain," served as a powerful reminder of the dangers of a strong executive. To contain power they set up an executive office that was constitutionally rather weak (Congress had, on paper at least, most of the power), based on the rule of law, with a separation of powers, ensuring a system of checks and balances.

To James Madison, the chief architect of the Constitution, a government with too much power was a dangerous government. A student of history, he believed human nature drove people to pursue self-interest, and therefore a system of government designed to have "ambition checked by ambition" set within rather strict limits was the only hope to establish a stable government that did not endanger liberty. Realizing that "enlightened statesmen" would not always guide the nation, Madison embraced a checks and balance system of separate but overlapping and shared powers. Madison's concern for a government that controlled and limited powers is seen throughout his writings, but nowhere is it more vivid than when he wrote in *Federalist Papers No. 51,* "You must first enable the government to control the governed; and in the next place, oblige it to control itself."[2]

Yes, government was to have enough power to govern, but no, it could not have enough power to overwhelm freedom. If one branch were empowered to check another, tyranny might be thwarted. There is scant concern in Madison's writings for the needs of a strong government. Thus, the Constitution was both an enabling and a constraining document.

Alexander Hamilton emerged as the chief defender of a powerful executive. An advocate of strong central government, Hamilton promoted, especially in the *Federalist Papers,* a version of executive power quite different from Madison's dispersed and separate powers. Where Madison wanted to check authority, Hamilton wished to enhance authority; where Madison believed the new government's powers should be "few and defined," Hamilton wanted to infuse the executive with "energy": A feeble executive implies a feeble execution of the government. A feeble execution is but another phrase for a bad execution; and a government ill executed, whatever it may be in theory, must be, in practice, a bad government.[3]

Hamilton wanted a strong president within a more centralized government. Yet such a system would undermine Madison's determination to check government power—the presidency, a unitary office headed by one person, would have no internal check. Thus Madison insisted on the need for strong external checks, that is, a strong Congress. While Madison may have won the day at the Convention, creating a presidency with fairly limited powers, history has been on the side of Hamilton.

Looming in the background was the influential presence of Thomas Jefferson. Friendly toward small government and democracy, Jefferson was suspicious of power. But Jefferson's vision of a decentralized government, an agrarian economy, and a robust democracy was given little attention at the convention.

The Madisonian model calls for both protections against mass democracy and limits on governmental power. This is not to say the founders wanted a weak and ineffective government—had that been their goal they could have kept the Articles of Confederation. But they did want a government that could not act too easily. The theory of government the Madisonian design necessitates is one of consensus, coalition, and cooperation on the one hand and checks, vetoes, and balances on the other.

The result is, as already stated, a rather weak presidency with few independent powers. The paradox thus created, especially in the modern period, is, again as already stated, how can a president bring Hamiltonian energy to this Madisonian system for Jeffersonian ends? The framers did not make it easy for the government to act or for presidents to lead. That was decidedly not their intent. And they left the powers and contours of the office somewhat vague—expecting Washington to fill in the gaps. This created, in Edward S. Corwin's words, "an invitation to struggle" for control of government.[4] A modern efficiency expert, looking at the framers' design for government, would likely conclude that the system could not work well: too many limits, too many checks, not enough power, not enough leadership. But this is the way the framers wanted it.

What did the framers create? The chief characteristics or mechanisms to control and empower the executive are: (1) *limited government*—a reaction against the arbitrary, expansive powers of the king or state, and a protection of personal liberty; (2) *rule of law*—only on the basis of legal or constitutional grounds can the government act; (3) *separation of powers*—the three branches of government are divided but sharing in power; and (4) *checks and balances*—each branch can limit or control the powers of the other branches of government.

In this system, what powers and resources has the president? Limited powers. In fact, the Constitution does not clearly spell out the power of the presidency. Article I is devoted to the Congress, the first and constitutionally the most powerful branch of government. Article II, the executive article, deals with the presidency. The president's power cupboard is, compared with that of the Congress, nearly bare or at least vague. Section 1 gives the "executive power" to the president, but does not reveal whether this is a grant of tangible power or merely a title. Section 2 makes the president commander-in-chief, but reserves the power to declare war for the Congress. Section 2 also gives the president power to grant reprieves and pardons, power to make treaties (with the advice and consent of the Senate), and power to nominate ambassadors, judges, and other public ministers (with the advice and consent of the Senate). Section 3 calls for the president to inform the Congress on the state of the union and to recommend measures to Congress; grants the power to receive ambassadors; and imposes on the president the duty to see that the laws are faithfully executed. These powers are significant, yet in and of themselves do not suggest a strong or independent leadership institution.

Thus, the president has two types of power: *formal,* the ability to command; and *informal,* the ability to persuade. The president's formal powers are limited and (often) shared. The president's informal powers are a function of skill, situation, and political time. While the formal powers of the president remain fairly constant over time, the informal powers are variable, dependent on the skill of the individual president. This is not to say a president's formal powers are static—over time presidential power has increased significantly—but the pace of change has been such that it was over one hundred years before the presidency assumed primacy in the U.S. political system.

The structure of government dispersed or fragmented power; there is no recognized, authoritative vital center; power is fluid and floating; no one branch could easily or freely act without the consent of another branch; power was designed to counteract power; ambition to check ambition. It was a structure developed by men whose memories of tyranny and arbitrary power by the King of England were fresh. It was a structure designed to force a consensus before the government could act. The structure of government created by the framers did not create *a* leadership institution, but three separate semiautonomous institutions that shared power. One scholar concluded, "the framers designed a system of shared powers within a system that is purposefully biased against change."[5]

The framers purposely created uncertainty about who holds power in the United States. A system of cross-powers and checked powers created a constitutional mechanism that prohibited one branch from exercising too much power on its own. Opportunities to check power abound; opportunities to exercise power are limited.

THE PRESIDENCY AS REDEFINED BY WASHINGTON AND HIS SUCCESSORS

As noted, the Constitution was the result of bargains and compromises. One area of dispute was the power the new president would be granted. Unable to come to terms with this question, the framers were forced to leave the powers of the president somewhat ambiguous. This was not especially troublesome because they knew who the first president would be: George Washington. The framers held Washington in such high regard that they were confident he would set the proper tone for the office.

George Washington loomed large in the early republic. Washington sailed in uncharted waters and is credited with establishing generally sound precedents. He lent dignity to this newly invented presidency, developed a version of consensus leadership where possible and executive independence where necessary, was guided by the rule of law and recognized the limits of his power, and generally remained true to the requirements of constitutional government. Above all, he established the legitimacy of the office of president.[6]

Keenly aware that as the first occupant of the presidential office his every act of commission and omission would be noticed, scrutinized, and perhaps established as precedent, Washington was careful to establish the independence of the office yet still respect the integrity of the Congress. Shortly after his inauguration he commented that "many things which appear of little importance in themselves . . . may have great and durable consequences from their having been established at the commencement of a new general government."[7] Washington also imposed his regal persona and his republican sentiments on the new office, and in effect, brought to life what the framers had merely invented.

Following Washington, the power of the presidency rose and fell depending on circumstances, the will of the president, and the demands of the time.[8] Thomas Jefferson (1801–9) stretched the powers of the presidency, using the nascent political party to aid in achieving his goals.[9] Andrew Jackson (1829–37) asserted the independence of the presidency and linked his power to the people. Jacksonian democracy represented a broadening of democracy and a connection between the president and the people about which the framers were suspicious. Jackson portrayed himself as the "tribune of the people" and asserted a brand of leadership linked to popularity.[10]

Abraham Lincoln (1861–65) demonstrated that during a crisis, the powers of the presidency could be expanded. Using a combination of claimed emergency and war powers, Lincoln took bold action during the Civil War, blockading Southern ports, calling up the militia, arresting persons suspected of disloyalty, suspending the writ of habeas corpus, all without congressional authorization. Lincoln knew he had intruded into areas of congressional authority, yet claimed the doctrine of necessity as his justification.[11]

During the presidency of Theodore Roosevelt (1901–9) the United States emerged as a world power, intent on making its mark on the international stage. TR aggressively asserted presidential power both at home and abroad, and reestablished presidential primacy. Roosevelt used the presidency as a "bully pulpit" and contributed to the rise of what is called the "rhetorical presidency,"[12] seeing the president as the "steward of the people." Viewing power expansively, TR asserted that "it was not only his right but his duty to do anything that the needs of the nation demanded unless such action was forbidden by the Constitution or by the laws. . . . Under this interpretation of executive power I did and caused to be done many things not previously done by the president and the heads of the departments. I did not usurp power, but I did greatly broaden the use of executive power."[13]

The presidency of Woodrow Wilson (1913–21) further established the United States as a world power and the presidency as a pivotal center or lever of American government. Wilson used powerful rhetorical skills in conjunction with a reliance on party and programmatic leadership to establish the presidency as the guiding force for the nation. Combined with his leadership during World War I, Wilson elevated the office of presidency to one of national and international leadership.[14]

Mixed in between the Jeffersons, the Lincolns, and the TRs were a series of lesser or even lackluster presidents who often diminished the office. For every Jackson, there was a Tyler; for every Lincoln, a Grant; for every Wilson, a Harding. Yet if there were presidential underachievers along with those who stretched the boundaries of presidential power, clearly the institutional trend was in the direction of growing and increased influence.

THE PRESIDENCY AS REDEFINED BY FDR AND THE MODERN PRESIDENTS

The institution of the presidency as we know it today was born in the 1930s. The stepchild of depression and war, it was during the Franklin D. Roosevelt era (1933–45) that the presidency became a truly modern institution.

The economic depression of 1929 and the world war a decade later contributed to the presidency-centered nature of our national government, and the leadership style and skill of FDR confirmed the United States (for better or worse) as a presidential nation.

FDR, considered by presidential scholars to be one of the nation's greatest presidents, was a powerful and effective chief executive. Under his leadership the presidency became the prime mover of the American system, and people began to look to the federal government and to the presidency as the nation's problem solver. The federal government had grown, and with it presidential responsibilities, ending the era in which presidents such as Calvin Coolidge could claim that his greatest accomplishment was "minding my own business."

Roosevelt was successful. In fact his success transformed both the presidency and public attitudes about it. He created expectations of presidential power and leadership that would be imposed on his successors. The "heroic" model of the presidency was established as a result of FDR's leadership, and for the next forty years, presidential scholars would often promote the model as good and necessary. FDR's successors would labor in his dominant shadow.[15] Elected four times, FDR got the cumbersome system moving. Increasingly, the presidency became more powerful and more personalized.

If Roosevelt was an important president, the myth of FDR took on even greater stature. An inflated view of Roosevelt passed for fact in popular and scholarly mythology of the presidency. He was, of course, not as powerful as many remembered him.

Truman assumed the presidency in the final days of World War II. In an effort to hasten an end to the war, Truman ordered atomic bombs to be dropped on Japan. After the war, it was Truman and his advisers who implemented the "containment" policy toward the Soviet Union, a policy each succeeding president would, more or less, follow until the end of the Cold War. Truman presided over the Marshall Plan for European recovery, helped establish the North Atlantic Treaty Organization (NATO), and led the United States back to a postwar domestic economic revival. It was also under Truman that the Korean War began.

To the surprise of most of his contemporaries, Truman became an effective president, though his popularity could never match that of FDR, and it would be only much later that his contribution was fully appreciated. This created a sense in which people said, "Surely if *he* can do the job, there must be something inherent in the office which brings out greatness in even the most common of men." Thus was born the "FDR halo," which could be passed down from president to president, a kind of magic that seemed to confer special powers on the occupant of the White House.

When Dwight D. Eisenhower became president in 1953, the Republican lent a bipartisan air to the majesty of the office. While not an activist president, Ike did manage to exert a hidden-hand type of leadership in an era where the public seemed anxious to take a break from the hurly-burly world of politics.[16] After all, the United States had been through a depression in the '30s, a world war in the '40s, and a nascent cold war in the late '40s and early '50s, and by the Eisenhower era the American people welcomed a low-profile leader. Ike, with a low-key, almost apolitical style, gave them what they wanted. He was popular, especially for a president who seemed to do little, and his popularity extended across the entire eight years of his tenure.

If the FDR halo seemed to be in limbo during the Eisenhower years, Ike's successor was determined to revive it. John Kennedy proposed an activist administration. Yet try as he might, President Kennedy's legislative proposals often fell prey to unresponsive leaders in Congress. Stymied by an intransigent Congress that took the system of checks and balances quite seriously, the Kennedy legislative record was at best mixed. The first Roman Catholic ever to be elected president, Kennedy barely won the presidency in 1960. Kennedy was responsible for the Bay of Pigs fiasco in Cuba, placed thousands of military advisers in Vietnam, and successfully led the nation through the Cuban missile crisis. His few progressive domestic initiatives, however, were blocked by a Congress controlled by conservatives in his own party. Kennedy did achieve tax cuts that stimulated economic growth, started the Peace Corps, and outlined the need for civil rights reform.

Scholars were as captivated as anyone at the prospect of a strong president leading the fragmented system to new heights, overcoming the chains of the checks and balances, and achieving mighty and good ends.

A sampling of quotes from the classic *The American Presidency* by Clinton Rossiter, first published in 1956, gives an indication of the status and esteem in which even the conservative Rossiter held the presidency.

- Few nations have solved so simply and yet grandly the problem of finding and maintaining an office or state that embodies their majesty and reflects their character . . .

- There is virtually no limit to what the President can do if he does it for democratic ends and by democratic means . . .

- He reigns, but he also rules; he symbolizes the people, but he also runs their government . . .

- The President is not a Gulliver immobilized by ten thousand tiny cords, nor even a Prometheus chained to a rock of frustration. He is, rather, a kind of magnificent lion who can roam widely and do great deeds so long as he does not try to break loose from his broad reservation.[17]

Rossiter wrote that the American presidency was "one of the few truly successful institutions created by men in their endless quest for the blessings of free government." He concluded:

> It is, finally, an office of freedom. The Presidency is a standing reproach to those petty doctrinaires who insist that executive power is inherently undemocratic; for, to the exact contrary, it has been more responsive to the needs and dreams of giant democracy than any other office or institution in the whole mosaic of American life. . . . The vast power of this office has not been "poison," as Henry Adams wrote in scorn; rather, it has elevated often and corrupted never, chiefly because those who held it recognized the true source of the power and were ennobled by the knowledge.[18]

Rossiter was not alone in his celebration of the presidency and presidential power. Plenty of other scholars joined in the chorus, viewing the president as the indispensible agent of reform. We can thus see the intellectual origins of what is called the "textbook presidency," a charismatic leader who was good, powerful, and essential. This textbook version plainly romanticized the office.

A presidency-centered model that came to dominate thinking was more than an operating style of government, it was also a philosophy of governing. It was an operating style in that it promoted a system of government in which the president was to direct or lead the people and the other branches of government from a perch of great power. It was a philosophy of government in that it legitimized a strong central government, which took power away from the other branches and, perhaps more important, took power from the people, and vested responsibility in the hands of government (the president) to solve problems. This diminished democratic responsibility in the people (a Jeffersonian goal), and promoted responsibility and power in the leadership class (a Hamiltonian goal). It also failed to recognize the potential danger of the heroic leadership model.

Lyndon Johnson's notable legislative achievements in the wake of the tragic assassination of John Kennedy confirmed for many the wisdom of the strong presidency model. The FDR halo was revived. Lyndon Johnson temporarily brought the strong presidency model back to life.

If the public suspended its skepticism about a strong presidency and placed its faith in the heroic model, why did scholars go along? There were voices in the wilderness, warning of the dangers of unchecked presidential power,[19] but in general scholars, like the public, were intrigued by or perhaps seduced by the strong presidency.

Just when the public was lulled into a false sense of complacency and security concerning the benevolence of presidential power, however, the war in Vietnam and the reaction against it caused everyone to rethink our assumptions.

U.S. involvement in Vietnam began quietly, escalated incrementally, and eventually led to tragedy.[20] By 1968, the United States was engaged in a war it could not win and from which it could not (honorably) withdraw. It was a "presidential war," and it brought the Johnson administration to its knees.[21]

As U.S. involvement escalated, and as victory seemed further and further away, blame was placed squarely on the shoulders of President Johnson. While the Constitution gives the power to declare war to the Congress, since the Truman administration, presidents had acted unilaterally in Vietnam. By the time Johnson came to office, presidents had been setting policy in Vietnam for twenty years, virtually unencumbered by the Congress. The tragedy of Lyndon Johnson is that after a positive start, after important legislative successes, the tragedy of Vietnam would overwhelm him and the nation. The nation was torn apart. The strong presidency, so long seen as critically important for the American system, now seemed too powerful, too unchecked, in short, a threat. After years of hearing calls for "more power to the president," by the late 1960s the plea was to rein in the overly powerful White House.

It was a rude awakening. All the hopes, all the trust, all the expectations that had been entrusted to the presidency had been jolted.

If Vietnam led to questions about the power of the presidency, Johnson's successor Richard Nixon would raise new concerns about that power. With the constitutional crisis known as Watergate, Richard Nixon, named an "unindicted co-conspirator" by a grand jury, became the nation's first president to resign from office.

As a reaction against the perceived abuses of power by Johnson and Nixon, Congress attempted to reassert its power. The Ford and Carter presidencies, rather than being imperial, seemed imperiled. It was a time of presidents being constrained.

In the 1980s, Ronald Reagan began to resurrect presidential power.

Yet his several successes were matched, if not overshadowed, by enormous budget deficits and the Iran-Contra scandal.

THE JOB OF THE MODERN PRESIDENT

Nowadays a president is asked to do countless things that are not spelled out with any clarity in the Constitution. We want the president to be a national renewer of morale as well as an international peacemaker, a moral leader as well as the nation's chief economic manager, a politician-in-chief, and a unifying representative of all the people. We want every new president to be everything, at least of virtue, that all our great presidents have been. No matter that the great presidents were not as great as we think they were. Rightly or wrongly, we believe our greatest presidents were men of talent, tenacity, and optimism, men who could clarify the vital issues of the day and mobilize the nation for action. We like to think our great presidents were transforming leaders who could not only move the enterprise forward but could summon the highest kinds of moral commitment from the American people.

Yet rarely is a president a free agent. He nearly always mirrors the fundamental forces of society: the values, the myths, the quest for order and stability, and the vast, inert, and usually conservative forces that maintain the existing balance of interests. Ours is a system decidedly weighted against radical leadership, a system that encourages most presidents, most of the time, to respond to the powerful, organized, and already represented interests at the expense of the unrepresented. Moreover, a president today must preside over a highly specialized and sprawling bureaucracy as well. A president can easily find himself sitting at the White House, "overworked and making the best of a bad situation, while all around him he has princes and serfs doing and undoing in thousands of actions the work of his administration without his having a clue."[22]

Recent presidents have often grown publicly frustrated in the job. Carter once likened it to an endless multiple-choice exam. They nearly always conclude that their responsibilities exceed their meager powers. Lyndon Johnson commented on his frustrations in achieving domestic progress this way: "Power? The only power I've got is nuclear . . . and I can't use that."[23]

Until recently it has been fashionable to say that a president wears many hats—a commander-in-chief hat, a chief-legislator hat, a chief-of-state hat, and so on. This simple metaphor of presidential hats belongs to a simpler past. Several years ago a prominent political scientist wrote that "the United States has one President, but it has two presidencies: one presidency is for domestic affairs, and the other is concerned with defense and foreign policy." Further, he added that presidents have "had much greater success in controlling the nation's defense and foreign policies than in dominating its domestic policies."[24]

By the 1990s, reality, as well as expectations, has expanded and recast the presidency, and has organized it around three major interrelated policy areas that we may call subpresidencies: (1) foreign affairs and national security, (2) aggregate economics, and (3) domestic policy, or "quality of life," issues. The president's time is absorbed by one or another of these competing policy spheres, and the staff and cabinet have come to be organized around these three substantive areas.

The Foreign Affairs Presidency

Modern presidents concentrate on foreign and national security policy, often at the expense of the other policy areas. This is understandable. Since World War II, the United States has been the hegemonic, or dominant, power of the Western alliance. Since the break up of the Soviet Empire in 1989, the United States has been recognized as the lone remaining superpower in the world. U.S. world leadership for the past sixty years has been linked to presidential power: a strong America necessitated, so many believed, a strong presidency. Thus, presidents focused increased attention on foreign affairs.

To be sure, an exclusively foreign, economic, or domestic problem is a rarity, and many issues intersect all three areas. Critical problems such as trade, inflation, energy development, drug abuse, or environmental problems, not to mention war, require planning and policy leadership that cut across the three presidencies. Still, a close examination of how presidents have spent their time in the past fifty years suggests that foreign policy matters (often crises) have driven out domestic policy matters. Economic matters usually come in second.

The founding politicians never intended a president to be the dominant agent in national policymaking, yet they did expect the president to be the major influence in the field of foreign affairs. In the eighteenth century, foreign affairs were generally thought to be an executive matter. The first task for a national leader is the nation's survival and national defense. Today, especially in the nuclear age, foreign policy responsibilities cannot be delegated: they are executive in character and presidential by constitutional tradition or interpretation. After the Bay of Pigs tragedy, President Kennedy vividly emphasized the central importance of foreign policy: "It really is true," he told a visiting Richard Nixon, "that foreign affairs is the only important issue for a President to handle. . . . I mean, who gives a s . . . if the minimum wage is $1.15 or $1.25, in comparison to something like this [the Bay of Pigs]?" [25]

Kennedy frequently said the difference between domestic and foreign policy was the difference between a bill being defeated and the country being wiped out. Both Kennedy and Nixon were personally more fascinated with foreign policy than with domestic or economic policy. Both wanted history to record that they had laid the foundation for peace not only in their own time but also for generations to come. President Carter spent more time on the Middle East issue than on any other matter during his first three years in office. Panama Canal politics and relations with the Soviet Union and China were preoccupations as well. He confessed he liked dealing with foreign policy because his capacity to act unilaterally seemed much greater in the foreign than in the domestic realm. President Reagan, after an initial flirtation with economic policy, spent most of his last six years concentrating on foreign affairs. George Bush had almost exclusively a foreign policy focus.[26] And Bill Clinton, who had a very ambitious domestic and economic policy agenda, found himself spending more and more time on foreign affairs as problems such as Haiti, Bosnia, Russia, the Middle East, and North Korea forced their way onto his agenda.

White House advisers from all the recent administrations agree that presidents spend at least half of their time on foreign policy or national security deliberations. In some instances, this emphasis on foreign policy and national security has occurred by choice, most notably for President Nixon, who said, "I've always thought this country could run itself domestically—without a President; all you need is a competent Cabinet to run the country at home. You need a President for foreign policy; no Secretary of State is really important; the President makes foreign policy."[27] President Bush also had a much greater personal interest in foreign affairs as opposed to domestic policy. President Johnson, on the other hand, was strongly disposed by experience toward domestic programs, but was unable to prevent his presidency from being consumed by military affairs. Aides, friends, and biographers all say that President Kennedy became far more engulfed by foreign policy matters than even he liked. "The President estimates that eighty percent of his first year in office was spent mulling over foreign policy."[28] After ten days in the White House, he talked as if he was overwhelmed by the flood of international crises and the ever-present possibilities of nuclear war:

> No man entering upon his office, regardless of his party, regardless of his previous service in Washington, could fail to be staggered upon learning—even in this brief ten-day period—the harsh enormity of the trials through which we must pass in the next four years. Each day the crises multiply. Each day their solutions grow more difficult. Each day we draw nearer the hour of maximum danger, as weapons spread and hostile forces grow stronger. I feel I must inform the Congress that our analyses over the last ten days make it clear that—in each of the principal areas of crisis—the tide of events has been running out and time has not been our friend.[29]

Presidents devote substantially more of their State of the Union addresses to national security matters than to any other topic. Through a quantitative analysis of these addresses, one scholar found "this policy is clearly the prime presidential concern . . . [and] that greater attention to international affairs results from the experience of being president. Attention to this policy area grows over time . . . in a pattern that can be related to the election cycle." Special focus on national security matters "mounts during the first, second and third years, then drops as a president faces re-election. During a president's second term, concern with international involvement grows again . . . substantially higher [in fact] than . . . during the first term."[30] Other scholars persuasively attribute this accentuated attention to more formal presidential powers in the national security area, better sources of information, and the weakness of Congress and outside groups in this sphere.

In addition, presidents naturally work hardest where they see both hope of success and leeway for significant personal impact. Former Nixon counsel Leonard Garment pointed to yet another reason when he explained Nixon's preference for foreign policy in this way to Theodore White: "In foreign policy, you get drama, triumph, resolution—crisis and resolution so that in foreign policy Nixon can give the sense of leadership. But in domestic policy, there you have to deal with the whole jungle of human problems."[31]

In any case, it seems vastly more "presidential" to be concerned with national security and world peace than with domestic problems. One notion holds that presidents should be involved in areas that allow them to make dramatic global choices. International travels and summit meetings with foreign leaders generate favorable short-term publicity and confer stature on an incumbent, whereas at home a president may be criticized as just another time-serving politician. International excursions are intertwined, of course, with domestic politics; they may even be undertaken to attract news coverage away from unsuccessful or embarrassing domestic initiatives. At home a president may be hamstrung by an opposition-dominated Congress, by a narrow electoral margin, by a hostile press, or by a scandal in his administration; but twelve miles offshore a president virtually *is* the United States. The purpose of some presidential trips is largely symbolic, yet the American people are tolerantly disposed. When the president uses the powers of his office to bring about better chances for peace, it is hard to resent his powers even if he does not succeed. The president will usually be rewarded in the opinion polls merely for trying.[32]

But presidents cannot focus on foreign affairs to the exclusion of the nation's domestic needs. George Bush is a case in point. After successfully winning the Persian Gulf War, Bush's popularity rose to historically unprecedented levels. Yet between January of 1991 and November of 1992, a slow economy and domestic drift led to electoral defeat. Presidents do not win by foreign policies alone.

Most scholars believe there are significant differences in the institutional balance of power when one compares the realms of domestic policy with foreign policy. Presidents, it is believed, have greater power in the foreign policy arena than they do in domestic and economic policy.

In the 1960s political scientist Aaron Wildavsky argued that since the end of the Second World War, presidents have been more successful at attaining their policy goals from Congress in foreign policy than in the domestic arena. Thus, Wildavsky concluded, a president is usually more effective in international affairs and foreign policy-making. Commenting on the power of the presidency in the international arena Wildavsky asserted, "there has not been a single major issue on which Presidents, when they were serious and determined, have failed."[33]

While Wildavsky's "two presidencies" notion may be oversimplified, it does highlight certain patterns. In foreign affairs, presidents can assert their will by being proactive, and force the Congress to be reactive. By setting policy, the president sets in motion events of his choosing. If Congress wishes to intercede, it does so in response to acts already taken. The initiative usually belongs to the president.

Constitutionally, the president may be on shaky ground, but in practical terms, it is difficult to stop a strong, assertive president in foreign affairs. Clinton's NATO intervention into Bosnia was a case in point. While all presidents claim their use of foreign policy power is grounded in the Constitution, such claims of independent authority are generally flimsy. It is the Congress, even in foreign affairs, that has the broadest and clearest constitutional mandate, but its power has atrophied due to lack of use.[34] Even the power to declare war, which is expressly granted to the Congress, has often slipped through the hands of the legislature. The *War Powers Act* and the constitution notwithstanding, it is the president who, by acting, holds primacy in war and foreign affairs.[35]

Much of foreign policy involves diplomatic activities that are generally under the purview of the executive, and in most areas, presidents simply presume authority to act. Likewise, much activity in foreign affairs does not require legislation,[36] and, to make matters worse (or better if you are a president), during a crisis presidents proceed with considerable authority and few checks.

Perhaps the chief power presidents have is the presumption that they speak for the nation to the world. This allows a president the opportunity to act, to make decisions, to announce policies, thereby preemptively eliminating most potential challengers. By fait accompli acts, the president may tie the hands of potential rivals on the congressional side. To defy the president on a matter of national security *may* appear unpatriotic, especially in moments of international crisis. Thus, presidents can both muffle potential critics and force the Congress to submit to their will. Examples of Teddy Roosevelt sending the fleet halfway around the world in defiance of congressional budget restrictions, then saying, "I sent the fleet . . . will they leave them there?"; George Bush sending U.S. troops to the Middle East following the 1991 Iraqi invasion of Kuwait, preparing for war, building public support, *then* challenging Congress to issue a vote of support, or Clinton sending U.S. troops to Haiti in 1994, serve as examples of how presidents, by acting and forcing the hand of Congress, can get their way.[37]

Over the course of a term, presidents spend more and more time dealing with foreign affairs precisely because they so often get frustrated being rebuffed in domestic and economic policy. Since presidents need to be seen as accomplishing things in order to maintain their reputation as "the leader," they almost naturally gravitate to the area in which their power is greatest. The irony, of course, is that the president has the greatest power in the area where he is probably most (potentially) dangerous and in need of controls and checks and balances, and is most controlled in the least dangerous area and the area most in need of leadership.

The Economic Presidency

The second largest portion of presidential policy time is spent on aggregate economic, or macroeconomic, policy. Aggregate economics refers here to issues of monetary and fiscal policy such as trade and tariff policy, inflation, unemployment, the stability of the dollar, and the health of the stock market, as opposed to the more explicitly domestic concerns of education, ecology, health, housing, welfare, social justice, the civil service system, and, more generally, the quality of life in the United States.

In the early days of the Republic, presidents had only peripheral input into the development of economic policies. The nation was young, not yet a world power, and Congress was in primary control of determining the budget and setting economic policy. Presidents began to get more involved in setting economic policy with passage of the Budget and Accounting Act of 1921. This law directed the president to formulate a national budget and created a Bureau of the Budget. The Pandora's box was opened. Beginning with Franklin D. Roosevelt, presidents began to expand their control over economic and budgetary matters.

During the 1930s, with the depression, the Supreme Court began to give judicial sanction to governmental and presidential intervention in the economy. The Constitution thus shifted from a limiting to an empowering document in presidential economic management. Since then, we have witnessed the development of a more "controlled" economy and the growth of presidential power in the economic sphere. In the post–World War II period, the passage of the Employment Act of 1946 further focused economic policymaking in the executive branch.

While the expectations of presidential responsibility and control of the economy have grown enormously in the past fifty years, the powers given to the president to manage the economy have not risen accordingly, creating yet another presidential paradox. We expect presidents to bring about economic breakthroughs, but do not give them the power necessary to work miracles.

In recent years, presidential involvement in economic policy has grown to the point where primary responsibility for economic management now rests in the White House.[38] Historical circumstance, presidential desire, legislative acquiescence, and public expectation have all contributed to the emergence of a powerful, but limited, presidency in economic affairs. While congressional efforts to recapture their constitutional duty over the "power of the purse" are evident (e.g., the Congressional Budget and Impoundment Control Act of 1974), the president remains the central, yet not controlling, figure in setting economic policy for the nation. The Congress in 1996 appeared to recognize this circumstance when it granted presidents, beginning in 1997, the right to veto specific appropriation items passed by Congress (the item veto), subject to override by the Congress.[39]

Presidents are held responsible for the overall "health" of the economy. They are praised or blamed for fluctuations in the inflation rate, unemployment, economic growth, and so forth. "Tinkering" with the economy has given way to attempts at more comprehensive economic management, orchestrated from the White House.

A president can scarcely hide from the hard, visible quantitative economic indicators: unemployment rates, consumer price and inflation rates, the gross domestic product, interest and mortgage rates, commodity prices, oil import price levels, and stock-market and bond-market averages. In the era of CNN and the internet these figures are available to everyone, at any time, and the American people increasingly judge and measure their presidents on whether they can cope aggressively with recession and inflation, can offer effective economic game plans (preferably without tax increases), and can use the nation's budget as an instrument for ensuring a healthy and growing economy. The complex issues of tax reform and income redistribution are always on the agenda of national politics. While presidents prepare and send to Congress a budget, often, as the old adage has it, "the president *proposes* and Congress *disposes.*" Congress most assuredly amends and modifies the budget before it approves it.

But a president's statutory responsibilities for maintaining full employment and pursuing stabilization policies are not matched by political resources or available expertise. In most areas of our economy the public is relatively weak or unorganized, but an uneven performance by the economy will call down on a president immediate pressure from wealthy businesspeople, unions, and farmers and from their friends in Congress.

A further paradox of the president's role as economist-in-chief of the nation relates to the contradictory demands we place on the government. We want to cut spending, cut taxes, and cut the deficit, but we don't want any of "our" favorite programs cut (and every line in the budget is somebody's favorite). Most people want the government to do more but spend less. In a robust economy, the hard choices become much easier, but in a staggering economy or one burdened with debt, all the choices become hard.

The Domestic Presidency

Presidents concentrate on those areas in which they think they can make the greatest impact, in which the approval of interest groups and the public can be most easily rallied. Yet, involvement in domestic policy is costly both financially and politically. Moreover, newly elected presidents find that budgets are virtually fixed for the next year and a half if not longer, and that in domestic matters they are dependent on Congress, specialized bureaucracies, professions, and state and local officialdom. Is it surprising, then, that the implementation of domestic policy can become the orphan of presidential attention? Is it possible presidents rationalize that they must concern themselves with foreign and macroeconomic policy as they lose heart with the complicated, hard-to-affect, divisive domestic problems?

From Roosevelt through Clinton, recent presidents have complained that progress on the domestic front was more difficult than they had imagined it would be. Presidents also regularly complain that there are greater limits on their ability to bring about favorable results in the domestic sphere than they had imagined. The Clintons' efforts to restructure national health care programs and policy is a notable example.[40]

Thus John Kennedy was once moved to quip about a relatively low-priority project, the architectual remodeling of Lafayette Square across from the White House, "Let's stay with it. Hell, this may be the only thing I'll ever really get done."[41] Lyndon Johnson was always disappointed by the slow pace of progress in first passing and then implementing Great Society programs. He often thought a significant portion of the programs he had fought so hard to pass had been sabotaged by indifferent or even disloyal bureaucrats. Nixon's contempt for the domestic bureaucracy was well known. His replacement of every one of his original cabinet officers from his first term is further evidence of his frustration in dealing with his executive branch "colleagues." Ronald Reagan rallied the ideological troops with promises of a domestic crusade, but treaded gingerly in the domestic arena, disappointing his most ardent followers as he failed to deliver on key domestic items.

Jimmy Carter learned early and often about the limits of his domestic leadership powers. One hilarious example of this lesson concerned a mouse that had climbed inside a wall of the Oval Office and died. The odor became offensive one day just as Carter was about to greet a visiting diplomat. An emergency call went out to the General Service Administration (the agency that maintains and oversees federal property), but GSA refused to respond, insisting that it had already exterminated all the mice in the White House. The dead mouse, the GSA officials reasoned, had obviously come in from outside of the building and was therefore the responsibility of the Department of the Interior. However, Interior officials objected; they contended that the mouse was not their concern because it was now inside the White House. An exasperated Carter finally ordered officials from both agencies to his office where he angrily told them, "I can't even get a damn mouse out of my office. . . ." A special task force representing the two agencies was established and they proceeded to get rid of the mouse!

Of course, the presidential litany about powerlessness can be self-serving: it can be conveniently used to suggest that presidents personally are more or less blameless for not seriously undertaking the painful domestic fights that need to be fought. It is also part of the rationalization for switching from exacting "home-repairman tasks" to the more glamorous role of world statesman.

THE MULTIDIMENSIONAL PRESIDENCY

Within each of these presidencies or policy subpresidencies, a president is asked to provide functional leadership in seven activity areas: crisis management, symbolic leadership, priority setting and program design, recruitment of advisers and administrators, legislative and political coalition building, program implementation and evaluation, and oversight of government routines and early warning of problem areas. These are not compartmentalized, unrelated functions but rather a dynamic, seamless assortment of tasks and responsibilities. This job description does not exhaust all presidential activity; rather, the examples in Table 5.1 attempt to classify the major functional as well as substantive responsibilities of the office. Of course, political activity solely for personal enhancement (such as reelection) should be acknowledged as a presidential preoccupation just as staying elected is a prime objective for most legislators.

In practice, no president can divide his job into tidy compartments. Instead, a president must see to it that questions are not ignored simply because they fall between or cut across jurisdictional lines. Presidents must act alternately, and often simultaneously, as crisis managers and as symbolic, priority-setting, coalition-building, and managerial leaders. Ultimately, all these responsibilities mix with one another. Being president is a little like being a juggler who is already juggling too many balls and is forever having more balls tossed into the game.

Crisis Management

A crisis is a situation that occurs suddenly, heightens tensions, carries a high level of threat to a vital interest, provides only limited time for making decisions, and possesses an atmosphere of uncertainty. Crisis management involves both precrisis planning and the handling of the situation during a crisis.

When crises and national emergencies occur, we instinctively turn to the president. Presidents are asked during a time of crisis to provide not only political and executive leadership but also the appearance of confi-

TABLE 5.1
The Presidential Job Description

Types of Activity	The three "subpresidencies"		
	Foreign policy and national security	Macroeconomics	Domestic policy and programs
Crisis management	Wartime leadership; missile crisis, 1962; Gulf War, 1991	Coping with recessions, 1982, 1992	Confronting coal strikes of 1978; LA riots, 1992; LA earthquake, 1992
Symbolic and morale-building leadership	Presidential state visit to Middle East or to China	Boosting confidence in the dollar	Visiting disaster victims and morale building among government workers
Priority setting and program design	Balancing pro-Israel policies with need for Arab oil	Choosing means of dealing with inflation, unemployment	Designing a new welfare program, health insurance
Recruitment leadership (advisers, administrators, judges, ambassadors, etc.)	Selection of secretary of defense, U.N. ambassador	Selection of secretary of treasury, Federal Reserve Board governors	Nomination of federal judges
Legislative and political coalition building	Selling Panama or SALT treaties to Senate for approval	Lobbying for energy-legislation package	Winning public support for transportation deregulation
Program implementation and evaluation	Encouraging negotiations between Israel and Egypt	Implementing tax cuts or fuel rationing	Improving quality health care, welfare retraining programs
Oversight of government routines and establishment of an early-warning system for future problem areas	Overseeing U.S. bases abroad; ensuring that foreign-aid programs work effectively	Overseeing the IRS or the Small Business Administration	Overseeing National Science Foundation or Environmental Protection Agency

dence, responsible control, the show of a steady hand at the helm. Popular demand and public necessity force presidents nowadays to do what Lincoln and Franklin Roosevelt once did during the national emergencies of their day, namely, whatever is required to protect the union, to safeguard the nation, and to preserve vital American interests.

In normal times, the checks and balances of the U.S. political system can be formidable, severely limiting presidential initiatives. But in a political crisis, most of these checks evaporate, and the president is given wide discretionary leeway in the exercise of executive prerogative or emergency powers. While the Constitution contains no explicit provisions for government during a crisis, in an emergency a president can invoke emergency statutes or merely assume power and become the nation's crisis manager in chief (as did Lincoln in the Civil War, Franklin Roosevelt during the Great Depression, and Kennedy during the Cuban missile crisis).[42]

The most significant factor in the swelling of the presidential establishment in the post-1939 era has been the accretion of new presidential roles during national emergencies, when Congress and the public have looked to the president for decisive responses. The Constitution neither authorized presidents to meet emergencies nor forbid them to do so. All strong presidents have taken advantage of this omission. The Great Depression and World War II in particular caused sizable increases in presidential staffs. After the Russians orbited Sputnik in 1957, President Eisenhower added science advisers; after the Bay of Pigs in 1961, President Kennedy enlarged his national security staff. The Cold War commitments as President Kennedy enumerated them in his Inaugural Address, to "pay any price, bear any burden, meet any hardship, support any friend, oppose any foe," and the presence of nuclear weapons fostered the argument that only presidents could move with sufficient quickness and intelligence in national security matters. When major crises occur, Congress traditionally holds debates but just as predictably delegates authority to the president, charging him to take whatever actions are necessary to restore order or regain control over the situation.

Primary factors underlying the transformation of the presidency of the Constitution to the modern crisis-management presidency are the invention of nuclear weapons, the emergence of the Cold War, the permanence of large standing armies, and the interdependent role of the United States in the world economy. Constant crises in national security have dominated American thinking throughout the twentieth century. Such crises include the Japanese bombing of Pearl Harbor in 1941, the North Korean invasion of South Korea in 1950, the Russian launching of Sputnik in 1957, the offensive missiles placed in Cuba by the Russians in 1962, the Viet Cong offensives in South Vietnam in the mid and late 1960s, and the various Middle East wars and incidents of the 1960s through the 1990s.

Symbolic, Morale-building, and Shamanistic Leadership

The American presidency is more than a political or constitutional institution. It is a focus for intense emotions. The presidency serves our basic need for a visible and representative national symbol to which we can turn with our hopes and aspirations. The president has become an icon who embodies the nation's self-understanding and self-image. Thus, we place the president on a high pedestal, yet we demand democratic responsiveness. The president is at once our national hero and national scapegoat.[43]

Presidents are the nation's number-one celebrity; most everything they do is news. Merely by going to a sports event, or a funeral, by celebrating a national holiday, or visiting a particular nation, presidents not only command attention, they convey meaning. By their actions presidents can arouse a sense of hope or despair, honor or dishonor.

Although Americans like to view themselves as hardheaded pragmatists, they, like humans everywhere, cannot stand too much reality. People do not live by reason alone. Myths and dreams are an age-old form of escape. People will continue to believe what they want to believe. And people turn to national leaders just as indigenous people turn to shamans—yearning for meaning, healing, empowerment, legitimacy, assurance, and a sense of purpose.[44]

Americans expect many things from their presidents—honesty, credibility, crisis leadership, agenda-setting and administrative abilities, and also certain of the tribal leader or priestly functions we associate with primitive or religious communities. A president's personal conduct affects how millions of Americans view their political loyalties and civic responsibilities. Of course, the symbolic influence of presidents is not always evoked in favor of worthy causes, and sometimes presidents do not live up to our expectations of moral leadership. Still, a great many people find comfort in an oversimplified image of presidents as *warrior-captains* firmly at the helm of the ship of state, as *emancipators or liberators* in the Exodus tradition, as *priests and prophets* of our civil religion, and as *defenders of the democratic faith* and *evocative spokesmen* of the American Dream.

United in this one institution are multiple roles that are on the surface confusing and conflicting. A president's unifying role as a head of state and symbol-in-chief, especially in times of crisis, often clashes with his advocacy of program initiatives and partisan responsibilities. Further, presidents invariably take advantage and borrow from their legitimacy as representative and symbolic head-of-state to expand their political and partisan influence.

The framers of the U.S. Constitution did not fully anticipate the symbolic and morale-building functions a president would have to perform. Certain magisterial functions such as receiving ambassadors and the granting of pardons were conferred. Yet the job of the presidency demanded symbolic leadership from the very beginning. Washington and his advisers would readily recognize that in leadership at its finest, the leader symbolizes the best in the community, the best in its traditions, values, and purposes. Effective leadership infuses vision and a sense of significant meaning into the enterprise of a nation.

George Washington was one of the few continental figures of his day. He was already a warrior-hero. He had commanded with distinction the revolutionary patriots for over eight years. In victory he became a prime symbol of that crusade. His integrity and judgment, his lengthy service both in and out of uniform to his country, and his devotion to his troops and his countrymen, set him apart from his fellow founders.

Washington wasn't a great philosopher, orator, lawyer, or even organizer. Even his military abilities have been questioned. He was not especially imaginative and, although his mind was keen, it was slow in operation.

But the nation needed a hero. America had no heritage of celebrated public servants other than those in England, and hence it was essential for national pride to endow our first hero with lavish praise. His countrymen did precisely this. And Washington understood their need and not only accepted it graciously, he used it as a means of legitimizing both his new office and the new national government he had labored so long to bring into being.[45]

No matter how ably others had explained the Constitution, and especially the provisions for presidential leadership, it was now up to President Washington to carry out the promise of the office and establish the precedents to be followed as long as the country survived. Washington was fully aware that the process of making the Constitution work had only just begun. He knew that written documents do not implement themselves. He appreciated that a living, real constitution includes customs, traditions, practices, interpretations, and precedent setting to fill out the vagueness of the written provisions.

Among President Washington's many legacies to our political system was his acting out superbly the head-of-state and symbolic functions we now regularly expect of presidents. The framers may not have spelled out the symbolic roles of the presidency for two reasons. First, they wanted to devise a system that was a government of laws rather than a government that depended on indispensable individuals. More than anything they were trying to invent new forms of government and constitutionalism that moved away from the British model. On the other hand, the availability of Washington and the expectation that he would serve as the first president doubtless also had much to do with their leaving these aspects of the presidency underdefined. For in Washington they realized they had a person of commanding presence and a person of optimism, vision, and self-confidence.

In some ways Washington became the nation's first secular priest or societal shaman. In helping his countrymen to transcend their ordinary definitions of reality he helped instill a new nationalism and a new sense of purpose. His combination of *warrior* and *priest, liberator* and *definer* of a national vision presaged additional extraconstitutional responsibilities for future American presidents.[46]

The warrior-symbol   In deciding to make the executive office also the place for commander-in-chief responsibilities, the framers fused together the roles of executive leadership and military or security leadership. This unification, which had been absent under the Articles of Confederation and especially during the American Revolution, guaranteed that Americans would expect presidents to know about war, be capable of wartime leadership, and, if possible, have military experience. Not surprisingly, then, we have elected several military heroes. Presidents Washington, Jackson, Grant, Teddy Roosevelt to a lesser extent, and Eisenhower all benefited from their military achievements, which enhanced their legitimacy as presidential candidates. And the presidency itself, because of these warrior-presidents, took on additional meaning. Each brought to the office some of the aura of legend and the charisma of reputation and authority earned in defense of the nation-state. While this never guaranteed effectiveness in presidential performance, particularly in the case of Grant, it expanded the public's acceptance of presidential authority—and in four of their cases it helped achieve for them rather high evaluations for their presidencies. By contrast, President Clinton has had difficulties establishing his legitimacy in part due to his lack of military experience.

Military heroes rise to leadership roles in every society and ours is no exception. On balance, the military veterans who have become president have generally bent over backward to not unnecessarily militarize the presidency. Yet the emergence of the nation as the leading military power in the world necessarily blurs the distinction between the job of the president as political and as national security leader. Whereas national security was once merely one of the jobs of the president and only occasionally demanding, it now compels the modern president to spend at least half of his time planning strategy and presiding over a sprawling military complex, often with as many as 500,000 troops in uniform stationed in all reaches of the world.

The emancipator-liberator symbol The Exodus or emancipator "paradigm" or theory of emergency leadership is one that is deeply embedded in the cultural consciousness of the West. One writer describes the Exodus as an account of deliverance not only in religious terms, but also in a secular, this-worldly account. The Exodus paradigm is one of liberation, of a march to freedom. Americans, like the Israelites, in their own way have had to march, and have had to have leaders and discipline, to free themselves from oppression and depression.

The American Revolution, although in many ways a conservative revolution, was a war of liberation and a determined, disciplined, focused striving to bring about a deliverance. If an aspiration, hope, and promise motivated that rebellion, forceful and committed leadership was needed to help realize the triumph. And the story and legend of George Washington are inextricably linked to this, our own story of liberation.

Yet perhaps it is the story of Abraham Lincoln that epitomizes the liberator-emancipator role of the American political leader. Political scientist Clinton Rossiter once wrote of Lincoln as the supreme American myth, the richest symbol in the American cultural experience. "He is," wrote Rossiter, "the martyred Christ of democracy's passion play. And who, then, can measure the strength that is given to the President because he holds Lincoln's office, lives in Lincoln's house, and walks in Lincoln's way?" Rossiter goes so far as to say the final greatness of the presidency lies in the truth that it is not just an office of incredible power, but a breeding ground of indestructible myth.[47]

Lincoln's special role in American civic and cultural life comes not from his revelation of God's will, but because he revealed and unlocked the higher aspirations and promise of America. In his ability to be a common man with uncommon instincts he preserved the nation, presided over our worst ordeal, and encouraged the rebirth and liberation the nation so urgently needed.[48]

So what if the nation often lavishes too much meaning and too much praise on the individual in the White House at that time? Even a free people need their heroes, and it is our fortune that our foremost heroes have been those who by empowering us helped us transcend our flaws and deficiencies. If the real Lincoln was not exactly a saintly Emancipator, neither was he a racist. He grew as he rose to power and he grew also once he attained power. The flawed, fatalistic, and politically ambitious Lincoln struggled with himself and in doing so helped his country struggle and resolve the haunting moral paradox of slavery in a nation based on the Declaration of Independence.[49] Ultimately, Lincoln provided transcending liberating leadership. Thus in late 1862 he would remind the Congress and the nation that the dogmas of the quiet past are inadequate to the stormy present. "The occasion is piled high with difficulty, and we must rise with the occasion. As our case is new, so we must think anew, and act anew. We must disenthrall ourselves, and then we shall save our country." "Fellow citizens," Lincoln went on, "we cannot escape history. . . . The fiery trial through which we pass, will light us down, in honor or dishonor, to the latest generation. . . . In giving freedom to the slaves, we assure freedom to the free—honorable alike in what we give, and what we preserve. We shall nobly save, or meanly lose, the last best hope of earth."[50]

Lincoln, caught up in the turbulent social forces of his time, helped point the way to a resolution of the nation's most perplexing moral problem. The Lincoln myth is not that he was a political saint, but rather that he acted so as to encourage the country toward self-improvement and the liberation of the nation's more generous impulses toward one another.

In his actions, and even more in the legends that developed around him after his death, the Lincoln story is rich in meaning and symbol. Lincoln helped his nation renew human hopefulness, he appealed to our better angels. Lincoln also added considerably to the legitimacy of the presidency and expanded the public's expectations of the possibilities of presidential moral leadership.[51]

Eighty years later, Franklin Roosevelt's determined personal leadership designed to cope with the devastation of the Great Depression also fit the emancipator-liberator symbolic role. Again the legend and myths that developed are often enlarged out of proportion. Yet in politics, the perception of the image is usually as important and more lasting than the reality. Today Roosevelt is remembered for his willingness to risk his political career to stake out bold new measures needed to rescue the nation from its worst economic disaster.

Similarly, his risking all to come to the aid of the Allies in defeating fascism is also interpreted as leadership of liberation. Somehow, whether during the aftershocks of Pearl Harbor, the invasion of Normandy, or Roosevelt's personal diplomacy and personal visits to the troops in North Africa, the president seemed to be engaged in bringing new meaning to the mission of America. While a wide variety of meanings were doubtless conveyed, the emancipator-liberator, the leader in defense of freedom, always loomed large and forms the FDR legend today—"a Soldier of Freedom" as one of his biographers puts it, and, we might add, the foremost architect as well of the modern American presidency.

Presidents as defenders of the faith   Every four years Americans elect a politician to serve as president. Yet we also yearn for a high priest of sorts, because despite our separation of church and state, we need performed for us some of the same functions that shamans, medicine men, and other practitioners of ritualistic arts perform in other societies. Having no national religion, we look nonetheless for spiritual meaning. And we often look to our presidents as high priests.

In our time, Ronald Reagan had an uncanny understanding of the symbolic roles a capable president must perform. He had a conscious appreciation of the need for the president to reaffirm our basic goals, to celebrate liberty and freedom, and to participate fully in the rituals that both give meaning to American life and help people understand the larger events of which they are a minor part. Few presidents have been better than Reagan at helping the nation observe its ritual occasions such as Memorial Day, July Fourth, Thanksgiving, and Veteran's Day or participate in the Olympics, summit diplomacy, and similar ceremonies. No one was better at performing national chaplain services—as Reagan did in comforting Americans after the Challenger disaster or as he grieved with and gave meaning to those who mourned the deaths of U.S. Marines in Lebanon. Reagan understood that Americans have a civil religion and semisacred symbols and that these reflect our human need to make sense of eclectic experiences and give meaning, form, order, and assurance.

President Clinton made several attempts to focus on values and what some have called "the politics of meaning."[52] Such efforts to reflect and become philosophical are a recognition that the presidency truly is, in FDR's phrase, a place of moral leadership, where—although no president can single-handedly create a social movement or implement a progressive reform—a president can help set a moral tone and place major issues in larger contexts of justice, fairness, and liberty.

Much of this commingling of the political and the culturally symbolic in presidential performance is an altogether understandable human response to societal yearnings. Leaders sometimes have no choice but to fulfill tribal roles, no matter how pragmatic, educated, sophisticated, or secular the society. Rituals are ceaselessly reinvented by the human heart.

It may be, in fact, that Americans need more symbolic or ritualistic leadership than many societies simply because we prohibited royalty and the establishment of a state religion. Even many of our Western democratic allies find it desirable to separate their head-of-state functions from their legislative and political leadership functions. Thus England has her queen and her prime minister. Germany has a president to perform head-of-state roles and a chancellor as politician-in-chief. Even the former Soviet Union differentiated many of its hierarchical functions, distributing them to the Communist Party's general secretary, a president, and others.

We do it our own way—and we may pay a price for this union of sometimes conflicting roles in the same office. Indeed, the rise and enlargement of the symbolic features of the American presidency create a number of problems.

First, presidents are sometimes tempted to engage a bit too much in symbolic leadership at the expense of political leadership. Understandably, presidents enjoy the symbolic and ceremonial duties. In the performance of these functions they represent the whole people. But, as political and partisan leaders, they often have to divide us. Much of the job of a president requires setting priorities, building new coalitions around these priorities, hiring and firing top personnel, bargaining with Congress, persuading interest group leaders, and, in general, negotiating compromises with other prominent leaders at home and abroad. These tasks require a president to take controversial stands and make tough political decisions. They also require a president to work closely with party leaders and implement party platform ideas. In short, they require a president, not just to preside, but to take risks; not just to attend ceremonies and give sermons, but to engage in conflict and division. As head of state and symbol-in-chief, a president seeks to unite us, reassure us, and emphasize order, stability, and continuity. As a political executive a president has to confront problems, antagonize opponents, and occasionally stir us from our complacency.

There is nothing wrong with the symbolic powers that come with the job. They can become a problem, however, when they lead the public to believe symbolism equals accomplishment, or when ceremonial and symbolic requirements keep presidents from performing their other demanding duties. This was one of the criticisms of the Reagan presidency.

Second, sometimes a president invokes the warrior, priest, defender-of-the-faith images that come with the job and help legitimize presidential authority to give credibility to decisions or actions that are not deserving of our approval. Perhaps Richard Nixon's enemies list, his deceptiveness in obstructing the judicial investigation of the Watergate break-in, and his extensive invocation of executive privilege best illustrate abuse of the office. Nixon portrayed himself as a leader, like Lincoln, embattled in crisis and needing more deference, loyalty, secrecy, and imperial authority.

Third, the amplification of the symbolic roles of the American president may do a disservice in diminishing competing forms of democratic leadership. Our system was not designed to achieve the acquiescence of the many in the rule of the favored few. It is sometimes said that people are ruled by their imaginations; yet it is perhaps more valid to suggest they are governed by the weaknesses of their imaginations.

Especially in the age of television, our presidents loom so large as to dominate much of the public discourse. It is difficult for political rivals or counter-leadership to get air time to present alternative interpretations of what might be desirable policy.[53]

No one, certainly not the founding politicians, ever intended that presidents would be the sole interpreter of the meaning of America. Most assuredly we want a nation that is capable of rich dialogue about our purposes, vision, and future aspirations. We also want a nation of leaders, not a nation dependent on a single leader and a single, centralized leadership institution. Our strength has always come from our diversity, our willingness and eagerness to debate, listen to alternative viewpoints, and nurture dissent. A top-heavy leadership structure with undue reverence and deference to the presidency would undermine much that is precious to the American experiment.

If the profound symbolism of the presidency has costly implications for the quality of the relationship between citizens and the presidency, it affects fully as much the ways in which presidents and their associates conceive of themselves and their jobs. The reverence and loyalty rendered a new president are a rich resource, yet an overindulgent citizenry can distort the president's psychological perspective and sense of right and wrong.

Fortunately, Americans have an excellent appreciation for humor, and our chief deflators of presidential pomposity or phony religiosity are sometimes our cartoonists, comedians, and humorous columnists. A president who goes too far must beware of Garry Trudeau, Jay Leno, David Letterman, Mark Russell, Russell Baker, Dave Bary, and Maureen Dowd, to name just a few potential adversaries.

On balance, we turn to presidents for more symbolic meaning and ritual leadership than was ever intended—and more than is probably desirable. Such dependence has consequences.

Vision, Priority Setting, and Program Design

Presidents, by custom, have become responsible for proposing new initiatives in the areas of foreign policy, economic growth and stability, and the quality of life in America. This was not always the case. But beginning with Woodrow Wilson and particularly since the New Deal, a president is expected to promote peace, prevent depressions, and formulate domestic programs to resolve countless social problems.

Reagan presents an interesting case of a president who shaped the national policy agenda, because he was a more ideologically committed president than most of his predecessor. Reagan's conservatism helped him develop a vision, a policy agenda, and a program of action. There is a good and a bad side to being an ideologue. It can be good in that a committed ideology can simplify complex problems; on the bad side, many problems do not lend themselves to simple solutions. On the good side, ideology can help get an administration marching in the same direction; on the bad side, people may march like lemmings, over a cliff. On the good side, ideology highlights certain variables in problem solving; on the bad side, it hides or obscures other equally important variables. On the good side, ideology gives passion to purpose; on the bad side it can also degenerate into crusading extremism.

By contrast, Bill Clinton, a self-proclaimed "New Democrat" and political moderate, must seemingly recreate himself on every issue. Being more moderate and pragmatic may avoid the mistakes of either extreme, but it creates problems of its own. It is difficult to develop a vision and generate a committed following around being moderate. While it may be easier for political moderates to get elected, and it was for Clinton in 1996, it may be harder for them to set visions and govern.[54]

Because power in the American system is not clearly given to any one branch, a policy entrepreneur can, on certain issues, capture control of the policy agenda. If this can be done, a president may be able to force an otherwise reluctant Congress to at least meet part of his policy goals. And no one is better positioned to attract attention than the president. If presidents are to lead and not merely preside, they must share and try to influence the agenda.

Presidents are often criticized for not being programmatic enough, for not being more passionately committed. The fact is, presidents, by force of habit, almost always want to extend their personal influence, not their ideology, over people and programs. They are generally suspicious of long-range planning. Politicians inherently fear being in advance of their time. They enjoy flexible processes and eschew explicit platforms forever visited on them by well-meaning expert advisers. Presidents are constantly being asked to plan, to forecast, to prevent us from going down the path to policy disasters, yet their instinct is "to leave their options open."

The essence of the modern presidency lies in its potential capability to resolve societal problems.

The effective president will attempt to clarify many of the major issues of the day, define what is possible, and harness the governmental structure so that new initiatives are possible. A president, with the cooperation of Congress, can set national goals and propose legislation. Close inspection indicates that in many instances a president's so-called new initiatives are measures that are or have been under consideration in Congress.

Recruitment Leadership

A president's most strategic formal resource is the ability to recruit able people who share the president's convictions to fill high-level positions in both the executive and judicial branches. According to what might be called the "good-person theory," a presidency is only as good as its staffs, cabinet, and counselors. Quality and loyalty are what is wanted. In practice, the view that a president dominates the recruitment and appointment process is misleading.

After running for the presidency for several years, the successful candidate often finds that the people courted during the campaign—delegates, press, financial contributors, machine leaders, advance aides, or political strategists—do not have the skills or experience needed to manage the executive branch. Kennedy repeatedly complained, "People, people, people! I don't know any people [for the cabinet and other top posts]. I only know voters! How am I going to fill these 1,200 jobs?"[55] Almost half of Kennedy's eventual cabinet appointees were unknown to him. Carter supposedly had almost 3,000 posts to fill, but in an effort to restore dignity to his cabinet officers he permitted them to allocate a large portion of the vacancies.

In the postelection rush many appointments are made on the basis of subjective judgments or ethnic or geographical representation and from too limited a field of candidates. Ironically, at the time when presidents have the largest number of jobs to fill and enjoy their greatest drawing power, they have less time and information available to take advantage of this major prerogative than at any other point. Later in their term, presidents have fewer vacancies and usually less prestige; candidates from outside the government are wary about being saddled with troubled programs and recognize that little can be accomplished in an administration's last year or two.

A president sometimes errs in recruiting only like-minded appointees; this can lead to an amiability and conformity in thinking. Social psychologist Irving Janis pointed out that this leads to a dangerous inclination to seek concurrence and groupthink at the expense of critical judgment.[56] The cult of loyalty to President Nixon doubtless encouraged the suspension of critical and objective thinking on the part of many of the people involved in the Watergate affair.

Many other circumstances limit the presidential appointment prerogative. The chief limit, of course, is that the members of Congress (primarily the Senate) usually view the recruitment process, especially the appointment of key policymaking officials, as a power presidents must share with them.

Some potential appointees do not want to live or even be near Washington, D.C. Others balk at the idea of having to disclose their financial background and income. Others do not care to live in the glare of an intrusive media, or have every detail of their past in print on the front page of the newspapers. Involvement with past administrators or too close an association with past scandals or major industries connected with a new assignment may be enough to occasion congressional hostility to a presidential nomination. Congress is somewhat more sensitive these days to any kind of potential conflict of interest, especially when there is divided government. Carter had to withdraw a number of his nominations to regulatory agencies, and his first choice of head of the CIA, Theodore Sorensen, withdrew under pressure. Both Bush and Clinton were forced to with-draw nominees when relatively minor past blemishes were brought to light by the opposition. In 1997, Clinton was forced to withdraw both CIA nominee Tony Lake and William Weld, his nominee for U.S. Ambassador to Mexico. Dozens of presidential nominees have had their chances ended when a key senator or senators refused to hold confirmation hearings.

Presidents also nominate federal judges. They are appointed for life. Here the Justice Department, the American Bar Association, and the members of the U.S. Senate play a large role. ABA rankings and evaluations are weighed heavily in most administrations. The tradition of "senatorial courtesy"—the right of a senator to veto a candidate from his home state for purely personal reasons—also constrains a president's appointment of federal district judges.

The selection of Supreme Court justices who have a constitutionally prescribed independence from the executive, as distinct from appointments nominally under a president's line of authority, can also lead to major disappointments. Historically, for example, the Senate has rejected nearly 20 percent of the presidential nominees to the Supreme Court, most recently including one each in 1968, 1969, 1970, and 1987. Political scientist Robert Scigliano suggested that one Supreme Court justice in four consistently rules quite differently from what the president who appointed him expected and that numerous other justices fail to conform to expectations in cases of particular importance.[57]

Effective presidents shrewdly use their recruitment resource not only as a means of rewarding campaign supporters and enhancing their ties to Congress but also as a vital form of communicating the priorities and policy directions of their administrations. One student of the appointment process described the importance of this function:

> A President's nominees are the primary link between him and the millions of men and women in the federal bureaucracy. Most of these men and women are located in the executive branch of the government and technically they work for the President. But he has little power to hire and fire them, he cannot control their political loyalties, and he lacks the time and resources to supervise their activities. His executive appointees must act as his surrogates in dealing with the federal bureaucracy. The quality and the character of the people he chooses to serve in executive positions will have a major impact on the ability of his transient administration to control the permanent government.[58]

Even when presidents have recruited able and loyal persons to their administration they run the risk of high turnover. The average cabinet member has stayed in the post for only about two and a half years, although President Clinton's cabinet averaged a bit longer.

Legislative and Political Coalition Building

To govern in the United States is to build coalitions, to form alliances and power networks. In relatively few areas can a president act unilaterally and not face challengers to his authority. Custom and the design of the U.S. system necessitate coalition building by political leaders.

The centrifugal force of American politics pulls the system apart and encourages independent entrepreneurship. Power is not fused as in parliamentary democracies, presidents and members of Congress are elected relatively independent of one another, parties do not develop a great deal of cohesion among the branches. Political scientists Benjamin Ginsberg and Martin Shefter agree that since elections fail to provide clear governing coalitions, the president and Congress end up resorting to "politics by other means" in order to influence policy.[59]

This dilemma highlights the essential difference in what it takes to get elected and what it takes to govern. Getting elected requires the development of an "electoral coalition." Governing requires the development of a "governing coalition." These may be very different, even contradictory things.

This difference between electoral and governing coalitions was noted by presidency scholar James MacGregor Burns: "There's an increasing disparity between the tests for winning office and what's required in governing. The kinds of quick, dexterous ploys, mainly public relations ploys, that are called for in public campaigning are a very far cry from the very solid coalition building that is needed to make this system of ours work."[60]

To get elected, Bill Clinton made appeals to a wide variety of traditionally Democratic interest groups, hoping to pull a sufficient number of these groups together to ensure his election. It was a successful electoral strategy. But was it good for governing? As soon as the election was over, all these groups approached Clinton demanding he keep his promises and deliver the goods. Clinton tried to deliver on his promises. For example, he broadened federal support for abortion rights (which garnered some backlash) and attempted to allow openly gay persons to serve in the military. This generated an incredible backlash—opposition from the military and fundamentalist religious groups all but paralyzed the new administration in the honeymoon period and did significant political damage to the new president precisely at the time he should have been at his power peak. In effect, Clinton's honeymoon was in part sacrificed at the altar of his electoral coalition.

A president's proposals may be resisted by a Congress even fresh on the heels of a triumphant election. Franklin Roosevelt, after his election in 1936 by the largest plurality in electoral history, suffered his most embarrassing congressional rebuff in the defeat of his bill to enlarge the Supreme Court. For a different set of reasons, immediately following his electoral landslide in 1972 President Nixon experienced a strikingly similar congressional stubbornness. Congress opposed Nixon more often in 1973 than it had opposed any president in the previous twenty years.

The skills and resources requisite for winning office are not necessarily those useful in dealing with legislators. John Kennedy, in spite of fourteen years in Congress, had little taste for effectively courting Congress. From the perspective of the presidency, he saw the collective power of Congress as a bloc far stronger than he had thought when a senator:

> It is very easy to defeat a bill in the Congress. It is much more difficult to pass one. To go through a subcommittee . . . and get a majority vote, the full committee and get a majority vote, go to the Rules Committee and get a rule, go to the Floor of the House and get a majority, start over again in the Senate, subcommittee and full committee, and in the Senate there is unlimited debate, so you can never bring a matter to a vote if there is enough determination on the part of the opponents, even if they are a minority, to go through the Senate with the bill. And then unanimously get a conference between the House and Senate to adjust the bill, or if one member objects, to have it go back through the Rules Committee, back through the Congress, and have this done on a controversial piece of legislation where powerful groups are opposing it, that is an extremely difficult task.[61]

One of the misleading indications of presidential power or success is the so-called presidential box score used by the *Congressional Quarterly* to indicate successes and failures in legislative programs. If Congress has approved a majority of a president's legislative program, the Congressional Quarterly may headline its story, "Congress Acts Favorably on President X's Requests." The impression often given is that a president is not only devoted to high principle but has also given independent, creative, galvanizing impetus to legislative progress. It is as if the president is the chief legislator; the Congress is mainly passive and has yielded its legislative authority.

However, box scores must be used with caution. First, they are often deceptive because legislative measures are by no means equal in importance. Moreover, scores fail to distinguish between measures that were central and those peripheral to presidential priorities. Further, they are skewed by the high percentage of presidential requests in the areas of defense and foreign policy, which Congress approves somewhat more readily than other matters. They show nothing of those programs a president wanted but, recognizing the overwhelming likelihood of defeat, never requested at all. Much of what a president does not achieve consists of what is never requested, rather than of what is proposed but does not get passed. Finally, sometimes a president merely anticipates what is going to pass Congress and adds an endorsement at a late stage in its legislative development.

Congress may not be designed to serve as a unifying leadership agency, yet it can play a relatively active role in determining or thwarting particular policies. It has clearly played such a role in tax matters, in policies determining freedom of information, and in the procurement; development, and evaluation of weapons systems. And as the country learned in the 1990s, it can occasionally provide important budget leadership.

Another area in which Congress influences presidential initiatives is in the shaping of legislation. Much of the recent writing about the presidency suggests that Congress merely delays and amends and is basically incapable of creating legislation or formulating policy. In fact, however, Congress sometimes plays the dominant role in initiating legislation; the formulation as well as the enactment of virtually all major legislation relating to domestic and economic policy is the result of extensive conversations between the presidency and Congress and between both of these institutions and pivotal interest groups.[62]

Much of the policy presidents supposedly formulate and propose as their own is derived directly from traditional party priorities, from previous presidents, or from Congress. Just as the celebrated New Deal legislation had a fairly well-defined prenatal history extending back several years before its espousal by FDR, so also recent investigations into the origins and enactment of most of the New Frontier and Great Society legislative programs, for example, broader medical-care programs, federal aid to education, and the more activist stance on civil rights, indicate they too were the fruition of past recommendations by the Democratic party.

Presidents have major resources with which they can enhance their political coalition and legislative lobbying roles. First, they can achieve a closer contact with the American public than any other politician in the country. The president can and often does appeal directly to the people for their support. This does not always work, yet when a president can define an issue and rally the public's concern, Congress and other power centers usually will take careful note.

A second resource is the use of patronage. Patronage today refers not only to jobs and favors rendered but also to campaign trips into districts and states, invitations to the White House, and countless considerations that a president can exchange in an effort to win friends and influence votes. The president's congressional liaison office, with a staff of sometimes up to twenty in recent years, works full time trying to employ these possibilities for maximum advantage.

The third resource is a president's role as a political party leader. Although a president has no formal position in the party structure, a president's influence over national policies and appointments and celebrity status command respect from party leaders. Presidents need the party's support to enact their program. Presidents find they must bargain and negotiate with party leaders just as they do with other independent power brokers.

Policy Implementation and Evaluation

Implementation, the carrying out and realization of presidential goals, is a crucial, if underappreciated, part of a president's job. The provisions made within the executive office to ensure presidential control over implementation simply do not guarantee this happens. Repeatedly, federal programs fail to accomplish desired goals. When such failures occur, it is relatively easy to blame the original legislation rather than examine what happened after a bill became a law. Of course, poorly written legislation can yield poor results. But students of national policymaking now realize even the best legislation can fail owing to problems encountered during implementation. Sometimes these problems can lead to the outright failure of a program; more often, however, they mean excessive delay, underachievements of desired goals, or costs far above those originally expected.

Implementation can be viewed as a process involving a long chain of decision points, all of which need to be cleared before a program can be successfully carried out. At each decision point is a public official or community leader who holds power to advance—or delay—the program. The more decision points a program needs to clear, the greater the chance of failure or delay. Special problems result if the successful implementation of a national program depends on the cooperation of state and local officials. The state or locality may be eager to help; another might be opposed to a program and try to stop it. The advent of new federalism and the vast growth in the number and scope of federal grant-in-aid programs have added to the problem of implementation.[63]

If anything has been learned by students of the presidency during the past fifty years, it is that policymaking does not end once a law is passed (or when a court decision or presidential executive order is handed down). The implementation of such laws, especially the development of specific guidelines and regulations, can have just as great an impact on public policy as the law itself. The ultimate responsibility for the implementation of the nation's governmental decisions rests with the president. "Implementing a public policy may include a wide variety of actions including issuing directives, enforcing directives, disbursing funds, making loans, awarding grants, making contracts, collecting information, disseminating information, assigning personnel, hiring personnel, and creating organizational units. Rarely are policies self-executing, i.e., implemented by their mere statement such as a policy not to recognize a certain government. Most policies require some positive action."[64]

We generally elect politicians as our presidents and then, almost as an afterthought, we hope they will be competent executives.[65] Executive skills are not the highest priorities people seek in presidential candidates. More often than not our presidents come from legislatures and law practices, which may prepare them well for passing laws but ill prepare them for policy-implementation leadership.

Presidential decisions transmitted to the bureaucracy often have a way of getting watered down or ignored. The routines of governmental bureaucracies are geared to maintaining the status quo. That's fine if a president merely wants to maintain the status quo. But most presidents want to change at least some policies. Yet policies persist from one administration to the next remarkably unchanged. Resistance to change is reinforced by several factors, chief of which are (1) inadequate communication and determination by a president, (2) the alliances between bureaucrats and the appropriate special interests and congresspersons, and (3) inadequate information and evaluation processes.

Presidents must be clear in their policy directives and persuasion efforts, for lack of clarity is a prime excuse for officials, even appointed political officials, to go their own way. An aide to Franklin D. Roosevelt emphasized how even cabinet members ignore a president's requests: "Half of a President's suggestions, which theoretically carry the weight of orders, can be safely forgotten by a Cabinet member. And if the President asks about a suggestion a second time, he can be told that it is being investigated. If he asks a third time a wise cabinet officer will give him at least part of what he suggests. But only occasionally, except about the most important matters, do Presidents ever get around to asking three times."[66]

Time and again all our recent presidents have been astonished at instances of noncompliance with their suggestions for implementation. But seldom is it a problem of outright defiance. Typically it is because of red tape and cautious, slow-moving officials who believe they know how to do their jobs. Sometimes, of course, this does involve differing policy preferences.

Too often the White House underestimates the importance of clear communications at the outset of an administration and later overestimates the uncooperativeness of career public servants. Caution is needed not to breed a we/they relationship or to breed hostility and disloyalty where none exists. Too often, also, the White House accepts the size and complexity of the large bureaucracies of the permanent government as insuperable and nonadaptive obstacles. What is needed is a White House strategy that consciously recognizes presidential dependence on the federal bureaus and field operations and takes into account the fact that presidential policy objectives will continue to lose clarity between the stages of legislation and implementation. What is also needed are innovative outreach and feedback strategies designed to keep organizations well informed of presidential intent, motivated to carry out such intents, and accountable for performance. These, of course, are far easier to talk about than to achieve.

Another major factor minimizing presidential influence in the implementation stage is the fact executive branch departments are more the creation of Congress than the White House. Special-interest groups often effectively capture administrative as well as legislative officials and succeed in fragmenting the organization to their own ends.

Congress, even when it is dominated by the president's party, often does not want to increase presidential discretion within executive agencies. The congressional committee structure in large measure parallels executive-branch organization, and what often appear to be structural absurdities in the executive branch may persist because of long-standing jurisdictional disputes within the Congress. Citing their responsibility of administrative review, these committees jealously protect what they consider to be their prerogative to determine how the departments and agencies they oversee are to be restructured. In addition, members of Congress guard those areas in which they have developed expertise and the close relations with government officials and extragovernmental clientele. A committee reorganization could diminish a member's sources of campaign finance or even jeopardize his or her chances for reelection.

Some observers suggest presidents have ignored the managerial side of their responsibilities because of a lack of incentives or rewards in this area.[67] In the words of one public administration expert, "No President has been able to identify any significant political capital that might be made out of efforts to improve management except for the conservative purpose of economizing or reducing costs."[68] Others feel that the way in which our civil-service system is organized encourages bureaucratic reluctance or resistance.[69] Still others remind us that the president frequently lacks authority to issue directions to operating departments. "The President's title as Chief Executive is misleading: there is very little that he can or does execute. . . . The work of the executive branch is carried out by operating departments granted specific powers and responsibilities by Act of Congress, and not by Presidential delegation."[70]

In sum, the array of formal powers and accumulated "chief executive" prerogatives that we commonly associate with the modern presidency do not, in fact, guarantee the achievement of very much.[71]

Oversight and Early-Warning System

Still another part of the job of the modern president is the design of oversight and early-warning systems and putting these to effective use. Often, little presidential time or stamina remains for inquiring into those routine activities that make up the great bulk of federal governmental work. Unfortunately, routine activities that are neglected or improperly monitored and evaluated can escalate to crisis proportion. But, for the most part, presidents must delegate large amounts of discretion to political subordinates and career professionals, and their influence over routine activities is felt only indirectly: through appointees, budgetary examinations, or legislative clearance. Presidents may view these activities as self-executing, or may even attempt to dissociate themselves from them, but much of the administrative and executive burden of the presidency consists of imaginative supervision of program implementation. The functional task of a president in the leadership of federal implementation activities is much more than merely issuing directives to the cabinet. It entails compromise, coalition building, education, and political leadership just as much as does the winning of congressional or public support. The quality of bread-and-butter service and assistance programs is important to the average American; a president, like it or not, is held responsible for the general quality of governmental performance.

Only through the skillful monitoring of routine governmental activities can a president know whether citizens are getting a fair return on their taxes. Only by a more imaginative use of presidential resources in overseeing these activities can the activities be prevented from becoming sources of crises themselves. To be sure, supervision and reform must come not only from presidents but also from other quarters as well. But how effective a president is in fashioning an executive oversight and review system is crucial. On the one hand, presidents must be able to delegate vast responsibilities to talented managers; on the other hand, they must have an early-warning system that alerts them through, among other means, their managers about inadequate government performance, about experimentation that yields negative results, and about progress in research and development that could provide corrective feedback.

In short, an effective organization is a learning organization—a system that ensures governmental activities concerned with implementation and routine would be brought to a president's attention when necessary and as they affected priority-setting and political-leadership tasks instead of only when they become matters of crisis management. Interior and transportation officials, for example, should detect evidence of impending fuel shortages sufficiently in advance to permit a president to commission studies and devise remedial strategies before the crisis occurs. White House supervision should have uncovered the extent and scope of illegal CIA domestic spying in the 1960s and the tragic dealings of CIA agent Aldrich Ames in the 1990s, long before they were discovered by the press and others. Similarly, data from government-run or -funded experiments or from program evaluations should be processed by the White House in order to fund more desirable or workable priorities. If one of the purposes of presidential power is to execute federal laws faithfully and to help avert costly crises, the presidency must be organized systematically as a learning agency, capable of anticipating and preventing system overload, communications failures, and corruption or ineptness in the operations of the executive branch.

CONCLUSION

Several underlying incentives help shape the performance of the presidential job. As they have operated in the recent past, these incentives ensure that certain responsibilities get special attention, whereas others become neglected. Preoccupation with problems of national security and macroeconomics leaves little time for leadership in the area of domestic policy. Crisis management, symbolic leadership, and priority setting also crowd out the tasks of lobbying and program implementation and supervision. Creative follow-through is seldom adequate; the routines of government often go neglected. Program evaluation and the imaginative recruitment of program managers for appropriate tasks never receive the sustained attention they merit. On balance, White House officials often do what is easy to do or what they perceive as urgent, sometimes to the neglect of doing what is important.

Presidents have concentrated in recent years on selected areas of the presidential job. In part this may be because we have created a nearly impossible presidential job description—we give our presidents too much to do and too little time in which to do it. In part, however, recent presidents have also been, or so it would appear, lulled into responding to those parts of their job that are more glamorous, more prominent. Presidential activity in symbolic, priority-setting, and crisis contexts conveys an image of strength, vigor, and rigor.

But measuring presidential strength or evaluating presidential leadership requires a more comprehensive look at what a president is doing and what a president can achieve not only in the three substantive subpresidencies but also in the seven functional leadership categories outlined in this chapter. The true measure of a president's effectiveness is his capacity to integrate multiple responsibilities in order to avoid having initiatives in one area compound problems in another or having problems go unattended merely because they defy the usual organizational boundaries. Another challenge of presidential leadership is the intricate balancing of each aspect of the presidential job with the others. A close examination of presidential performance in recent years relative to the whole matrix of the job suggests that presidents are strong in some areas and weak in others, and that the overarching job of synthesis and integration is seldom performed adequately.

The job of the modern presidency was forged in the World War II and Cold War eras. It called for a powerful presidency to stand up to enormously challenging problems. But in the post–Cold War era, we need to rethink the place of the presidency within our constitutional system of separate powers.

We are not likely to redefine in any measurable way the presidential job, although altering our expectations of the office in a post–Cold War age could help. Nor will structural, institutional reforms that seek to make the presidency more efficient and manageable be able to resolve most of the paradoxes and dilemmas compounding the job of the presidency, for almost all of the complexities of the presidential condition in America revolve around political problems and the diversity of American values. Those who devise structural solutions to what are essentially political questions are likely to be disappointed. In the twenty-first century we shall get improved presidential performances only when we have a clearer idea of what we want to do as a nation and when we understand better the mix of incentives that now shape the way presidents respond to their political and executive duties.

CHAPTER 6

The President and Congress

All the public business in Congress now connects itself with intrigues, and there is great danger that the whole government will degenerate into a struggle of cabals.

John Quincy Adams, *Diary,* January 1819

(Congress is) functioning the way the Founding Fathers intended—not very well. They understood that if you move too quickly, our democracy will be less responsible to the majority. . . . Exhaustion and exasperation are frequently the handmaidens of legislative decision. . . . I don't think it's the function of Congress to function well. It should drag its heels on the way to decision.

Congressman Barber B. Conable, Jr., *Time,* October 22, 1984

Our republic, we know, was designed to be slow-moving and deliberative. Our founding fathers were convinced that power had to be entrusted to someone, but that no-one could be entirely trusted with power. They devised a brilliant system of checks and balances to prevent the tyranny of the many by the few. They constructed a perfect triangle of allocated and checked power. . . . There could be no rash action, no rush to judgment, no legislative mob rule, no unrestrained chief executive.

The difficulty with this diffusion of power in today's cyberspace age is that everyone is in check, but no-one is in charge.

Sen. William S. Cohen, (R-Maine) (later secretary of defense), *Washington Post National Weekly Edition,* January 28–February 4, 1996, p. 29

171

Major Pierre L'Enfant, the principal designer of the nation's capital, had more than design in mind when he located the president and Congress at opposite ends of an avenue. Congress, he decided, would be housed atop Jenkins Hill, giving it the high ground. The president would be housed about one mile away, at the end of a long street that would serve as a threadlike link connecting these two institutions. Not only did the Constitution separate the executive and legislative branches, but geography would as well.[1]

Yet for the U.S. system to work, both branches must frequently find ways to bridge the gap, and join what is separated. The framers saw the separation of powers not as a weakness but as a source of strength, as a way to ensure deliberation and as a way to prevent tyranny. Cooperation between president and Congress was what the framers required if energetic government was to be achieved. But how was a president to develop the collaboration of the branches necessary to make the system of separate institutions sharing power work?

The relationship between president and Congress is the most important one in the American system of government, and while presidents spend considerable energy courting the media and appealing to the public, they do so primarily to gain leverage with the Congress. For, in the long run, presidents must get Congress to accept their proposals. Only Congress can allocate resources, and presidents who consistently attempt to go around the Congress cannot long succeed. A president may not like it, but sustained cooperation with Congress is a necessity.[2] Presidents may act semiautonomously of the Congress in a few areas, but most of the president's goals require a partnership with Congress.

This is no easy matter. As one long-time presidential adviser noted, "I suspect that there may be nothing about the White House less generally understood than the ease with which a Congress can drive a president quite out of his mind and up the wall."[3]

THE PRESIDENT'S CONSTITUTIONAL PLACE

In matters requiring legislative authorization Congress ultimately has the upper hand. While a president may be seen as leader of the nation, agenda setter, vision builder, and legislator-in-chief, it is often Congress that has the final say. And since there are multiple veto points in Congress, any of which may block a president's proposal, the forces wishing to prevent change almost always have the upper hand. The American system has many roadblocks, yet few avenues for positive change.

Many people expect presidents to lead the Congress. But public expectations notwithstanding, a president's legislative powers are constitutionally (Article II, Section 3) thin. The Constitution grants Congress "all legislative power" but establishes a relationship of mutual dependence and power sharing. The more ambitious a president's agenda, the more dependent that president is on congressional cooperation. Presidents have the veto power, and as of 1997 the item veto. They report on the state of the union, and extraconstitutionally they may suggest legislation, lobby Congress, set the agenda, and build coalitions. Still, overall presidential powers over Congress are limited.[4]

The peculiar paradox is that in a real sense, as Secretary William Cohen noted, no one is in charge. Responsibility and power are fragmented. This often creates a fascinating blame game, with presidents blaming the Congress for the ills of the nation and the Congress blaming the president. This was illustrated during the Reagan presidency when Reagan, who never once in eight years submitted a balanced budget to the Congress, railed against the big-spending Democrats in Congress, and Democrats countered, attacking the president as a budget buster. Who was right? Both were. Because we separate power, both could lay blame yet neither had to accept responsibility. Accountability in such a system is difficult to maintain.

In this the U.S. system stands in contrast to parliamentary systems such as that of Great Britain. Rather than having a separation of powers, parliamentary systems have a *fusion* of power. The prime minister is elected from and by the majority party in the Parliament. Prime ministers are responsible to, yet also have power over, the majority party, and are thereby granted more legislative power than a U.S. president possesses. Fusing the executive and legislative powers together creates more power and greater accountability. If things go well or ill, the voters know who is responsible.

HISTORICAL EVOLUTION

Congress was established as the dominant branch of government, and the first few presidents were only minimally involved in the legislative process. But the founders' vision soon gave way to political reality, and slowly presidents began to pull power into the White House. Congress was often a willing participant in giving additional power to the presidency. The rise of the legislative presidency did not follow a clear unobstructed path. Some of the more ambitious chief executives such as Jefferson used the newly forming political parties to exert influence in Congress. Others such as Jackson exploited popular opinion to gain influence. Still others such as Lincoln and FDR gained additional powers during crisis.

Sprinkled between these power-aggrandizing presidents were less assertive or weaker chief executives. Thus, an institutional ebb and flow characterizes power relations between president and Congress. Over time, however, this tug of war has moved toward a stronger president in the legislative arena. Today, "What was designed as a congressionally centered system has evolved into a presidentially centered . . . system."[5] Still, Congress has a way of frustrating even the most skilled of presidents, especially when there is divided government, with the control of the presidency and Congress in the hands of vying political parties.[6]

While the popular expectation is that the president is the "chief legislator" who "guides Congress in much of its lawmaking activity,"[7] reality is often different. The decentralized nature of power in Congress, the multiple access and veto points within the congressional process, the loosely organized party system, the independent/entrepreneurial mode of the legislators, the weakness of the legislative leadership, all conspire against presidential direction and leadership.[8] This means that generally presidents impact Congress only "at the margins," and are more "facilitators" than leaders.[9]

Lest the system break down into hopeless gridlock, the president and Congress must find ways to work together. The theory on which the U.S. government is based hinges on a reasonable amount of cooperation between the two branches. As Justice Robert Jackson wrote in 1952, "while the constitution diffuses power the better to secure liberty, it also contemplates that practice will integrate the dispersed powers into a workable government."[10]

WORKING WITH CONGRESS

Immediately on a new president's taking office, members of Congress take the measure of the person. Will the president stand firm? Will the president bend? Will the new president work with us or act independent of Congress? Will the president be highly partisan or bipartisan? Will the president be able to intimidate us, or can we lead the White House? Who will set the tone?

From early August of 1990 through much of January 1991 President George Bush deftly maneuvered the United States in the direction of what became known as Operation Desert Storm to liberate Kuwait. Bush said initially his decision to send troops to Saudi Arabia was a "wholly defensive" mission. Yet he increasingly enlarged our military presence, and he rallied both the United States and a twenty-eight–nation coalition to prepare for war against Iraq's Saddam Hussein. In December of 1990 Bush signaled cabinet officials that he wanted to go to war to liberate Kuwait. He is reported also to have told foreign leaders in the coalition that if he decided to go to war, he would do so whether Congress agreed or not.[11]

Congress eventually, after heated arguments, passed resolutions supporting the use of military force to liberate Kuwait. There was, however, no formal declaration of war. Critics assert that American presidents now have more personal power to make war than do the leaders of any other major democracy. And they lament that this is precisely what the framers of the Constitution set out to prevent.[12]

Consider another example of executive-congressional conflict. In his 1991 memoir *Under Fire,* Lieutenant Colonel Oliver North, who served as a national security aide in the Reagan White House, admitted he deliberately misled members of Congress when they asked him about covert operations on behalf of Nicaraguan rebels. North was later indicted and convicted for lying to Congress, although ultimately the decision was overturned. North knew what he had done was wrong, but he did not consider it a crime. It was, he now says, just part of the "two hundred years of sparring between the Executive Branch and Congress over the control of foreign policy."[13]

The tumultuous Senate confirmation hearings of Clarence Thomas are yet another example of the politics of shared powers. Presidents nominate individuals to fill vacancies on the Supreme Court, but the Senate has the power to confirm or deny such nominations. The Clarence Thomas hearings raised issues of sexual harassment and integrity as well as debates about liberal and conservative court policy. Thomas eventually won confirmation in a 52–48 roll-call vote—the closest for a confirmed Supreme Court nominee in more than one hundred years.[14] Judge Thomas termed the last segment of his hearings "a high-tech lynching" and said, near the end of the proceeding, that he was no longer interested in whether he got the job; he just wanted to clear his name. The hearings angered a lot of people, and "senators in both parties admitted to a common embarrassment at the institution's failure to deal better with the divisive issues before them."[15]

The celebrated sparring between President Clinton and House speaker Newt Gingrich in 1995 over who would control the nation's public policy agenda may have seemed unusual, but in historical terms it merely reflects the ongoing struggle between the president and Congress.

These are just examples of skirmishes between the president and Congress as they try to adapt the Constitition of 1787 to today's realities. In this chapter we begin by looking at our system of separate branches and the conflict between the branches. We then treat two interpretations of the relationships between the presidency and Congress. Finally, in the second half of the chapter, we examine some of the measures Congress has taken since the mid-1970s to try to reassert its authority in its continuing struggle with the White House.

THE POLITICS OF SHARED POWER

The framers anticipated that the president and Congress would often disagree over policy, for they gave the president veto power over legislation but gave the Congress the power to override that veto. They gave the power to nominate top personnel to the president but gave the Senate power to confirm top personnel. They authorized presidents to negotiate treaties but required the Senate's approval by a two-thirds vote before the president could ratify a treaty and make it part of the law of the land. The framers also knew from their study of history that heads of government were more prone to going to war than were the people's representatives, so they vested the final authority for war in the legislature.

Constitutional democracy in the United States was designed to be one of both shared powers and division. The framers wanted disagreement as well as cooperation because they assumed that the checks and balances within the government would prevent a president and Congress from "ganging up" against the people's liberties. The framers actually made such disagreements inevitable by providing a president, Senate, and House of Representatives elected by different constituencies and for different lengths of service.

The United States is unique among major world powers because it is neither a parliamentary democracy nor a wholly executive-dominated government. Our Constitution plainly invites both Congress and president to set policy and govern the nation. Leadership and policy change are encouraged only when two, and sometimes all three, branches of government concur on the desirability of new directions. At one time Congress was the dominant partner, but recently, as we have noted earlier, presidents seem to hold the upper hand, despite constitutionally specified restraints.

The politics of shared power has often been stormy, as the Persian Gulf War, Iran-Contra affair, Clarence Thomas confirmation hearings, and Clinton health care proposal illustrate. The politics of shared powers is characterized by changing patterns of cooperation and conflict depending on the partisan and ideological makeup of Congress, the popularity and skills of the president, the strength of the political parties, and various events that shape the politics of the times. As we discuss later, the politics of shared powers in an era of divided government can aptly be described as the politics of Congress's varying success at asserting itself in relation to the initiatives and leadership provided by presidents.

Divided government has become the norm rather than the exception. Since 1952 there has been a split in partisan control of the presidency and Congress for about two-thirds of the time. Only Presidents John Kennedy, Lyndon Johnson, Jimmy Carter, and for a brief time Bill Clinton, all Democrats, enjoyed majority control by their own party in Congress, and even they had considerable trouble getting support for many of their legislative programs.

The opposition party in Congress regularly mounts its own programs. It will, when possible, defeat a president's policy initiative and substitute its own. This effort becomes all the more troublesome, of course, when Congress is controlled by the opposition party. Certainly Presidents Bush, Reagan, and Clinton have had their troubles with a House of Representatives controlled by the opposition.

The politics of shared power necessitates the development of agreement between a president and Congress. The model, or theory of government, on which the American system was founded is based on consensus and coalition building. Consensus means agreement about *ends;*

coalitions are the *means* by which the ends are achieved. Since power is fragmented and dispersed, something (crisis) or someone (usually, a president) has to pull the disparate parts of the system together.

Power can only be legitimately exercised when the different parts of the system act together. Power, which in many ways floats in the American system, has to be harnessed and channeled. For this to happen, a working consensus on what to do has to be developed, and a coalition of forces has to be formed.[16]

A consensus can be formed if the president has a clear, focused agenda; if the president can forcefully and compellingly communicate a vision; and if the public is ready to embrace that vision. If a president can develop a consensus he can then muster the power to form the coalitions necessary to bring the vision to fruition.

Simply placing a legislative package at the doorstep of Congress is not enough. Presidents must work to build support within and outside Congress. They must build coalitions. The office does not guarantee political leadership; it merely offers incumbents an invitation to lead politically. It is in this sense that those best suited to the job are those who can creatively shape their political environment and savor the rough-and-tumble give-and-take of political life.

Of course, building coalitions is easier said than done. Institutional combat, not power sharing, characterizes this relationship, as both branches attempt to usurp power and govern autonomously.

The political party can serve as an aid in coalition building. In 1981, Ronald Reagan relied on a unified Republican party to help push his program through Congress. President Clinton spent a good deal of time lobbying congressional Democrats on behalf of his agenda, but the Democrats are a fractured and combative lot, and Clinton had a difficult time keeping his majority together in Congress.[17]

Because Clinton had an ambitious agenda yet only a thin congressional majority, and because the opposition party was unified against the President, Clinton was forced to rely on the Democrats alone for votes. This emboldened some Democratic legislators who felt they had leverage with the President and could trade favors for their votes.

THE PRESIDENT IN THE LEGISLATIVE ARENA

Although separation of powers and divided government are obstacles, they are not insurmountable barriers to good public policymaking. Presidents and Congress can legislate when the leaders of both institutions bargain and compromise in ways that overcome the roots of division. In fact, while the Constitution disperses power and invites a continuous struggle between these two branches, it also requires the two branches to collaborate. And usually they do. Even when the relationship is regarded as hostile, "bills get passed and signed into law. Presidential appointments are approved by the Senate. Budgets are eventually enacted and the government is kept afloat. This necessary cooperation goes on even when the White House and the Capitol is divided between the two major parties."[18]

Under what conditions are presidents likely to establish their agenda and get congressional support? Put another way, when is the Congress most likely to follow a president's lead? And when is it likely that a president will follow the lead of Congress?

Several factors lend themselves to presidential success in dealing with Congress. First, in a crisis, the president is accorded a great deal of deference, and the Congress usually supports the president. Second, when a president has a clear electoral mandate (when the campaign was issue oriented, the president won by a wide margin, and the president's party has a sizable majority in Congress), the Congress sometimes will follow. Third, a president can exert pressure on members of Congress when he won the election by a landslide and ran ahead of the member in his or her own district (usually referred to as "presidential coattails"). Part of President Clinton's problems with Congress can be traced to the fact that he ran behind most members in their districts, did not face an emergency (as of late 1997 at least), and did not win a clear electoral mandate in 1992 or 1996.

Fourth, presidential popularity is often said to be a source of power over Congress. But how easily can a president translate popularity into power? Many social scientists are dubious, believing that popularity has only a marginal impact on presidential success within Congress.[19] Others argue that popularity does translate into power.[20] Political scientist Richard Neustadt placed considerable emphasis on presidential prestige as a source of power and influence, and his view was echoed by another political scientist who wrote that

> Presidential poll ratings are important because they are thought to be important. They are thought to be important because political leaders look for indications of when it is safe or dangerous to oppose their policy interests or career ambitions to those of the President and because indications of political support—which in other political contexts might be preferred—are too limited in scope to be relied upon in this context.[21]

Fifth, skill does make a difference, but how much of a difference? As is the case with popularity, scholars disagree, some arguing that skill is of little importance,[22] others that it is important.[23] One scholar warns that

> it is important to depersonalize somewhat the study of presidential leadership and examine it from a broader perspective. In this way there are fewer risks of attributing to various aspects of presidential leadership consequences of factors largely beyond the president's control. Similarly, one is less likely to attribute incorrectly the failure of a president to achieve his goals to his failure to lead properly. Things are rarely so simple.[24]

High skill levels may give a president greater leverage to win in the congressional process. Skills such as knowledge of the congressional process and needs, good timing, bargaining, deal-making, persuasion, and coalition-building skill; moving the public; setting the agenda; self-dramatization; arm twisting; trading; consultation and co-optation, and even threats can all be used to advance the president's goals.

Lyndon Johnson, one of the more skilled of the modern presidents vis-à-vis Congress, emphasized that a White House had to work continuously to build bridges with Congress. He urged his aides and cabinet members to get to know the members of Congress as well as they knew their family.

President Johnson perfected what some called "the Johnson treatment." Journalists Rowland Evans and Bob Novak described it as follows:

> Its tone could be supplication, accusation, cajolery, exuberance, scorn, tears, complaint, the hint of threat. It was all of these together. It ran the gamut of human emotions. Its velocity was breathtaking, and it was all in one direction. Interjections from the target were rare. Johnson anticipated them before they could be spoken. He moved in close, his face a scant millimeter from his target, his eyes widening and narrowing, his eyebrows rising and falling. From his pockets poured slips, memos, statistics. Mimicry, humor, and the genius of analogy made The Treatment an almost hypnotic experience and rendered the target stunned and helpless.[25]

Lyndon Johnson could be relentless, but he was afforded that luxury only partly due to personal skills. He had other significant advantages that made his power more robust and his threats more credible (compare the resources Johnson had with the more meager power resources of a Bush or Clinton). One of the most important of these power resources is party support.

Partisan support in Congress is the sixth major factor shaping presidential success with Congress. Parties can serve as a bridge linking the institutional divide between the president and Congress. Lyndon Johnson had such a large majority of party members in Congress that even if several dozen Democrats abandoned the president, he could still get his majority in Congress. George Bush on the other hand faced a Congress in control of the opposition, and was thus stymied.

Linked to this is the seventh factor: the nature of the opposition in Congress. How many votes do they have? How ideologically driven are they? How cohesive are they? How willing are they to work with the president? In 1981, President Reagan faced a Senate controlled by his own party but a House controlled by the Democrats. Reagan was able, via skill, luck, and circumstance, to win over enough Democrats (referred to as the "Boll Weevils") to win several significant legislative battles. By contrast, while President Clinton had majorities in both houses during his first two years in office, the Republican opposition was so unified against the president (on Clinton's 1993 economic package, not one Republican senator supported the President) that Clinton had a difficult time with Congress.[26] Yet, in spite of these difficulties, Clinton achieved a lofty 86.4 percent bill passage rate in his first two years in office. In 1995, the year after the Republican takeover of Congress, Clinton's success rate plummeted to 36.2 percent, the lowest score of any president in modern times, and the biggest drop (over 50%) ever. In 1996, Clinton achieved a 55.1 percent rate, the single biggest jump, since 1953. This jump reflects both Clinton's moving to the center, and the Republicans, stung by the blame they received for the 1995 government shut-down, learning to work with Clinton on several key issues.

Due to the nature of partisan politics, the eighth factor that shapes presidential success or failure in Congress is the nature of consultation between the two branches. One of the lessons President Clinton learned is that a president must consult, not only with the president's own partisans but with the opposition as well. Attempting to gain cooperation and agreement must be the first step. A president also needs to set up an effective legislative liaison office that listens and works closely with the leadership on the Hill. Then too, a president needs to appreciate the policy values and political needs of the national legislators. Presidents sometimes need to follow as well as lead, especially when there are better ideas to be found in Congress (see Table 6.1).

Finally, the type of agenda a president pursues has a significant im-

TABLE 6.1
Success Rate History of Modern Presidents
with Congress

Eisenhower	Nixon	Reagan
1953 89.0%	1969 74.0%	1981 82.4%
1954 82.8	1970 77.0	1982 72.4
1955 75.0	1971 75.0	1983 67.1
1956 70.0	1972 66.0	1984 65.8
1957 68.0	1973 50.6	1985 59.9
1958 76.0	1974 59.6	1986 56.1
1959 52.0	**Ford**	1987 43.5
1960 65.0	1974 58.2%	1988 47.4
Kennedy	1975 61.0	**Bush**
1961 81.0%	1976 53.8	1989 62.6%
1962 85.4	**Carter**	1990 46.8
1963 87.1	1977 75.4%	1991 54.2
Johnson	1978 78.3	1992 43.0
1964 88.0%	1979 76.8	**Clinton**
1965 93.0	1980 75.0	1993 86.4%
1966 79.0		1994 86.4
1967 79.0		1995 36.2
1968 75.0		1996 55.1

Source: Updated from *Congressional Quarterly Reports,* December 19, 1992, p. 3896.

pact. George Bush had a very thin legislative agenda, and he pursued it only half-heartedly. Bill Clinton had a large, ambitious agenda, making it more difficult to get the Congress to go along. Given the antipolitics mood of the times, it is surprising that Clinton did as well as he did in his first term. An ambitious agenda is difficult to pass in good times, but in tough times, the tough choices appear to be impossible choices.

In recent decades the inside bargaining skills necessary to cut deals have been replaced by or supported by major marketing or self-promotion efforts. In the loose bargaining regime of the Congress, presidents think their time is better spent appealing directly to the public for support, which may translate into clout in Congress. But ultimately presidents must cut deals in Congress.[27]

Congress can, of course, play a major role in setting and sometimes shaping the national public policy agenda. It may be difficult for a plural institution to lead, yet Congress sometimes acts as a leadership and law-making as well as representative institution. When is this the case? This happens when a party enjoys strong majorities in both chambers, when a president is vulnerable, such as at the end of a term, or is politically wounded (as Nixon was in 1973 and 1974), and when Congress has strong leaders, as when Sam Rayburn and Lyndon Johnson exercised impressive leadership toward the end of the Eisenhower presidency.

But the central question under consideration here is: Can a president lead Congress? The answer remains: Yes, but not often, and not usually for long. A mix of skill, circumstances, luck, popularity, party support, timing, and resources need to converge if effective collaboration of these two highly political branches is to occur.

What follows is a consideration of some of the resources and realities that can affect presidential-congressional leadership.

THE PRESIDENTIAL VETO

A president can veto a bill by returning it, together with specific objections, to the house in which it originated. Congress, by a two-thirds vote in each chamber, may override the president's veto (see Table 6.2). Another variation of the veto is known as the pocket veto. In the ordinary course of events, if the president does not sign or veto a bill within ten weekdays after receiving it, it becomes law without the chief executive's signature. But if Congress adjourns within the ten days, the president, by taking no action, can kill the bill.

The veto's strength lies in the failure of Congress to get a two-thirds majority of both houses. Historically Congress has overridden less than 7 percent of presidents' general vetoes. Yet a Congress that could repeatedly mobilize a two-thirds majority against a president can almost take command of the government. This is a rare situation, but such was the fate of Andrew Johnson in the 1860s.[28]

TABLE 6.2
Presidential Vetoes, 1789–1997

	Regular vetoes	Pocket vetoes	Total vetoes	Vetoes overridden
Washington	2	—	2	—
Madison	5	2	7	—
Monroe	1	—	1	—
Jackson	5	7	12	—
Tyler	6	3	9	1
Polk	2	1	3	—
Pierce	9	—	9	5
Buchanan	4	3	7	—
Lincoln	2	4	6	—
A. Johnson	21	8	29	15
Grant	45	49	94	4
Hayes	12	1	13	1
Arthur	4	8	12	1
Cleveland	304	109	413	2
Harrison	19	25	44	1
Cleveland	43	127	170	5
McKinley	6	36	42	—
T. Roosevelt	42	40	82	1
Taft	30	9	39	1
Wilson	33	11	44	6
Harding	5	1	6	—
Coolidge	20	30	50	4
Hoover	21	16	37	3
F. Roosevelt	372	263	635	9
Truman	180	70	250	12
Eisenhower	73	108	181	2
Kennedy	12	9	21	—
L. Johnson	16	14	30	—
Nixon	26	17	43	7
Ford	48	18	66	12
Carter	13	18	31	2
Reagan	39	39	78	9
Bush	29	17	46	1
Clinton (first term)	17	0	17	1

Sources: Updated from Statistical Abstract of the United States, 1986, p. 235; Senate Library, *Presidential Vetoes* (Washington, D.C.: Government Printing Office, 1960), p. 199.

Presidents can also use the veto power in a positive way. They can announce that bills under consideration by Congress will be turned back unless certain changes are made. They can use the threat of the veto against some bills Congress wants badly in exchange for other bills that the president may want. A presidential veto can also protect a national minority from hasty, unfair legislation passed in the heat of the moment. But the veto is essentially a negative weapon of limited use to a president who is pressing for action.

The presidential veto power has stirred little controversy. Carter vetoed only thirty-one bills, Nixon vetoed forty-three, Ford sixty-six, and

Reagan seventy-eight. Bush vetoed forty-six bills, and Clinton vetoed seventeen bills in his first term.

In short, there is little Congress can do when confronted with a veto. It must either get enough votes to override the veto or modify the legislation and try again. Presidents are able to make the vast majority of their regular vetoes stick.

THE PRESIDENTIAL ITEM VETO

In 1996 Congress passed and President Clinton signed into law a measure authorizing an item veto for presidents. This now enables a president to veto or delete from a larger bill a discretionary spending item, new direct spending, or items of a limited tax benefit. A president, according to this law, has five calendar days after the passage of the spending measure to notify Congress of the decision to veto that part of the measure. Congress then has thirty days to act and does not require a two-thirds vote to override, merely a majority vote of each chamber of Congress.

Presidents had long wanted this additional power to help them fight deficit spending and to cut expenditures when the primary reason for the expenditure was to help members of Congress get reelected. Such "wasteful" items were often added to vital appropriation bills at the last minute, and presidents had felt forced to sign the bill, the wasteful along with the crucial. Now with the item veto it will be fascinating to see whether presidents will exercise this new power vigorously and how they will act when a colleague's pork barrel items are isolated from needed major spending bills.

Some members of Congress and some constitutional scholars believe enacting an item veto provision, if such was deemed desirable, should have been done through the complex process of amending the Constitution rather than being decreed by legislative statute. Others believed Congress had the authority to voluntarily delegate this kind of power to the president. The Supreme Court ruled in 1997 in a 7 to 2 decision that Clinton could exercise this power, though they reserved the right to review the item veto's constitutionality later after it was used.

THE LEGISLATIVE VETO

Since the 1930s and most especially since the 1970s, Congress has exercised what came to be known as the legislative veto. Using this device, Congress would draft a law broadly but incorporate a provision allowing Congress to review and veto the executive branch's implementation of the law. The legislative veto could be put into effect by a majority vote of one house, both houses, or sometimes even by a single congressional committee, depending on how the law was written.

Whatever the form, the legislative veto allowed Congress to delegate general power to the executive branch and then take it away without having to secure presidential approval. In effect, it permitted Congress to legislate without exposing its handiwork to a presidential veto, as the framers intended. The Constitution stipulates that every bill, resolution, or vote for which the agreement of the Senate and the House of Representatives may be necessary, shall be presented to the president for approval or veto. Joint resolutions were regularly submitted to the president, but concurrent resolutions were not. In the past this made little difference, because concurrent or simple one-house resolutions were mainly used to express congressional opinion and had no force of law.

But the legislative veto was also used to keep presidents in check. Arms sales had to be submitted to Congress for its scrutiny. Presidential use of military troops abroad had to be reported to Congress and was subject to recall by unilateral congressional action. In short, angered by presidents who had either lied to or ignored Congress, assertive legislators tried to use the legislative veto to recapture congressional power.

The legislative veto originated in 1932 when Congress passed a resolution allowing President Herbert Hoover limited authority to reorganize the executive branch agencies. The resolution stipulated that the president's proposals would not be put into effect for ninety days, during which time either house of Congress, by a simple resolution, could veto the proposal. For the next fifty-one years, the legislative veto became a standard practice.[29]

Between 1932 and mid-1983 more than two hundred pieces of legislation included some provision for a legislative veto. The device was used to ensure that bureaucratic regulations conformed to congressional intentions. This was important in the 1970s because of the rapid increase in the number of such regulations, which often have the same or nearly the same force as laws. In any given year in the late 1970s or early 1980s, Congress might have passed a few hundred public laws; but the administrators in about 708 executive branch agencies were responsible for twenty times as many regulations.

Then, in June 1983, Chief Justice Warren Burger, speaking for a Supreme Court majority in a 7–2 decision *(INS v. Chadha)*, said that to maintain the separation of powers, the carefully defined limits on the power of each branch must not be eroded. The legislative veto was found unconstitutional. Said Burger: "With all the obvious flaws . . . we have not yet found a better way to preserve freedom than by making the exercise of power subject to the carefully crafted restraints spelled out in the Constitution." In effect, the Court told Congress that to obtain more influence over an agency or the presidency, Congress should pass a law that accomplished this explicitly. In the words of dissenting Justice Byron White, the Court's decision "strikes down in one fell swoop provisions in more laws enacted by Congress than the Court has cumulatively invalidated in its history."[30]

In this historic decision, the Court tried to curb a weapon that Congress had sometimes used effectively to intimidate executive branch officials. Clearly, the existence of the legislative veto stimulated compromise and understandings between executive and legislative officials. Some observers viewed the decision as just one in a long series of Supreme Court rulings that generally approve and encourage an assertive and increasingly powerful presidency.

Because of this decision, Congress has been a little more explicit about the policy directions it sets. It is also exercising some options that make it clear to executive departments that if congressional committees are displeased by the actions of the departmental officials, congressional retribution in the form of reduced appropriations is likely. Congress has found ways to exercise the functional equivalent of a legislative veto.[31] Dozens of such quasi-legislative vetoes have been signed into law in the past decade. Further, there have been scores of informal agreements between the branches, such as the Bush-Congress agreement on aid to the Nicaraguan rebels.

These quasi-legislative vetoes and similar agreements survive, despite the Court's ruling, because they serve a purpose. They give the executive branch the flexibility it desires and that Congress might otherwise not provide, while allowing the Congress to retain a certain amount of ongoing control. "The legislative veto procedure represented a classic quid pro quo," says Louis Fisher. "It attempted to reconcile the interests of both branches: the desire of (executive) agencies for greater discretionary authority and the need of Congress to maintain control short of passing another public law."[32] Hence legislative vetoes or agreements very similar to them are likely to be with us for the near future.

PRESIDENTS AND CONGRESS IN FOREIGN AFFAIRS

Although the framers of the U.S. Constitution may not have anticipated the degree to which presidents would become so centrally involved in the making of domestic public policy, scholars have justified the need for strong dynamic presidents who would help overcome the fragmentation of power in the country and in Congress. The creaky machinery of our government, they contend, can be made to work only if we give a president the proper amount of help and authority.

The advent of radio and television greatly increased the visibility of the president, so that Americans now look to the president to cope with the full range of domestic and economic problems, such as recession and pollution. Considering that presidents get so much more publicity than do members of Congress, this is hardly surprising.

If the constitutional framers expected presidents to play a relatively minor role with respect to domestic policy, they had different expectations with respect to foreign and national security matters. That is why they made the president the agent in making and ratifying treaties, with senatorial consent, and also gave the president control over the military forces.

Although foreign affairs in the eighteenth century were generally thought to be an executive matter, the framers did not want the president to be the only or even the dominant agent.[33] Various powers vested in Congress by the Constitution were explicitly designed to bring the national legislature into the making of foreign and military policy. Indeed, a good part of the Constitution was written to deprive the executive of control over foreign policy and foreign relations, which under the English system were so dramatically vested in the king. Thus the framers gave Congress as a whole the sole power to declare war, and they plainly intended the Senate to serve as a partner in the shaping and making of foreign policy. Constitutional scholar Leonard W. Levy goes even further: "The framers meant, at the most, that the president should be a joint participant in the field of foreign affairs, but not an equal one."[34]

But as the world leadership role of the United States has increased, so also has the foreign policy role and the power of the president. From 1940 or so onward, "the Oval Office is where foreign policy for better or worse has been made."[35] Policies that have succeeded in recent decades—the Marshall Plan, the Truman Doctrine, the Panama Canal Treaty, and the Camp David Accords—have been those in which presidents encouraged careful collaboration and debate in Congress. When presidents have involved Congress and the people in the shaping of new foreign policies, those policies have generally won legitimacy and worked. "The policies that have failed have tended to be those adopted by presidents without meaningful debate—Roosevelt's Yalta policy toward Poland, Kennedy's intervention in the Bay of Pigs . . . Reagan's Iran-contra policy, among others."[36]

THE "IMPERIAL PRESIDENCY" ARGUMENT

No matter what the circumstances, though, most people expect Congress to exercise fully the system of constitutional checks to rein in the president, and vice versa. Late in the 1960s and in the early 1970s, a number of events made many fear that Congress had become too passive and the presidency too powerful.

In the early 1970s, the Vietnam War was finally winding down, but not without leaving the nation reeling from the protests, civil unrest, and loss of lives associated with that unpopular and largely unsuccessful war. The nation's grief and outrage were fueled, too, by revelations that both Johnson and Nixon had misled Congress and acted in some instances without congressional approval in making and conducting the war. These disconcerting events were compounded by the Watergate scandals and the growing realization that a president of the United States had acted to obstruct justice. These developments led Arthur M. Schlesinger, Jr., a historian and former adviser to John F. Kennedy, to write *The Imperial Presidency,* charging that presidential powers were so abused and expanded by 1972 that they threatened our constitutional democracy.[37]

Schlesinger and other proponents of the "imperial presidency" view contend the problem stems in part from ambiguity concerning the president's power as commander-in-chief; it is a vaguely defined office, not a series of specific functions. Schlesinger and others acknowledge that Johnson and Nixon did not create the imperial presidency; they merely built on some of the more questionable practices of their predecessors. But some observers contend there is a distinction between the abuse and the usurpation of power. Abraham Lincoln, Franklin Roosevelt, and Harry Truman temporarily usurped power in wartime. Johnson and Nixon abused power, deceiving Congress, misusing the Central Intelligence Agency, and manipulating public opinion and the electoral process.

Secrecy has often been used to protect and preserve a president's national security power. Nixon, it is said, pushed the doctrine beyond acceptable limits. Before Eisenhower, Congress expected to get the information it sought from the executive branch, and instances of secrecy and executive privilege were the rare exceptions. By the early 1970s, however, these practices had become the rule. And a Congress that knows only what the president wants it to know is not an independent body.

Political scientist Theodore Lowi contends presidents have little choice but to be at least semi-imperial, given the role of the executive in the development of the American national economy and its related regulatory and bureaucratic machinery. He suggests Schlesinger's interpretation exaggerates the case of personal abuse of power by Nixon and others and underestimates the fact that the modern presidency is largely the construction of the Congress with the cooperation of the federal courts. The vast growth of presidential power, he claims, began with the coming of New Deal domestic programs and cannot be linked solely with the expansion of the president's foreign policy powers. Although "there may be many specific cases of usurpation by modern presidents, these are extreme actions in pursuit of powers and responsibilities by and large willingly and voluntarily delegated to the president by Congress."[38]

Still, Schlesinger's views are a useful point of departure for discussing the alleged too-powerful presidency. The chief complaints involve such presidential activities as war making, emergency powers, diplomacy by executive agreement, and government by veto.

PRESIDENTIAL WAR-MAKING POWERS BEFORE 1974

The Constitution delegates to Congress the authority to declare the legal state of war (with the consent of the president), but in practice the commander-in-chief often starts the fighting or initiates actions that lead to war. This power has been used by the chief executive time and time again. In 1846, President James K. Polk ordered American forces to advance into disputed territory. When Mexico resisted, Polk informed Congress that war existed by act of Mexico, and a formal declaration of war was soon forthcoming. William McKinley's dispatch of a battleship to Havana harbor, where it blew up, helped precipitate war with Spain in 1898. The United States was not formally at war with Germany until late 1941, but prior to the Japanese attack on Pearl Harbor, Franklin Roosevelt ordered the navy to guard convoys to Great Britain and to open fire on submarines threatening the convoys. Since World War II presidents have sent forces without specific congressional authorization to Korea, Berlin, Vietnam, Lebanon, Grenada, Cuba, Libya, Panama, Kuwait, and Haiti—in short, around the world.

From George Washington's time until today, a president, by ordering troops into battle, has often decided when Americans will fight and when they will not. When the cause has had political support, the president's use of this authority has been approved. Abraham Lincoln called up troops, spent money, set up a blockage, and fought the first few months of the Civil War without even calling Congress into session.

It became obvious that presidents needed the power to respond to sudden attacks and to protect the rights and property of American citizens. One State Department official once described this enlarged mandate as follows:

> In the twentieth century the world has grown much smaller. An attack on a country far from its shores can impinge directly on the nation's security. . . . The Constitution leaves to the President the judgment to determine whether the circumstances of a particular armed attack are so urgent and the potential consequences so threatening to the security of the U.S. that he should act without formally consulting the Congress.[39]

Congress was angered when it learned (several years after the fact) that in 1964 President Johnson won approval of his Vietnam initiatives on the basis of misleading information. Under Nixon in 1969 and 1970 a secret air war was waged in Cambodia with no formal congressional knowledge or authorization. The military also operated in Laos under Nixon without formally notifying Congress. It was to prevent just such acts as these that the framers of the Constitution had given Congress the power to declare war; many members of Congress believed that what happened in Indochina in the 1960s and 1970s was the result of the White House's bypassing the constitutional requirements. They also agree, however, that presidential excesses came about because Congress either agreed too readily with presidents or did little to stop them.

During hostilities, especially if the military action is not an all-out war with the nation's vital interests clearly at stake, such as World War II, the country and Congress typically rally behind a president. As casualties mount and fighting continues, support usually falls off. In both Korea and Vietnam, presidential failure to end the use of American ground forces led to increased political trouble. Eisenhower swept into power in 1952 saying, "I shall go to Korea," thus arousing hope among voters that he would end the Korean conflict. Nixon won in 1968 when Johnson was forced out over Vietnam. But even though Congress may have been misled during the Vietnam War, it enthusiastically supported the president and went along with his actions. Not until the war turned sour did senators and representatives begin to charge misrepresentation. Why, then, were they so easily talked into approving funds for the war? (They continued to pass appropriations for it right up to April 1975). The general lesson appears to be that the country, Congress, and the courts tend to go along with a president's judgments about military actions overseas. It is also true that presidents who involve the nation in a limited war for limited stakes do so at great political peril; unless they get the war over quickly and with few casualties, they face the prospect of defeat in the next election.

There are additional reasons why no formal congressional declaration of war has been issued since 1941. During a state of war the president assumes certain legal prerogatives that Congress might not always be willing to grant. There are also international legal consequences of a formal declaration of war regarding foreign assets, the rights of neutrals, and so on, which our allies would not always be willing to recognize and which would be difficult to insist upon. Moreover, there is the psychological consequence of declaring war, compounded by the fact that, according to Article II, Section 2, of the United Nations Charter, war is illegal except in self-defense.

The end of the war in Vietnam, the 1974 impeachment hearings, and the resignation of President Nixon gave Congress new life.[40] It set about to recover its lost authority and discover new ways to participate more fully in making national policy. We now examine Congress's more notable efforts to reassert itself.

THE CONTINUING DEBATE OVER WAR POWERS

The dispute over war powers arises because of some seemingly contradictory passages in the Constitution, which state that Congress has the power to declare war, that the executive power shall be vested in the president, and that the president shall be the commander-in-chief of the army and navy.

In 1973 Congress overrode Richard Nixon's presidential veto and enacted the War Powers Resolution, which declared that henceforth the president can commit the armed forces of the United States only: (1) after a declaration of war by Congress; (2) by specific statutory authorization; or (3) in a national emergency created by an attack on the United States or its armed forces. After committing the armed forces under the third circumstance, the president is required to report to Congress within forty-eight hours. Unless Congress had declared war, the troop commitment must be ended within sixty days. The president is allowed another thirty days if the chief executive claims the safety of U.S. forces requires their continued use. A president is also obligated by this resolution to consult Congress "in every possible instance" before committing troops to battle. Moreover, at any time, by concurrent resolution not subject to presidential veto, Congress may direct the president to disengage such troops. A concurrent resolution is passed when both chambers of Congress wish to express the "sense" of their body on some question. Both houses must pass it in the same form. These resolutions are not sent to the president and do not have the force of law. Because of a 1983 court ruling, the question of whether Congress can remove the troops by concurrent resolution or legislative veto is now in doubt.

Not everyone was pleased by the passage of the War Powers Resolution of 1973.[41] Nixon vetoed it because he said it encroached on presidential powers. Purists in Congress and elsewhere said it was clearly unconstitutional, yet cited reasons different from Nixon's. They said it gives away a constitutional power plainly belonging to Congress, namely, the war-making or war-declaring power, for up to ninety days.

Still other observers, however much they may have thought this resolution was defective, believed nonetheless that war powers legislation was of symbolic and institutional significance because it reflected a new determination in Congress at the time. Presidents had been put on notice that the commitment of American troops is subject to congressional approval. According to the resolution, presidents have to persuade Congress and the nation that their actions are justified by the gravest of national emergencies.

Even so, all recent presidents have opposed the War Powers Resolution as unwise and overly restrictive. They claim it gives Congress the right to force them to do what the Constitution says they do not have to do, withdraw American forces at some arbitrary moment. The War Powers Resolution has not really been tested in the courts because it raises political questions judges generally seek to avoid.

In 1982 President Reagan reported to Congress, along the lines suggested in the War Powers Resolution, after he sent troops into Lebanon. Congress arranged a compromise favorable to the White House and sent it to Reagan for his signature. He made clear in a written statement that his compliance did not "cede any of the authority vested in me under the Constitution as president and as commander in chief. . . . Nor should my signing be viewed as any acknowledgement that the president's constitutional authority can be impermissibly infringed by statute."[42]

At other times, as in Grenada and Libya, when Reagan used the military in overseas operations the president did not invoke the War Powers Resolution. He generally acted as though he did not need any authorization. How did Congress react? Members grumbled but took no action.

President Bush insisted, like Reagan before him, that he had an obligation to conduct national security and military affairs as he saw fit—unrestricted by the War Powers Resolution. Bush pointed out that presidents throughout history have had to order military actions without congressional authorization. His sending of troops to Panama in 1989 was in this tradition.

Bush maintained he sent troops to Saudi Arabia in 1990 in response to a request from the Saudis. But when it became clear that Iraq's Saddam Hussein would not pull out of Kuwait, Bush marshaled the support of the United Nations and then sought the support of Congress for offensive military action in the Persian Gulf. "The current situation in the Persian Gulf, brought about by Iraq's unprovoked invasion and subsequent brutal occupation of Kuwait, threatens vital U.S. interests," wrote Bush to Congress. He requested both chambers to adopt a resolution stating that Congress supports the "use of all necessary means to implement U.N. Security Council Resolution 678."[43]

Bush's letter to Congress was the first time since 1964 that a president had asked Congress for a show of support for military engagement. After a soul-searching debate, the House approved the requested resolution 250–183; the Senate by a more narrow 52–47. A majority of the Congress, controlled by Democrats, acted to support a Republican president. It is not clear what Bush would have done if Congress had not given him his support. He is reported to have told journalists that he had the authority to act without such a vote, but plainly he wanted to have their political support.

Most constitutional experts agree that the January 1991 vote by Congress authorizing Bush to use force qualified as an exercise of its constitutional authority to declare war. Congress does not have to say in so many words that it declares war. What these resolutions did was to authorize the use of the military and an invasion of Kuwait to expel the Iraqi troops. That was about as much declaring of war as Bush needed. Senate Majority Leader George Mitchell (D-Maine) said Bush took the correct approach "when he requested authorization" from Congress. "His request clearly acknowledged the need for congressional approval."[44]

In his first term as president, Bill Clinton sent U.S. troops to Somalia, Haiti, and, under NATO auspices, Bosnia. In all cases, American soldiers were placed in danger of attack by adversaries. But President Clinton acted on his own, without direct congressional authorization, although Congress rather grudgingly did pass both supportive and restricting resolutions with respect to Clinton's sending troops on a Bosnian peacekeeping mission.

What are the lessons of the War Powers Resolution of 1973? On the one hand, Congress reasserted itself, at least in theory, and tried to get tough about unilateral presidential war making. On the other hand, presidents have mostly ignored the resolution and viewed it as a nuisance. In the 1990s many members of Congress recognized that the 1973 approach was not effective and perhaps not wise. Those who want to strengthen the resolution and force presidents to comply with it to the letter do not have the votes to get their colleagues to confront the White House. Many Republicans would prefer to scrap the resolution altogether, saying it has not worked and it is not proper for Congress to undermine a president who has to act fast in today's military emergencies, never mind what the Constitution may imply on the matter.[45]

Another group in Congress would like at some point to modify the War Powers Resolution to make it workable. They would have Congress establish a special consultative group of some eighteen congressional leaders who would meet with the president before decisions are made commiting American troops to situations where hostilities are probable.

Constitutional and political questions will continue to surround the war-making powers as long as our constitutional democracy survives. The constitutional debate over the precise character of this power continues even as practical accommodations in our constitutional system evolve.[46]

CONGRESS AND THE INTELLIGENCE AGENCIES

Presidents have also been charged with abusing the intelligence and spying agencies. The Central Intelligence Agency (CIA) was established in 1947, when the threat of "world communism" led to a vast number of national security efforts. At that time, Congress recognized the dangers to a free society inherent in such a secret organization. Hence it was stipulated that the CIA was not to engage in any police work or to perform operations within the United States.

From 1947 to the mid-1980s, no area of national policymaking was more removed from Congress than CIA operations. In many instances Congress acted as if it really did not want to know what was going on. Said one senator: "It is not a question of reluctance on the part of CIA officials to speak to us. Instead it is a question of our reluctance, if you will, to seek information and knowledge on subjects which I personally, as a Member of Congress and as a citizen, would rather not have."[47] There is much evidence that both Congress and the White House were lax in supervising intelligence activities. By 1973, the CIA was accused of plotting assassinations, experimenting with mind-altering drugs, carrying out extensive foreign paramilitary operations, and perhaps most important, spying on American citizens during the Watergate era.

Congress has tried to reassert control over the CIA. It now requires the agency to report to two committees, the House and the Senate oversight committees, any plans for clandestine operations. In 1976, in an unprecedented exercise of power, Congress amended the Defense Appropriations Bill to terminate American covert intervention in Angola.

Presidents have criticized Congress for weakening the CIA, for going too far in making covert operations too difficult. President Carter especially pressed this case during the Iranian and Afghanistan crises of 1980. President Reagan gave the CIA a new era of prominence and enhanced powers; its role in coordinating American assistance to the Nicaraguan Contras is perhaps the most striking example. President Bush, a former CIA director himself, generally continued to be highly supportive of the CIA, yet the end of the Cold War and the dissolution of the Soviet Union understandably caused the CIA budget to be reduced in the 1990s.

Former CIA Director Stansfield Turner says congressional oversight of the CIA is useful. It forces intelligence officers to exercise greater judiciousness, he contends, and to maintain a healthy sense of the national temper. He also believes congressional oversight strengthens the hand of the CIA director in controlling what has always been a notoriously independent agency.[48]

After the Reagan administration's covert selling of arms to Iran was revealed, Congress considered a variety of proposals to place additional curbs on the CIA. None became law. President Bush, for his part, worked with congressional leaders to keep them informed about CIA covert activities, as he was required to do under existing law. President Clinton's CIA directors James Woolsey, John Deutch, and George Tenet had to report to Congress on a more regular basis.

Still, for all the talk of greater congressional control, the CIA and recent presidents have not been much hampered in carrying out the operations they have deemed necessary.[49] This is not to say Clinton or Congress has been especially pleased by the CIA in recent years. The CIA has plainly not had an easy transition from the Cold War to the post–Cold War era.

THE 1974 CONGRESSIONAL BUDGET AND IMPOUNDMENT CONTROL ACT

During the Nixon administration, some members of Congress used to joke that ours was a system of checks and balances all right. Congress wrote the checks, and the White House kept the balance. They were referring to President Nixon's frequent use of the power to impound funds appropriated by Congress.

By impounding funds a president forbids an executive branch agency to spend money even though the money has been appropriated by Congress. Impoundment may be necessary to accommodate a change in events (if a war ends, for example) or to alter a managerial approach (to carry out a project more efficiently). Before Nixon, impoundments were infrequent and usually temporary; they generally involved small amounts of money. Nixon stretched the use of impoundments to new lengths. He claimed that the Democratic Congress was spending too much and causing huge deficits. Congress responded that Nixon was using impoundment to set policy and that he was violating the Constitution, which states: "No Money shall be drawn from the Treasury, but in Consequence of Appropriations made by Law." Congress took this to mean it had the final say in fiscal policymaking. But Congress not only complained, it acted as well by passing the 1974 Congressional Budget and Impoundment Control Act.

This Act repealed the 1921 language used by Nixon to justify his impoundments. It also stipulated two new procedures, rescissions and deferrals, by which a president, at least temporarily, can override appropriations decisions or delay spending. A president may propose to cancel, or rescind, enacted appropriations or subsections of a larger appropriations bill, but unless Congress agrees (with a majority vote in both houses) to the rescission within forty-five days, the money must be spent by the executive branch. The new item veto certainly is a strengthened variation of this same authority.

Recent presidents have used this provision, which has cut billions from the budget, and Congress has mostly gone along with presidential suggestions for rescission. Some have complained, however, that this provision creates too much paperwork. Reports need to be sent to Congress even when a few thousand dollars are not spent for simple managerial and efficiency purposes. Others complain about the vagueness of the law. Still, this involvement by Congress and the need for presidents to win congressional approval for impoundment have reclaimed some of the diminished power of the purse for Congress.

The second provision, permitting the deferral of spending by the executive, has caused considerable confusion. According to the law, a president may propose to defer spending funds already appropriated for up to a year. The law requires only that the executive notify Congress of these deferrals and that new notification be filed with Congress to continue a deferral into a second year. The law permits either the Senate or the House to overturn a deferral. But in 1983 the Supreme Court invalidated legislative vetoes and in effect said that the only way to counter a president's deferral of funds was to pass a law doing so.

Before 1986, President Reagan, until then the only president to be affected by the 1983 Supreme Court ruling, was able to defer some spending and work out informal agreements with Congress to make the system work in a reasonably acceptable manner. Later, Reagan and his new budget director began using the deferral more frequently and for larger spending projects. Both Congress and the federal courts protested that this deferral was a violation of the law. When a federal district court ruled that the White House was wrong to halt an expenditure of more than $5 billion for housing and other matters, it in effect affirmed Congress's authority to shape federal spending. Meanwhile, some members of Congress proposed the repeal of the entire deferral provision, saying the deferral process is "a mess." But no change has taken place.

By the 1990s national concern had turned from fears about the president's failure to spend money appropriated by Congress to fears that Congress would appropriate too much money, thereby adding to the nation's huge deficits. Nowadays questions are more likely to be raised about how to strengthen presidential control over spending rather than how to limit such authority.

The Congressional Budget and Impoundment Control Act of 1974 also responded to the fact that since the days of President Roosevelt congressional influence over federal spending had diminished, while that of the Office of Management and Budget in the executive office of the president had increased. With no budget system of its own, only many separate actions and decisions, Congress had become dependent on the president's budget proposals. Members of Congress grew to appreciate the old saying, "The one who controls the purse has the power."

The 1974 act created a permanent budget committee for each chamber of Congress and a Congressional Budget Office (CBO). This office provides budgetary and fiscal experts and computer services to give Congress technical assistance in dealing with the president's proposals. Some members of Congress hoped the CBO would provide hard, practical data to guide the drafting of spending legislation. Others saw it as a potential "think tank" that might propose standards for spending and national priorities. Optimists hoped this budget reform act would help Congress react more effectively and in a timely fashion to the president's budget proposals. They hoped, too, it would tie separate spending decisions to fiscal policy objectives.

In fact, the CBO is most frequently used to provide routine cost estimates of spending and tax bills and to keep track of the overall budget level. Its budgetary timetable gives Congress three additional months to consider the president's recommendations. By May 15 of each year, Congress adopts a tentative budget that sets target totals for spending and taxes. These targets serve as guides for the committees considering detailed appropriations measures. By September 15 Congress is supposed to adopt a second resolution that either affirms or revises the earlier targets. If necessary to meet the final budget totals, this resolution must also dictate any changes in expenditures and revenues.

How has the "reformed" budget process worked in recent years? The quality of information produced by the CBO has improved congressional deliberation on the budget; the new budget committees in each house have worked reasonably well; and the budget resolutions have provided a vehicle for certain helpful debates on key economic issues. But overall the new budgetary process has not diminished the budgetary powers of the president. Nor has it reversed the spiraling national debt. Confusion still surrounds the budgetary process. The whole point of this new budget process was to force Congress to make choices, to put together in one place the spending claims and the revenues, and to decide what it wants. Yet Congress only reluctantly applies cuts to the "sacred cows" of medicare, defense, and senior citizen entitlement programs.

Many people believe the problem is structural and not the result of personal faults of members of Congress. Congress reflects many competing local pressures, while presidents reflect national pressures. Congress may well have to restructure itself to strengthen places where overall consensus can be built. But this would mean weakening subcommittees and curbing the recent tendencies to disperse power—something Congress is unlikely to do. Yet if it cannot do this, Congress will very likely have to respond to the choices and priorities set by presidents.[50]

BUDGET REFORM REVISITED: GRAMM-RUDMAN-HOLLINGS

After several years in which virtually no one believed the budget process had worked well, Congress, with White House support, voted to approve the Balanced Budget and Emergency Deficit Reduction Act of 1985 (popularly known by the names of its sponsors in the U.S. Senate, Phil Gramm, Warren Rudman, and Ernest "Fritz" Hollings). After years of being unable to cure the problem of deficit spending, Congress opted for more surgery. This legislation set maximum allowable deficit levels on a declining basis from 1986 to 1991, when the deficit was supposed to be at zero; Congress later set 1993 as the date to be at zero. In 1996, they set a seven-year target. The reduction of deficits was intended to be achieved through across-the-board spending cuts. Initially Gramm-Rudman (as it is usually called) had given the comptroller general of the General Accounting Office the final word in ordering the president to trim spending and reduce deficits. But a 1986 Supreme Court ruling (*Bowsher v. Synar,* 478 U.S. 714) said that giving this authority to the comptroller general would violate the separation of powers doctrine because the comptroller general is an officer of Congress. Congress reverted to a fallback position that requires both houses to pass a resolution making the spending cuts. Then Congress sends this resolution to the president to order the cuts. Certain programs, such as for Social Security or interest on the national debt, are exempt from this process, but almost all domestic and military spending is subject to these cuts.

Even some of the sponsors of this budget-balancing measure called it a "bad idea whose time had come." They insisted it was needed to bring a dose of reality that would force hard choices about domestic spending, military spending, and taxes. It is far preferable, proponents argued, than the processes of the previous decade, during which no choices were made and annual deficits soared.

Critics of Gramm-Rudman called it the worst form of congressional posturing and fear it will bring about an unprecedented shift of power from the legislative to the executive branch. "With this additional power," said Senator Bill Bradley (D-N.J.), a president, "if he plays hard ball, could dismantle the nondefense portion of the budget and wreak havoc with America's poor."[51] Some critics have noted other aspects of this measure that transfer power to the executive. Thus, Gramm-Rudman legislation requires the president to bring future federal budgets into line with the deficit-reduction schedule by reducing, or even eliminating, cost-of-living allowances and similar automatic spending increases previously enacted in entitlement programs.

Today, as we enter the twenty-first century, we still have an enormous debt. Congress and the president found countless ways to evade the Gramm-Rudman deficit targets. The savings and loan bailout, for example, was handled outside these targets. Too often Congress simply voted to exclude spending outlays from Gramm-Rudman calculations. Cuts in defense spending are due to the end of the Cold War, not to these budget reforms. Brookings Institution political scientist Thomas E. Mann concludes, "The charade of Gramm-Rudman has to end. It has only served to introduce deceit into budget policy."[52] Still, the Gramm-Rudman debates and related earlier efforts led to serious deficit reduction efforts in the late 1990s.

A major goal of all recent budget reforms was to force policy and priority consensus where none had existed. "Budgeting rules, however, have not changed the desire of presidents and members of Congress to represent and to be reelected," writes James Thurber. "It is clear that the executive-legislative divisions over budget policy remain as entrenched as ever. Increased budgetary discipline by Congress and the President is not a simple matter of legislation."[53]

CONFIRMATION POLITICS

The framers of the Constitution regarded the confirmation process and its advice and consent by the Senate as an important check on executive power. Alexander Hamilton viewed it as a way for Congress to prevent the appointment of "unfit characters."

Presidents have never enjoyed exclusive control over hiring and firing in the executive branch. The Constitution leaves the question somewhat ambiguous: "The President . . . shall nominate, and by and with the Advice and Consent of the Senate, shall appoint Ambassadors, other public Ministers and Consuls, Judges of the Supreme Court, all other officers of the United States." The Senate jealously guards this right to confirm or reject major appointments. During the period of congressional government after the Civil War, presidents had to struggle to keep their power to appoint and dismiss. Presidents in the twentieth century have gained a reasonable amount of control over top appointments. Presidents have won more influence, in part, because public administration experts warned that chief executives without compatible cabinet level appointees of their choice cannot otherwise be held accountable.

Today the U.S. Senate and the president often struggle over control of top personnel in the executive and judicial branches. The Senate has taken a somewhat tougher stand. Time spent evaluating and screening presidential nominations has increased. "Our tolerance for mediocrity and lack of independence from economic interests is rapidly coming to an end," said one senator. Another summed it up this way: "Surely, we have learned that one item our government is short on is credibility." Senators are especially concerned about potential conflicts of interest and about related character and integrity concerns.

The Senate rejected several nominations of Presidents Bush and Reagan, including turning down Bush's choice for secretary of defense, John Tower. Tower, a prominent former senator, was accused of having a drinking problem and perhaps a too-cozy tie to several defense contractors. Tower denied the charges, promised not to drink liquor if he was confirmed, yet still failed to win confirmation. He was the first cabinet nominee to be rejected since 1959, when a rejection came late in Eisenhower's second term, not during the "honeymoon" first year as was the case with Bush's defeat with Tower.

President Clinton often watched in frustration after several of his nominees for attorney general, CIA director (Tony Lake in 1997), surgeon general, and judges and ambassadors withdrew after their nominations raised stiff opposition in the Senate.

The Senate's role in the confirmation process was never intended to eliminate politics but rather to use politics as a safeguard. Some conservatives in recent years object that the Senate has rejected occasional nominees because of their political beliefs and thus has interfered with the executive power of presidents. In such instances, so this complaint goes, the Senate's decision is not a reflection of the fitness of a nominee but rather of the political strength of the president.

The Senate's participation in recent years has become more thorough and more independent. "Unfortunately, the process has also become more tedious, time-consuming, and intrusive for the nominees," according to one study. "For some, this price is too high, particularly in conjunction with the requirements of the Ethics in Government Act of 1978. For others, the process is annoying and distasteful but not enough of a roadblock to prevent them (nominees) from going forward."[54]

Because of the tradition called "senatorial courtesy," a president is not likely to secure Senate approval for appointments against the objection of the senators from the state where the appointee is to work, especially if these senators are members of the president's party, even if his party does not control the Senate. Thus, for nearly all district court judgeships and a variety of other positions, senators can exercise what is in fact a veto. This veto can be overridden only with great difficulty. Further, it is usually exercised in secret and is subject to little accountability. But this form of patronage is sufficiently important to senators that senatorial courtesy is likely to continue.

It is useful to note a distinction between judicial appointments, especially those to the Supreme Court, and administration appointments. The Senate plays a greater role in judicial appointments because of both the life terms judges and justices serve and the fact that they constitute an independent and, as we discuss in Chapter 8, vital branch of the government.[55] There is, as noted, an argument that when it comes to cabinet-level positions in the executive branch, a president ought to be able to choose those who share the president's views and will carry out the general views of the White House. In contrast, a president is not expected to have or to enjoy partisan loyalty from those nominated to the bench. Thus Presidents Bush and Reagan had more difficulty in winning confirmation for their nominations to the Supreme Court than they had with their cabinet nominations. Court nominees regularly endure aggressive political and policy scrutiny.[56] More on this in chapter 8.

The confirmation provisions in the Constitution have fulfilled most of the intentions of the framers. The Senate has been able to use its power to reject unqualified nominees. It has sometimes also been able to prevent those with conflicts of interest from taking office. In addition, senators have been able to use the confirmation process to make their views known to prospective executive officials. Indeed, the very existence of the confirmation process generally deters presidents from appointing weak, questionable, or "unfit characters." Yet, by and large, presidents have still been able to appoint the people they want to important positions.

FUSING WHAT THE FRAMERS SEPARATED

A few reformers believe relations between the president and Congress are so bad that what we have is a "constitutional disorder"[57] and only major reforms can right the wrong the framers inflicted upon us. The "wrong" in this case is said to be the separation of powers that fragments and shares political power between the different branches. These commentators see the separation as a weakness that often leads to divided government, which brings about divisive politics, which ends up in deadlock.

This separation is a structural problem, and the proposed cure is the *fusion* of executive and legislative power. The model is British parliamentary democracy.[58] In *Constitutional Reform and Effective Government,* James L. Sundquist offers a parliamentary critique of American government.

Failed government is the problem for which Sundquist suggests the U.S. system has no real cure. What happens "when a president fails as a leader?" Stalemate. But would a system organized along more parliamentary lines solve this problem? Sundquist believes there is much we can learn from the Canadian and British parliamentary systems, yet warns against wholesale grafting of the parliamentary model onto the U.S. system. He writes that "for most constitutional reformers the parliamentary system represents only a source of ideas for incremental steps that might bring more unity to the American government, each step to be considered on its own merits in terms of its adaptability to American tradition and institutions. Parliamentary democracy is not a model to be adopted in its entirety, supplementing the entire U.S. constitutional structure with something new and alien."[59]

Throughout this work, Sundquist is sensitive to how proposed parliamentary reforms might fit into the American system. He focuses his concern on five areas: divided government, length of terms of office, reconstructing failed governments, fostering interbranch collaboration, and altering the checks and balances.

Divided government often leads to stalemate except under unusual circumstances. "When government is divided, . . . the normal and healthy partisan confrontation that occurs during debates in every democratic legislature spills over into confrontation between the branches of the government, which may render it immobile."[60]

In recent years, divided government has been the norm, leading to a higher probability of deadlock. The solution? Sundquist examines four proposals: (1) a presidential-congressional team ticket, (2) changing the ballot format to make ticket splitting more difficult, (3) separating the presidential and congressional elections by a short time period, and (4) arbitrarily assigning the president's party enough seats in the Senate and House to gain a majority (a bonus seat plan).

Sundquist also looks at plans to lengthen or adjust the terms of office for the president and Congress, examining the four-eight-four plan, the eight-year Senate term, the six-six-three plan, repeal of the Twenty-second Amendment, and eliminating the mid-term elections. The goal in all of these plans is to link the electoral fate of the legislature more closely with that of the president, thereby promoting opportunities for true mandates and presidential coattails.

How to reconstruct a failed government? All too often in recent years, the United States has faced, for various reasons, failed governments. In a presidential system there is little short of impeachment that can be done to remedy this situation. Says Sundquist:

> All governments descend into periods of ineffectiveness, when for any of a wide range of causes leadership fails, public confidence is lost, and conflicts within the government deepen and remain unresolved. When this occurs, parliamentary systems possess a safeguard; their governments can be dissolved at any time and new elections can then install new leaders with a fresh mandate from the people. Or weak leaders can be replaced with stronger ones even in the absence of new elections, without provoking a constitutional crisis . . . the United States, in contrast, is in bondage to the calendar. The nation's leader is elected for four years, and no matter what his failures the office is his as a kind of property right until his term's scheduled expiration—with two exceptions. He may be removed under the impeachment clause of the Constitution if he is convicted of "treason, bribery, or other high crimes and misdemeanors." He may also be relieved of his duties under the Twenty-fifth Amendment, adopted in 1967, if the vice president and a majority of the cabinet

declare him to be unable to perform the functions of his office and, in the event the president disputes that finding, two-thirds of both houses of Congress confirm his incapacity.[61]

To overcome this problem, Sundquist looks at several reforms: special elections, no confidence votes, and other "safety valves."

In an effort to promote greater collaboration between the branches, Sundquist offers several proposals, all designed to modify the separation of powers. Among these are such ideas as allowing members of Congress to serve in the president's cabinet, giving cabinet officers a place in the Congress, and, most important of all, strengthening political parties. He notes: "Whatever can be done to strengthen political party organizations will serve to improve cohesion between the executive branch and the President's party in the Congress. The party is still the web that infuses the organs of government with a sense of common purpose."[62]

But in the long run, Sundquist laments, "proposals for formal institutional linkages between the branches . . . hold little promise."[63]

Still, Sundquist offers more than an exercise in intellectual calisthenics. After examining a multitude of reform proposals, he gives, in order of importance, his own formula for constitutional reform: (1) a team ticket; (2) four-year House terms, eight-year Senate terms; (3) a method of special elections to reconstitute a failed government; (4) removal of the prohibition against dual officeholding (to permit members of Congress to serve in the executive branch); (5) a limited item veto; (6) restoration of the legislative veto; (7) a new war powers amendment; (8) approval of treaties by a majority of both houses; and (9) a national referendum to break deadlocks.

While these are not a panacea to break us out of the deadlock of stalemated government, and while Sundquist is not optimistic about the chances for implementation, he does offer an integrated package of reform proposals designed to marry some aspects of the parliamentary model to the idiosyncrasies of America's traditions and institutions.

James MacGregor Burns, in his *The Power to Lead,* takes a similar reformist approach, examining the symptom of failed government and prescribing the cure of a parliamentary alternative. Burns, like Sundquist, sees antileadership tendencies intentionally built into the Constitution, and this leads to either deadlock or the overpersonalization of politics.

Burns' structural critique of leadership failure finds two systems of government at work, "One had infinite tendencies toward gradualism, delay, compromise, deadlock, the other toward decision, action, miscalculation, and catastrophe. Thus the original source of leadership failures, as well as leadership potential, lies in the tendencies built into the system constitutionally toward an oscillation between statis and spasm."[64]

The cycles of leadership, "stalemate to crisis," are, in Burns's analysis, dangerous and need to be overcome. In exploring the various reform proposals he divides reformers into "process changers," those who seek various individual changes in the process of government, and "structuralists," those who seek major changes along variations on the British parliamentary or French presidential models.

Burns advocates several of the same reform proposals as Sundquist, hoping again for "real teamwork" and "genuinely collective leadership." Yet even more central to Burns's critique is the need to develop more responsible political parties.

> Political parties can serve as the mainspring of democracy, as the vital link between voters and rulers. They organize and focus public opinion. They aggregate "special" interests. They provide meaningful cues to voters, present them with alternatives at election time, propose programs, develop among party followers and party leaders in office support for such programs. They unify legislators and executives among themselves and with one another. They help hold government officials accountable to voters. They mobilize popular support for candidates and officeholders and hence are indispensable to democratic leadership.[65]

To accomplish the goal of party regeneration, Burns suggests "formalizing party membership, drastically reducing the number of binding presidential primaries, and enlarging the role of all party cadres and leaders in presidential conventions."[66]

Burns urges us to: (1) strengthen party and collective leadership in the House and Senate, and between Congress and the president; (2) broaden the impeachment process to include loss of confidence; (3) make major parties more organized, disciplined, programmatic, and principled; (4) create a team ticket; and (5) enable presidents to choose members of Congress for the cabinet.[67]

Like Sundquist, Burns does not see much support for his reformist proposals: "Our only chance," he argues, "lies in beginning with modest efforts to strengthen both the constitutional and party structures in the hope that gradual renewals and reforms simultaneously in both the party and constitutional spheres would set up a symbiotic relationship out of which might come major changes." [Such is] "a strategy for gradual structuralists."[68]

Mainstream political scientists disagree strongly with the prescriptions put forward by Sundquist and Burns. They say a president who enjoys both a working majority in Congress and a strong party might be too free to impose his or her own program on the nation. And if the Twenty-second Amendment were voided, this vigorous president might, over time, be able to dominate the judicial branch as well. Critics also say these schemes are designed almost entirely to strengthen the White House and weaken Congress.

Defenders of our presidential system say it generally serves us well. It is a system, they say, that reflects the cautious temperament of the American people. The vitality of the system depends on new ideas and new people, not on importing parliamentary features. In parliamentary systems the parties are so tightly organized that they often become rigid. The

American system was intended to have tension in it. It sometimes swings a bit too far in one or another direction, yet it generally works. It is, moreover, especially well designed to protect the liberties of individuals and minorities and it has succeeded in preventing authoritarianism.

Finally, opponents of the Sundquist-Burns proposals say the principle of separation of power is no less valid today than it was in 1787. The national legislature was designed to prevent tyranny of any one leader and tyranny of the majority. Yes, the system is often slow and cautious and hostile to bold leadership initiatives. Yet isn't this appropriate to its responsibilities and primary function? Most Americans want a certain amount of deadlock and delay on policy initiatives that emerge out of presidential or party brainstorming. The genius of the American Congress, then and now, is its capacity for deliberation, debate, and prudent reflection.[69]

THE CONTINUING STRUGGLE

Most observers now believe that Congress has not won back many of its allegedly lost powers, and they are skeptical of Congress's ability to match the president's advantages in setting the nation's long-term policy direction over the long run. Ronald Reagan demonstrated that a popular president who knew what he wanted could not only influence the national policy agenda but also win considerable cooperation from the Congress. George Bush demonstrated that a president could win considerable public and congressional support for his foreign policy and military initiatives, even with Congress controlled by the opposition party.

The American public may be skeptical and critical about its political leaders, but it has not really lost a belief in the efficacy of strong, purposeful leadership. Whether or not people believed in Ronald Reagan's policy priorities, many supported his view that the country needed a strong president who would strengthen the presidency and make the office a more vital center of national policy than it had been in the years immediately following the Watergate scandals. Whether people were critical of Clinton's character or not, Americans generally recognized his leadership efforts to encourage trade, promote new jobs, and advance the cause of civil rights and environmental protection.

A central question during the 1970s was whether, in the wake of a somewhat diminished presidency, Congress could furnish the necessary leadership to govern the country. Most people, including many members of Congress, did not think Congress could play that role. The routine answer as we enter the twenty-first century is that the United States needs a presidency of substantial power if we are to solve the trade, deficit, productivity, and other economic and national security problems we currently face. We live in a continuous state of emergency. Terrorism or nuclear warfare could destroy our country. Global competition of almost every sort highlights the need for swift leadership and a certain amount of efficiency in government. Many people realize, too, that weakening the presidency may, as often as not, strengthen the vast federal bureaucracy and its influence over how programs are implemented more than it would strengthen Congress.

Congress simply is not structured for sustained leadership and direction. Power in Congress is too fragmented and dispersed. Congress can, on occasion, provide leadership on various issues, yet it is far less able to adapt to changing demands and national or international crises that arise than is the presidency. The presidency is a more fluid institution and thereby can usually more quickly adjust and adapt.

Congress's efforts to reassert itself were more a groping and often unsystematic, if well-intentioned, attempt to be taken seriously than a concerted effort to weaken the presidency. It did not take long for observers to appreciate that when a president is unable to exercise authority and leadership, no one else can usually supply comparable initiative. Not only did we reaffirm the pre-Watergate view that the presidency is usually an effective instrument for innovation, experimentation, and progress, but a majority of Americans concurred that to the extent the country is governable on a national basis, it must be governed from the White House by a president and top advisers who can figure out a way to gain the cooperation of various coalitions.

WORRIES ABOUT AN "IMPERIAL CONGRESS"

Friends of whoever is in the White House sometimes contend it is not an imperial presidency but rather an imperial Congress that all of us should worry about. They said that since a president is the only elected officeholder answerable to the nation as a whole, the president is uniquely qualified to pursue foreign policy on behalf of the entire nation. They said the framers saw the wisdom of giving responsibility for foreign policy to the chief executive because Congress is poorly suited to micromanage U.S. foreign policy. In addition, Congress, because it is made up of 535 members tied to parochial interests, the argument went, is too slow, too unwieldy, and too easily captured by sectional interests or political action committees when it tries to carry out what properly are executive functions. Congress, these critics add, is more like a lawyer representing special clients than it is like a judge weighing the larger picture and the longer-term interests of the entire nation.[70]

Former Reagan national security aide Oliver North lamented that Reagan too often gave in to this all-powerful Congress when he should have ignored Congress's restrictions on war powers, support for the Nicaraguan resistance, and similar obstructions to the president's authority. Future presidents may well be "constrained by an imperial Congress, because Ronald Reagan, one of the most popular presidents in American history, did not reclaim the original powers of his office."[71]

Other critics of a too-powerful Congress say presidential nominees should not have to endure the critical confirmation hearings that have become commonplace in recent years. After all, they say, the president won the national election and should enjoy the power of appointment. The Senate, they add, should reject nominees only if they are manifestly unfit for office, not because certain senators differ with the nominee on policy matters.

There are those also who believe there is no need for independent counsels or special prosecutors. Let the president be judged by history, or by the voters at the next election. If a president does something unconstitutional, Congress has the power to impeach and convict a president. But if this is not the case, Congress should not interfere with presidents who are trying to lead, govern, and negotiate on behalf of the country. In short, Congress should seek to help rather than hinder a president. Better, they say, to follow the steady leadership of the president than the ever-changing whims of Congress.

There is something to be said for this argument, and it is a useful counterpoint to some of the exaggerations in the "imperial presidency" argument. Yet what is missing in the "imperial Congress" reasoning, argue critics of this view, is that a too-powerful presidency can also pose dangers to the kind of society and the kind of government we want. We are bothered, and rightly so, when an occasional president such as Richard Nixon says: "When the president does it, that means it is not illegal." We are concerned, as well, that shortcuts that bypass the checks and balances of our constitutional system and excessive secrecy by those who serve the president do not, in the long run, strengthen the presidency. Such evasions usually weaken the presidency and the constitutional system of government, for rarely will a foreign policy be maintained for long unless the American people are behind it. And if Congress does not support a policy, the American people usually are not going to support it either.

The challenge that confronted the framers—how to reconcile the need for executive energy with republican liberty—is still with us. The history of constitutional democracy has always been the search for limitations on absolute power and for techniques of sharing power. Our American style of constitutionalism and separation of powers, especially in the absence of a major crisis, often means a slow-moving and sometimes inefficient decision-making system. It is a system that often hinders rather than facilitates leadership, that invites contention, division, debate, delay, and political conflict.

The framers in 1787 knew well they were creating a unique, necessary, yet always potentially dangerous institution when they invented the presidency. This remains true. What is needed, of course, is a strong presidency, a strong Congress and a vigilant citizenry.

CHAPTER 7

The President and
Political Parties

No free country has ever been without parties, which are a natural offspring of Freedom.

James Madison, from a speech at the Constitutional Convention 1787

The spirit of party serves always to distract the public councils, and enfeeble the public administration. It agitates the community with ill-founded jealousies and false alarms; kindles the animosity of one part against another; foments occasional riot and insurrection.

George Washington, Farewell Address to the People of the United States, September 19, 1796

[The president] is the party nominee and the only party nominee for whom the whole nation votes. . . . He can dominate his party by being spokesman for the real sentiment and purpose of the country, by giving direction to opinion, by giving the country at once the information and the statements of policy which will enable it to form its judgments alike of parties and of men. . . . He may be both the leader of his party and the leader of the nation, or he may be one or the other. If he can lead the nation, his party can hardly resist him.

Woodrow Wilson, *Constitutional Government in the United States* (Columbia University Press, 1908), pp. 67–69

The Constitution deliberately created a separate legislative branch and a separate executive. Yet making our Madisonian system of separate powers work effectively requires some help to encourage meaningful cooperation. One of the devices that has often served this intermediary function is the American political party.

A political party at its best brings like-minded people together, that is, people who share similar policy perspectives. Thus members of Congress and a president who are members of a common party not only can talk more easily across the branches, but can devise collaborative mechanisms to produce agreed on legislation, treaties, budgets, and similar political decisions. One of the great paradoxes of the president-party relationship is that while the party can be one of the useful tools of presidential leadership, it is less developed and less appreciated than should be the case.

It is often said that presidents are leaders of their political parties. In fact, however, a president has no formal position in the party structure. In theory, the supreme authority in our parties is the national presidential convention. More directly in charge of the national party, at least on paper, is the national committee (and each national committee has a national chairperson).

In practice, successful presidents usually control their national committees and, often, their national conventions as well. Although the national committee picks the national party chair, a president almost always lets the committee know whom he wants. Modern presidents hire and fire national party staff almost at will. Several of our recent presidents have ignored their national party committees. Some have treated them with contempt.

Political parties once were a prime source of influence for a president. It used to be said, for example, that our most effective presidents were effective in large part because they had made use of party support and took seriously their party leadership responsibilities. But as our parties have declined in organizational importance, there has been more of an incentive for presidents to "rise above" party or "go public."

Today, the presidential-party relationship is strained. National party chairpersons come and go with embarrassing regularity and regular embarrassment. Few party chairpersons of the president's party have enjoyed much influence. And many were regarded at the White House as little more than clerks. Clinton's Democratic National Committee's fund-raising operation, especially some of its Asian contributions, proved especially embarrassing in 1996.

The central concern in this chapter is the awkward alliance between presidents and their parties. Both president and party need each other. Yet each sometimes becomes frustrated and even annoyed with the other. What has been the role of the president as party leader? What of presidents' use of party as an appeal to Congress for support of their programs? Why the apparent growing divorce between presidents and their parties?

What are the limits on presidents as party leaders and the limits of the party as a check on presidents? Could something be done to encourage more cohesion between presidents and parties?

PRESIDENT AS PARTY LEADER?

Our earliest presidents vigorously opposed the development of political parties. They viewed them as factions and divisive—something to be dreaded as the greatest political evil. The Constitution makes no mention of political parties. In his Farewell Address of 1796, George Washington warned against the "spirit of party generally," and said the nascent parties were "the curse of the country." But if the inventors of the presidency bemoaned the emergence of parties, they soon came to accept them as inevitable, and many even attempted to use parties to their advantage.

The president's role as party leader developed over time; it was a role grafted onto the presidency. The president serves as the party's "most eminent member," titular leader of the national committee, and most visible spokesperson for the party, and he initiates the party's program.

But the president is also, in many ways, quite independent from the party. Today, the party does not control the nomination; it is comparatively weak and relatively undisciplined; there are separate national, state, and local organizations; presidential involvement can be sporadic; and ordinarily, presidents need support from the opposition to pass their proposals and have them accepted as legitimate.

However, the president needs the party. The grids usually lock if the president, with help from the party, cannot grease the machinery of government: "To be successful, to be powerful the president needs the lubricating of party. Since Washington, all the presidents considered great or near great were active and powerful party leaders. And failures neglected or failed to adequately use party." [1]

One of the chief reasons presidents seek ways other than legislation to make policy is that they are so often unable to get Congress to pass their proposals. This sometimes leads to use of administrative tactics, or engaging in *fait accompli* acts, going public, or moving beyond the law. But if the president has a strong party on which to rely (as long as this party has a majority in Congress), the temptation to achieve results by other means is not as great. Parties can help presidents govern while helping maintain democratic accountability.

By 1800, the contest for the presidency had become a battle between political parties. The Jefferson-led Republicans took on and defeated the John Adams-led Federalists. But even Jefferson, sometimes called the founder of what later became the Democratic party, had doubts about party contests for the presidency. He disliked the Federalist party and its narrow, elitist constituency. He hoped the Federalist party would shrivel and collapse, thus allowing the more representative Republicans to remain in permanent control. Jefferson's party was in fact dominant for two generations.

Parties just naturally arose. They represented fundamental and enduring ideological and cultural divisions in American society. Out-of-power interests coalesced to put their own candidates in contention for the presidency. Party clashes became routine. The evolution of political parties had much to do with the successful functioning of both presidency and Constitution. They solved in part the problem of presidential recruitment. They served also as a means of checking presidents, of keeping them responsive to concerned grass-roots citizens, for to remain in office a president would have to win renomination from his political party.

As they evolved, parties helped narrow down the number of candidates, prepared platforms, and creatively mediated among the diverse interests that were pressing claims and pushing ideological views on future officeholders. Parties usually were able to find the common middle ground among more or less hostile groups so that agreement could be reached on general principles.

Moreover, the parties enabled officeholders to overcome some of the limitations of our formal constitutional arrangements. Political parties facilitated coordination among the branches. President Jackson especially used his resources to promote partisan control of government. Jackson's achievements as party leader transformed the office, so much so that he is sometimes called the "first modern president."

> The President became both the head of the executive branch and leader of the party. The first six Presidents usually acted in a manner that accorded Congress an equality of power. However, starting with Andrew Jackson the President began more and more to assert his role not simply as head of the executive branch but as leader of the government. By the skillful use of his position as head of the party he persuaded Congress to follow his lead, thereby allowing him to assume greater control of the government and to direct and dominate public affairs.[2]

Many historians and political scientists hold that the effective presidents have been those who, like Jackson, strengthened their position by becoming strong party leaders. Cooperation and achievement could be achieved through party alliances. "Since the office did not come equipped with the necessary powers under the Constitution, they had to be added through a historical process by the forceful action of vigorous Presidents whose position was strengthened by the rise and development of political parties."[3]

Few presidents have been able to duplicate Jackson's success. Most have found it exceedingly difficult to serve as an activist party builder and party leader while trying to serve also as chief of state and national unifier. President William Taft lamented that the longer he was president, "the less of a party man I seem to become."[4] President William McKinley said he could no longer be president of a party for "I am now President of the whole people." [5] Others complain that they cannot simultaneously be faithful party leaders and serve the nation impartially.

What, at least in theory, are the obligations of a president as a party leader? These vary, of course, depending on one's conception of the presidency and the party system. The textbook model has generally held that the president should promote party platforms, reward party loyalists, punish party mavericks, run proudly with the party ticket, and heed the interests and advice of party leaders. The president should be a party builder and strengthen the party ranks by communicating the party's purpose and positions. Presidents, it is believed, should be as much the product of their parties as they are the leader. It should be a two-way street, with parties serving to check ambition and to ensure accountable leadership. A president is expected to consult regularly with local, state, and national party committee officials.

Presidential practice is distinctly different. Few of our recent presidents have spent much time working with party officials save as it was necessary for their renomination. Most presidents of late have mistreated their national party committees.

The Nixon presidency in many ways was the ultimate in presidential hostility toward its own party. Nixon dumped Republican Party Chairman Ray Bliss, who was acknowledged as a brilliant party builder. Nixon time and again sought to divorce himself from the Republican label and from supporting Republican candidates. Nixon regularly ignored Republican Party officials and set a similar tone for his White House staff. Once, when National Chairman Robert Dole had been trying for some time to see the president, a White House aide is alleged to have said to Dole: "If you still want to see the president, turn on your television set tonight at 7:00. The president will be on then." Nixon yearned to be above partisan responsibilities. He wanted to be a bipartisan foreign policy leader. Accordingly, Nixon's definition of presidential leadership was that a president is there to make global decisions that no one else can make. He also felt that party officials do not know enough to clarify foreign policy issues, not to mention to make foreign policy decisions. Thus Nixon would be the peacemaker, the statesman, the globe-trotting diplomat.

But when a president strives to be above party, critics say, he and his aides may grow dependent on secretive and covert political operations. The campaign abuses of 1972 are one result. A president who divorces himself from his party is potentially an arbitrary president, one who may be too much in the business of self-promotion at the expense of party and public interests. A president out of touch with party politics usually feels little accountability to the people who are close to the grass-roots realities of political life.

The record of recent presidents as party builders is mixed. Ronald Reagan benefited from the efforts of Bill Brock who served as Republican party chair from 1977 to 1981. In the aftermath of the Watergate crisis,

Brock rebuilt the party. Under his leadership Republicans gained seats in Congress, raised large sums of money, and aided in the campaigns of Republicans. His efforts helped recapture the Senate in 1980, and paved the way for Ronald Reagan.

As president, Reagan was sensitive to party needs. Since Reagan was not actively involved in the day-to-day workings of his presidency he had time to cultivate party members and satisfy the partisan needs of members of Congress. He was especially willing to have photo sessions with Republican members of Congress, big donors, and the party faithful. This, along with Bill Brock's efforts, greatly enhanced the Republican party, and helped Reagan govern. But in his 1984 reelection bid, Reagan eschewed party and ran a more personalized campaign. He won in a landslide but failed to help his party, and thus had rough going in his relations with Congress during his second term.

George Bush ran a slashingly partisan campaign in 1988, but tried to govern in a less partisan manner, calling for a "kinder, gentler" America. This may reflect the split personality of George Bush, who tended to see politics as a dirty business, but government as an honorable enterprise.[6] It also reflected Bush's weak political position as he faced a Congress controlled by the opposition. Bush's position was further complicated by the fact that he was not trusted by the far right wing of his party, and even faced a nomination challenge from the right by Pat Buchanan in 1992.

The split within the Republican party hounded Bush, and this family fight went painfully public during the Republican national convention when the "religious right" clashed with party regulars. The self-proclaimed religious right presented strident, even hostile speeches that alienated much of the public, and hurt Bush in the general election.

Bill Cinton came to office with something unusual: his party in control of both houses of Congress. But the Democrats were divided and undisciplined. Clinton, who was governor of a one-party state, initially tried to govern in a "Democrats only" manner. This proved unsuccessful as the Democrats soon came to realize they could pressure the president to give them special favors in exchange for votes and the Republicans, using the filibuster to an unprecedented degree, stalled and delayed the president's program. In effect, Clinton was held hostage by both parties.

The mid-term elections of 1994 as well as the 1996 election changed the political climate. For the first time in forty years, the Republicans gained control of both the House and the Senate.

LIMITS ON PRESIDENTIAL PARTY LEADERSHIP

Once in office, presidents often bend over backward in an attempt to minimize the partisan appearance of their actions. This is so in part because the public yearns for a "statesman" in the White House, for a president who is above politics. Presidents are not supposed to act with their eyes on the next election; they are not supposed to favor any particular group or party.

Herein lies another of the paradoxes of the presidency. On the one hand, presidents are expected to be pure and neutral public servants, avoiding political and party considerations. On the other hand, they are supposed to lead their party, help cooperative members of Congress get reelected, and work closely with party leaders. Also, they must build political coalitions and drum up support, including party support, for what they believe needs to be done.

To take the president out of partisan politics, however, is to assume incorrectly that the president will be so generally right and the leaders and rank and file of their party so generally wrong that the president must be protected from the push and shove of political pressures. But what president has always been right? Having a president constrained and informed by party platforms and party leaders is what was intended when our party system developed. How often this was actually done is difficult to estimate.

Another profound limitation on the modern presidents as party leader is the fact that they must now communicate to citizens by television. Television is the main source of information for the citizen. Television is also a major weapon in the arsenal of presidential leadership. But it also apparently forces the president to bypass party structures—the nonpartisan direct television appeal has replaced the party rally. As a result, the party has lost one of its main functions, namely, being a source of information and communication between citizens and their government.

Presidents know that appeals to party during television addresses are not politically wise. Presidents are instructed by their pollsters and marketing managers to rely on popular appeals and encourage popular leadership rather than party leadership. Party organization thereby becomes subordinate. Time and again, a personalized entrepreneurial politics emphasizing the president triumphs over a politics emphasizing party purposes and party issues.

Television has also provided a means for third-party and extraparty interests to communicate their views to the public and bypass traditional party structures. Ross Perot and George Wallace were beneficiaries of the availability of television time. Television shows and news coverage of people like Pat Robertson, Rush Limbaugh, and Ralph Nader have helped to promote yet other alternatives to the regular two-party process. Little known "partyless" candidates who are wealthy, such as Steve Forbes, can spend millions of dollars in primaries, bypassing the party by going directly to voters, and become serious candidates without the party.

Another related but more subtle factor in lessening the role presidents play in party activities is the fact that for at least a generation now, ours has been a candidate-financed election system, not a party-financed system. A few states, such as Wisconsin, are an exception to this pattern.

Most candidates now raise their own funds and organize their own staffs and campaign committees. Generally speaking, the ablest organizers, campaigners, and media consultants have worked for candidates—candidates for the presidency, candidates for Congress, and so on. As the parties have grown weaker and as candidate-based organizations have become the routine, people who have remained as workers with party organizations are often less talented, or so at least is the perception of many people now involved in elective politics. The most talented campaign people now are usually in the White House or in Congress, either in office or in staff positions. So much is this the case that elected officeholders and their staffs often do not take "the party people" seriously.

During the Carter presidency, the Democratic National Committee staff was controlled by the White House staff. They took their instructions from Hamilton Jordan and Tim Kraft, two of the people who masterminded the Jimmy Carter primary victories in 1976. Neither Jordan nor Kraft were party professionals or party leaders. On the contrary, they were candidate loyalists who proved themselves and won their reputations as candidate promoters. Similarly, top Reagan strategists went to the White House after 1980, not to the Republican National Committee.

Not surprisingly, these candidate-oriented professionals look down on party officials, and especially party staffers. What have they ever won? This may not be as it should be, but it is the way attitudes get formed in the world of the practitioners.

Every recent White House has had an aide, usually a senior aide, assigned the function of liaison with the national committee staff of the president's party. Such persons (Kenneth O'Donnell for Kennedy, Marvin Watson for Johnson, Tim Kraft for Carter, Ed Meese and others for Reagan, and Harold Ickes in Clinton's first term) meet with the chairperson of the national committee, arrange for an occasional presidential visit to state and national party fund-raising functions, and host visits from visiting party delegations. Sometimes, for example, a state party chairperson or even a state party committee will be invited to the White House for staff briefings and perhaps even a lunch with the president. The liaison aide also oversees sensitive party patronage decisions and handles suggestions and complaints coming from national committee members and state chairpersons.

But the work of this White House aide has been overshadowed by the gradual development in the contemporary White House of an office of interest-group liaison. With a notable institutionalization in the Nixon White House, this staff (sometimes numbering as high as twenty-nine or more staffers) has sought an outreach to ethnic, professional, labor, business, religious, and every conceivable citizen-interest organization. Aides such as William Baroody, Jr. (Nixon and Ford), Midge Costanza and Anne Wexler (Carter), Elizabeth Dole (Reagan), and Rahm Emmanuel (Clinton) have pioneered this effort of working directly with groups that once enjoyed their access to politics primarily through the political parties. These aides bring delegations of interest-group leaders to the White House for briefings. They provide an opportunity for the groups to bring their views and grievances directly to top White House aides. In addition, these White House aides try to provide information and inside know-how to group leaders, with the hope they will back the president's programs in Congress and try to get their organization to support the president. It is estimated that the Carter interest-group liaison operation worked with as many as eight hundred groups and organizations by 1979. Groups and officials that once may have worked through party leaders (or bosses) now are organized on a national scale and as often as not have a lobbying office in Washington ready and willing to deal directly with White House aides and cabinet officials. Mayors, governors, county executives, labor leaders, ethnic group advocates, and so on—they no longer need or want to go through the party bosses to be heard. Plainly, this direct access and direct consultation operation has removed yet another function that party officials often believed was theirs.

Presidents in recent years have also believed, rightly or wrongly, that many of the toughest problems they face are policy controversies that defy party clarification or traditional party problem-solving approaches. They develop the view that many of the great issues of the day require study by blue-ribbon bipartisan presidential advisory commissions or by White House task forces. In order to get a prominent elder statesman or top professional of some standing to accept the chairpersonship of such a commission, the president usually has to promise that partisanship will play no role in the selection of personnel for the commission. Time and again, the reliance on presidential advisory system mechanisms has further minimized the role that party organizations have played as agenda setters and problem definers, not to mention as problem solvers.

In sum, these are several, but by no means all, of the limits that make it difficult for a president to serve aggressively as a party leader. We have not bothered to elaborate on the well-worn observation that our parties become truly national only for the purposes of electing a president and organizing the national legislature. Although national committees now enjoy greater influence in the conduct of presidential nominations, they still have little influence over our decentralized party structure between elections. Hence, a president's practical influence over the party is sharply limited. A president's power as party leader comes to an end when the president needs it most. The president is virtually unable to recruit party candidates for Congress or other offices and has little or no influence in the selection of state and local party officials. A president's once vast influence over patronage appointments is now considerably circumscribed. Presidents and cabinet members must weigh patronage claims of the party against their own need for expertise and talented assistance. Patronage has its limits. President after president is amazed at the ingratitude of those who have been on the receiving end (and, of course, for every one person who wins a position, a lucrative contract, or some other presidential favor, there are several more who felt they should have won it instead).

THE "NO-PARTY" PRESIDENCY

Are we moving to a no-party presidency?[7] Parliamentary regimes such as Great Britain have strong disciplined parties, and the prime minister can rely on the party in the Parliament to gain passage of his or her proposals. But in the United States, with weak, undisciplined parties, presidential power is more personalized. While parties matter in the United States, they do not matter as much as in Britain or other European democracies.

These factors undermine party in this country:

- An antiparty bias exists in the political culture.
- Our history is generally one of weak parties.
- The nomination process is not in the control of party regulars.
- The party does not run campaigns.
- Office seekers are entrepreneurs first, partisans second.
- Parties are not the prime fund raisers, individual office seekers are.
- Parties in Congress are relatively undisciplined.
- Presidents have short coattails.
- Presidents do not often rely on party, choosing other governing strategies such as
  - direct appeals to the public
  - bipartisanship
  - administrative or managerial strategies
  - taking action on their own
- Patronage has shifted from partisan to personal systems as presidents seek ideological and personal loyalty over party loyalty.

Presidents encounter a number of disincentives when determining what approach to take toward party leadership. In the nominating and governing processes, it is often more useful to eschew party than embrace party. As Robert Harmel writes,

> the context of presidential politics provides few incentives for presidents to lead their party organizations, and most recent presidents have acted accordingly. While none has totally abandoned his party, it can be said without exaggeration that the record has been one of neglect more than active party leadership.[8]

Given the difficulties presidents have in exerting effective party leadership, it may seem surprising that presidents *do* spend some of their precious resources on party. But as difficult as party leadership may be, the alternative, to allow the parties to deteriorate, is far worse. The demise of party has a number of adverse consequences. If party is of little and only sporadic use to presidents, it is at least of some use. In a process where the president's resource cupboard often seems relatively bare, party *may* serve as an aid in moving the machinery of government. It may not be much, but at least it is something.

PRESIDENTS AND USE OF PARTY APPEAL IN CONGRESS

Presidential control of party supporters in Congress has seldom been great. "Party discipline," wrote Arthur Schlesinger, Jr., "far from being a novelty of our fallen times, is one of the conditions that American democracy has endured from the start."[9] Of course, the situation varies from time to time. If Ronald Reagan enjoyed surprising party unity on his 1981 economic package, he enjoyed it less on social issues and defense spending matters. President Carter assuredly experienced a lack of party discipline among the Democrats in Congress. He found members of his own party were strong-minded and independent. Appeals to party were of marginal benefit; Carter increasingly found he needed Republican votes as well to secure most of his victories.

George Bush, facing Democratic control of both houses of Congress, presented an unambitious legislative package to the Congress and failed to push his proposals through. Bill Clinton, after early success in Congress, faced a legislature controlled by the Republicans in 1995 and was forced to alter both his agenda and his operating style. His attempt to pressure an unfaithful Democrat, Senator Richard Shelby of Alabama, led the senator to switch parties after the 1994 mid-term elections.

When a president's approval rating falls in the polls, the president's ability as a party leader in the Washington community withers. A sag in presidential popularity worries members of Congress of the president's party, for they fear they may be more vulnerable to opposition attacks back home.

Presidential attempts to unseat or purge disloyal members of Congress in the president's party have not worked. Roosevelt's celebrated "purge" of nonsupportive Democrats in the congressional elections of 1938 was mainly in vain. Anti–New Deal Democrats won reelection or election for the first time in most of the places where he tried to wield his influence.[10]

Presidential coattails, once thought to be a significant factor in helping to elect members of a president's party to Congress, have had little effect in recent years (see Table 7.1). Members usually get reelected because of the quality of their constituency services and the fact they can take advantage of incumbency, not on whether they have worked cooperatively with the White House. Congressional races are not notably affected by national issues or national trends (1994 was the exception). Time and again, presi-

TABLE 7.1
Presidential Coattails, 1932–1996

Year	President	Party	Gains or losses of president's party	
			House	Senate
1932	Roosevelt	Democrat	+90	+9
1936	Roosevelt	Democrat	+12	+7
1940	Roosevelt	Democrat	+7	−3
1944	Roosevelt	Democrat	+24	−2
1948	Truman	Democrat	+75	+9
1952	Eisenhower	Republican	+22	+1
1956	Eisenhower	Republican	−3	−1
1960	Kennedy	Democrat	−20	+1
1964	Johnson	Democrat	+37	+1
1968	Nixon	Republican	+5	+7
1972	Nixon	Republican	+21	−2
1976	Carter	Democrat	+1	+1
1980	Reagan	Republican	+33	+12
1984	Reagan	Republican	+14	−2
1988	Bush	Republican	−3	−1
1992	Clinton	Democrat	−10	0
1996	Clinton	Democrat	—	−2

Source: Adapted from Harold W. Stanley and Richard G. Niemi, *Vital Statistics on American Politics,* 3rd ed. (Washington, D.C.: Congressional Quarterly Press, 1992), pp. 124–25.

dents have found in mid-term congressional elections that they can do little to help members of their own party who are in trouble. Ford campaigned vigorously in 1974 for dozens of members of Congress only to see most of them defeated. Carter experienced similar disappointments in 1978, especially in Senate races. In the 1994 mid-term elections President Clinton's party lost fifty-two seats in the House and eight in the Senate.

Presidents today have little retaliatory leverage to apply against uncooperative legislators. Members of Congress, as a result of various congressional reforms, have more and more resources (trips home, large staffs, more research facilities, more home offices and office staffs in their districts, and so forth) to help themselves win reelection. With the dramatic growth of government programs and governmental regulation, members of Congress are in a good position to make themselves nearly indispensable to local officials and local business people, who need to have a Washington "friend" to cut through the red tape and expedite government contracts or short-circuit some federal regulation. These kinds of developments have enhanced reelection chances for most members while at the same time making them less dependent on the White House and less fearful of any penalty for ignoring presidential party appeals.

Note too that many members in Congress today are the product of "movement" or "new politics" experiences. Many of them had to buck local party establishments to win election in the first place. In this age of television and direct mail campaigns, candidates for Congress have often run virtually as independents or as outsiders. Their loyalty to party is thin to begin with. Their base back home is much more tied to professional, business, new politics, or consumer interest groups; to religious activists; and to political action committees than to old-line political party operatives. Moreover, they have often built their own personal organization. Often, the local party apparatus is moribund anyway.

From the president's vantage point, it is seldom helpful to punish party mavericks. In this reformed Congress, with power more dispersed and decentralized and where nearly everyone has a piece of the action, there is just too much risk for a president to single out a few party "disloyalists" for retribution. White House congressional relations aides know all too well that it is best to abide by the motto of "no permanent allies, no permanent enemies." You may lose someone on a vote today, but his or her vote may be crucial on some other measure next week. Then, too, a president has fewer patronage plums or perquisites these days with which to persuade a wavering member of Congress. More and more patronage jobs of the past are now civil service positions or are selected on the basis of merit. More and more governmental contracts or so-called "pork-barrel" expenditures come under close congressional or press scrutiny or are subject to some funding formula.

In short, a president's appeal to his fellow party members in Congress is effective only some of the time. It sometimes will help, but this kind of appeal is more unpredictable today than in the past. Party caucuses may be a bit stronger in Congress, but legislators know that neither the White House nor fellow members in Congress will penalize them if they can claim that "district necessities" forced them to differ with the party on a certain vote—even a key vote. Presidents doubtless will contrive to encourage party cohesion, but just as clearly, party support will vary on the kinds of measures the president is asking them to support and on whether or not parties continue to lose importance in national political life.

PROPOSALS TO INCREASE COHESION BETWEEN PRESIDENT AND PARTY

In spite of the revival of the Republican Party in the past decade, scholars still see the party system as weak and offer proposals that might revitalize the parties and help them recapture some of their historic mediating and moderating influence in U.S. political life.

Readers should be warned that most Americans could not care less. Most Americans are indifferent, at best, to the purposes of our parties. Moreover, even some informed observers say they do not know whether there is much realism in trying to resuscitate the parties. Critics of party renewal efforts sometimes contend there is too much nostalgia for a romanticized two-party system that seldom or never really existed.

Still, rethinking public policy toward the parties is much needed. In the campaign to revive and protect them, the following are the measures, at least bearing on the presidency, one hears most often. We will examine some of their merits and defects.

The most ambitious proposal seeks to replace the Madisonian system of separated powers with a parliamentary system of fused executive and legislative powers similar to that of Great Britain. Most observers note that parliamentary systems are more efficient than presidential types of democracies. Parliamentary governments tend to have strong, programmatic, disciplined parties.

A bipartisan committee of scholars and practitioners calling itself the Committee on the Constitutional System, founded in 1981, calls the weak party structure in the United States the chief flaw of the American system. The committee promotes parliamentary reforms so as to bridge the wide gap between the president and Congress, and party plays a prominent role in strengthening that bond. Among the proposals of the committee relating directly to party are:

1. Grant all party nominees for the House and Senate seats as uncommitted voting delegates at the national convention.

2. Make it easier to vote straight-line party tickets by having one line or lever for a party vote.

3. Create a Congressional Broadcast Fund to be used by the party to air TV spots.

4. Have strong party discipline in Congress.

Political scientist James MacGregor Burns, a proponent of these reforms, offers a slightly different list of party reforms.[11] (See chapter 6.) Parties have changed a lot since the 1960s, Burns says, mostly for the worse. To strengthen coherent, unified national parties, controlled from the grass roots, Burns would make major changes in the presidential selection process: participation in primaries would be based on active party membership and statewide primaries would be discouraged; instead delegates would be selected in local caucuses. State party committees could select at-large delegates to reward party officers and party loyalty, and Burns would make certain elements of the party platform mandatory for the party nominees to uphold.

Some proponents of party renewal say federal funding of presidential elections should be abolished or at least significant portions of this public financing should be channeled through the parties and not go directly to the candidates. They say federal funding bypasses parties too much and encourages autonomous, entrepreneurial political adventurers who are unaccountable to anyone. Campaign funds, these purists argue, should go to the party organization. Party regulars or party establishments would then have some control and, thereby, more of a function. Candidates would have to demonstrate their loyalty and their willingness to run with the ticket and on the platform as a quid pro quo of getting adequate financial help. This would presumably tie a candidate, and perhaps later the officeholder, to a closer relationship to the national committee. Moreover, it is pointed out by veteran observers that the two national committees already have experienced accounting offices and these staffs could help reduce the staff work, start-up costs, and probable errors of a newly formed candidate organization. Campaign financing legislation encourages separate candidate committees to be set up to receive and account for public funds, at least in primary elections.[12] If candidates trusted their national committees and national committee staffs, the funds could, at least in theory, come through these party officers to the candidates (this could be done at both the primary and general election stage).

However, the abuses of the pre-1974 system of private financing to candidates had to be remedied. It was predominantly a private candidate-financed system in which candidates and their campaign teams raised enormous sums of money from wealthy donors. If the party got involved, it was to serve the needs of the candidate and not to call the shots itself. The party fundraising apparatus had been reduced to financial middle-man (if it was used at all) for the candidate. In the 1972 Nixon campaign, things got so out of hand that Congress was forced to confront the abuses of this system.[13]

In response to these kinds of situations and the Watergate revelations, Congress passed the Campaign Reform Act of 1974, which substituted a publicly financed candidate system for a privately financed candidate system. National party committees receive some public funds for running the quadrennial national conventions.

Why did this campaign finance reform ignore and bypass the parties? Most members of Congress just did not want to strengthen party committees; members of Congress have little interest in giving party officials more power over who gets money in national elections. Congress obviously was thinking of itself as well. If public financing is extended to congressional campaigns, the last thing a member wants is to have to submit to local and state party officials for campaign funds. As things stand now, the candidate raises his or her own funds. It is a highly individualistic arrangement, and few members of Congress will entrust so important a career lifeblood factor to party bosses. Especially is this the case with officeholders who had to buck the party to get elected merely a few years ago. Further, in its weakened state, party machinery sometimes gets taken over by extremist factions.

If public funds are made available to the national committees, it will probably be done through indirect side payments of some kind or another. Thus, Congress in 1979 approved a lower postal rate for parties to use in fund-raising efforts. Another subsidy, which would make sense, is the provision of free television time to national party committees to present their views and purposes to the American people. Modest assistance of this kind may help to prop up the major parties. Too much funding of the parties, however, might cement into permanence the present parties at the expense of some future party that would be deserving of a chance to catch on.

Another general proposal to enhance cohesion between presidents and their parties calls for giving more power to party regulars. One variation calls for holding fewer direct primaries on the grounds that party conventions or party caucuses would allow party loyalists to have more of a say in the presidential nominating process. (Many people remain unconvinced, however, that a return to the smoke-filled room technique for candidate nominations will measurably strengthen our parties.) The reasoning here is that these kinds of persons have long-standing commitments to the preservation of the party and have a stake in the future of the party, not just in who the candidate will be this time. It is thought that these party professionals know better how to bring disparate factions together and how to compromise on explosive issues that might divide the party.

PROSPECTS

One of the many reasons politicians rely on the media and go public rather than use the party organization is that the media route is easier. Keeping a party organization intact or rebuilding a political party in a community is an exacting undertaking. It is no accident that more and more groups in America are single-issue organizations. It is difficult to hold a multiissue group together for long, and that is what a party is. We need to have a better understanding of why this is and why there appear to be patterns of party organization and practices of local parties that apparently ensure decay. Could it be otherwise? Might different practices or different incentives help overcome the ruts and the rot?

There have been some healthy signs in the past years. The party caucuses and party conferences in Congress have shown signs of life. Indeed, the Democratic Party caucus in the House of Representatives in the mid-1970s was instrumental in revitalizing that institution. The House Republicans Conference was instrumental in fashioning strategies that led to the Republican takeover of Congress in the mid-1990s. Party caucuses were used to help modernize and democratize the House of Representatives. The party caucus is also a place to discuss party commitments and adapt old commitments to new policy problems. Reagan benefited from extraordinary party caucus loyalty from Republicans in the early 1980s.

Parties need to be preserved: they need to be made constructive and a vital part of the political process. Those who are aware of the chaos and paralysis produced by thousands of new interest groups, each having an impact directly on the government in narrow public policy areas, long for almost any instrument that might prevent direct parochial impact and encourage intermediary processes of trade-off, mediation, and so forth. Stronger parties would obviously help here.

Parties help to give the electorate reasonable choices at the election booth. They can provide a forum for candidates of somewhat common perspectives to agree on broad purposes. They certainly will continue to be important in organizing the legislative branch. Perhaps, too, they sometimes will formulate strategies that can help us overcome some of our toughest economic policy questions.

Perhaps most important to the preservation of parties is a commitment to avoid further weakening of them. To move to the direct election plan might encourage splinter and single-issue parties and further weaken our major parties. Similarly, a national primary might be a severe blow to our party system. Political scientist Austin Ranney views the national primary proposal with alarm, saying it not only would greatly weaken state parties but would mean "a virtual end to the national parties as anything more than passive arenas for contests between entrepreneurial candidate organizations."[14]

Can parties as institutions be made to serve as a more effective check on presidents? This might be desirable, but again, the weakened condition of the parties and the history of party fragmentation and lack of discipline suggests that the prospects are not inviting.[15] One should not overlook, however, the role party officials and party stalwarts have played in checking and balancing presidents of their party. Johnson was forced to quit his 1968 bid less by Eugene McCarthy's insurgency than by Robert Kennedy and party regulars in Wisconsin and elsewhere making it clear he was unacceptable to broad segments within the Democratic Party. So also, Nixon was forced to resign when it was clear that stalwarts in his own party—people like Senators Barry Goldwater, Hugh Scott, and others—could no longer support him.

Parties need to be protected and preserved even if they are not likely to be much strengthened. Presidents in the future, as in the past, will find party leadership a necessary yet exacting task. The incentive is to ignore the party. A six-year nonrenewable presidential term would encourage that tendency, and for that reason, among others, should not be approved. The four-year term with the necessity of renomination at least ensures that most presidents in the first term will recognize the importance of the party to their political survival. A political party should retain the threat of dumping a president who has turned his back on his party's pledges or has ignored the party platform. We shall have some presidents in the tradition of the Jacksons, Wilsons, and Roosevelts who will serve as party leaders, but just as likely, we will have presidents in the Eisenhower mold, who eschew party responsibilities.

We really have not had a president in recent decades who has gone the extra mile and tried to be a party leader. The incentive system may seem stacked against it, yet such a counterintuitive strategy might just pay some surprising dividends. Parties will endure, but in the future as in the past, they will serve us in direct proportion to our taking them seriously and our willingness to make them work.

In the end we need to remember political scientist Clinton Rossiter's warning that there can be "No America without democracy, no democracy without politics, no politics without parties."[16] Likewise we will rarely have effective presidencies without healthy political parties that have integrity.

CHAPTER   8

The President and the Supreme Court

[The Supreme Court] has been in angry collision with
the most dynamic and popular Presidents in our
history. Jefferson retaliated with impeachment. Jackson
denied its authority; Lincoln disobeyed a writ of the
Chief Justice; Theodore Roosevelt, after his Presidency,
proposed a recall of Judicial decisions; Wilson tried to
liberalize its membership; and Franklin D. Roosevelt
proposed to "reorganize" it.

> Robert H. Jackson, *The Struggle: Judicial Supremacy* (New
> York: Vintage Books, 1941), p. x

Packing the Supreme Court just doesn't work . . .
whenever you put a man on the Court, he ceases to
be your friend. I'm sure of that. I've tried it and it
won't work.

> President Harry Truman, Speech at Columbia University,
> New York, April 28, 1959

For most practical purposes the President may act as if
the Supreme Court did not exist. . . . The fact is that
the Court has done more over the years to expand
than contract the authority of the Presidency. . . . In
the nature of things judicial and political, the Court
can be expected to go on rationalizing most
pretensions of most Presidents. It is clearly one of the
least reliable restraints on presidential activity.

> Clinton Rossiter, *The American Presidency* (New York:
> Harcourt, Brace and World, 1960), pp. 56, 58–59

The history of presidential–Supreme Court relations is primarily one of the nation's highest court aiding and abetting an expansive view of presidential power. Although the Supreme Court has occasionally halted presidential action or declared a presidential act unconstitutional, the Court has more frequently labored to approve or legitimize the growth of presidential power.

The framers of the Constitution intended the Court to serve, along with the presidency, as a major check on the anticipated or at least potential excesses of the national legislature. The Court, they hoped, would comprise wise, virtuous, and well-educated statesmen who would interpret and preserve the Constitution, especially from legislative encroachment. Most of the framers believed the Court would work hand in hand with the executive. Some even viewed it as part of the executive department or at least engaged in the same functions, namely executing, interpreting, and applying the laws.

Both president and Court have a national constituency. But where the president serves many masters, the Court serves only one: the Constitution. Both are expected to restrain Congress when needed.

Presidents and the Court have had their share of clashes. This is to be expected. The Constitution stands as the supreme law of the land in a nation governed by laws, not individuals. The rule of law has a key limiting and empowering function for our government. Yet the Constitution is sometimes ambiguous, sometimes silent on matters of grave concern; the law is flexible and changing, open to interpretation. Differences of opinion, interest, and interpretation are inevitable. In most but not all encounters between presidents and courts, presidents emerge victorious. The exceptions are notable. Jefferson was told by the Marshall Court that he had acted wrongly. Lincoln, after his death, was rebuffed for using military courts in areas where they were not justified. FDR saw several New Deal initiatives struck down by the Court as unconstitutional. Truman was ordered by the Court to release steel mills from federal control. Richard Nixon's policy and political intentions were overturned by the federal courts on at least seventy-five occasions, most notably in the summer of 1974 when the Supreme Court directed him to release White House tapes for use in criminal investigations.

Here are some basic characteristics of more than two hundred years of presidential–Supreme Court relations.

• There is no getting politics out of a nomination to the Supreme Court. Picking justices is a political act. Presidents usually want to "pack" or at least shape the Court, but they possess no special legal or constitutional right to impose their views on a whole branch for a generation. The Constitution entitles a president only to try, not necessarily to succeed.

• Presidents nominate Supreme Court justices and their nominees are usually confirmed by the U.S. Senate. Most justices conform to the general intentions of the presidents appointing them. However, about 25 percent of the justices deviate and become "wayward justices." That is, they vote and write decisions contrary to the views and values of those who appointed them. Several factors encourage justices, once on the bench for awhile, to grow independent from the presidents who nominated them.

• The Supreme Court is generally a friend of the presidency and supports the use of presidential power, especially in wartime or in the conduct of foreign affairs. Indeed the judiciary has played a significant role in the trend toward executive primacy in foreign policymaking.

• The justices can thwart presidential initiatives and are most likely to do so if evidence exists that Congress would agree with the Court, and if there is clear absence of any authority for the presidential action either in the Constitution or in congressional statutes. They are also likely to weigh in against a president when a president's public approval is reasonably low, at the end of a term or after the president has left office or died.

• Presidents take an oath of office, typically administered by the Chief Justice, pledging to uphold the Constitution and not necessarily what the Supreme Court says about the Constitution. Presidents reserve the right to interpret the Constitution for themselves. They may, as they usually do, defer to the judgment of the Court, yet they do not have to accept or agree with the judicial reasoning.

This chapter explores these realities and seeks to answer the following questions and paradoxes. How do presidents select people for the Supreme Court? Why do some nominees get rejected? Why do some justices grow noticeably independent once on the Court? What were the Lincoln and FDR experiences with the court and why did Roosevelt try to "pack the Court?" When, and in what circumstances, has the Court acted to expand the prerogatives of the chief executive? When and in what instances has the Court restrained the presidency? Has the Court been a reliable check on presidential power? Can we expect the judiciary to constrain a powerful and popular president bent on exercising his or her will? Finally, do national security considerations justify virtually any means, constitutional or otherwise, a president wishes to adopt to ensure U.S. sovereignty and survival?

PRESIDENTIAL NOMINATIONS TO THE COURT

At the federal convention of 1787 in Philadelphia many delegates believed the national legislature should appoint the justices. The important Virginia plan had suggested this method. But it was opposed as impractical. It had not worked well in the states. The legislature would be too large to make such personnel decisions. And there were other objections. Yet strong opposition to legislative appointment was met with equally strong opposition to appointment by the chief executive.

Alexander Hamilton, in an effort to break the deadlock, proposed that all major officials be nominated by the president with the Senate having the right to approve or reject them. Although Hamilton's proposal was an ingenious compromise, delegates rejected the concept of two-branch participation in the appointment process twice before approving it only in the last month of their deliberations.

In modern times, presidents have to look at the judicial selection process as a partnership venture between the White House and the Senate. A judicial nominee must win a majority confirmation in the Senate. Presidents usually get their way with cabinet and top executive department nominations. The widely accepted view is that presidents are entitled to the assistance in the executive branch of people they respect and trust and whose views are compatible with their own. The Supreme Court, however, is different. It's a separate branch and most members of the Senate nowadays think they have a co-equal responsibility for who should sit on the nation's highest court.

The U.S. Constitution is silent as to the criteria that should guide a president in selecting a Supreme Court nominee. As Terry Eastland, a former Reagan and Bush administration aide, writes: "This is a matter clearly within the president's discretion." However, "the considerations that have decisively influenced presidents in judicial selection have included political patronage, geographical balance, judicial philosophy, and—especially in recent years—race and gender."[1] Judicial values and judicial philosophy are plainly now the dominant criteria.

Most presidents understandably seek to nominate to the bench persons who share their policy preferences. In effect, presidents make predictions about the likely future performance and policy opinions of their potential nominees. If a president can nominate two or three members to the Court, that president's political philosophy can extend beyond his term in office. Thus presidents pay a great deal of attention to whom they shall nominate. In many instances this has led to nomination of close friends, advisers, and loyal partisans. Yet party label alone is not enough or even, in some instances, a key factor. President Theodore Roosevelt wrote to his friend Henry Cabot Lodge and outlined some of his views about what was important as he considered Massachusetts Chief Justice Oliver Wendell Holmes, Jr. as a Supreme Court nominee. To Lodge he wrote, "I should like to know that Judge Holmes was in entire sympathy with our views, that is, with your views and mine . . . , before I would feel justified in appointing him. . . . I should hold myself guilty of an irreparable wrong to the nation if I should put . . . [upon the Supreme Court] any man who was not absolutely sane and sound on the great national policies for which we stand in public life."[2]

In 1956 Dwight Eisenhower nominated a Democrat, William Brennan. He did so in part for political reasons. Brennan was a Catholic as well as a Democrat and the month was October, at the tail end of Eisenhower's reelection efforts. Eisenhower and his political advisers plainly wanted to attract Catholic votes. Eisenhower was blunt about it. His Republican National Committee chairman recalls Ike saying: "Maybe you'll disagree with me but I told [the Attorney General] that I didn't give a damn whether it's a Republican or Democrat, I just want to get a good Catholic for a judge."[3] Eisenhower and his aides appreciated that as the minority party in the nation they needed Democrats as well. "Eisenhower Democrats" had helped win their 1952 victory, and subsequently the 1956 victory as well.

Herbert Hoover also nominated a Democrat, indeed a man whose views were contrary to many of his own. Benjamin Cardozo, a prominent legal scholar who had served for years as a distinguished member of the New York Court of Appeals (that state's highest court) won nomination from Hoover because he was the virtually unanimous choice of law school deans, legal scholars, editorial and countless other public opinion leaders. Hoover was reluctant yet finally agreed to appoint him. Hoover later acknowledged this had been one of his best decisions.

President Nixon was less concerned with party affiliation than with getting Court members who were conservatives who would exercise judicial restraint. His was of course a reaction to the judicial interventionism of the Warren Court. Nixon also wanted some justices who would be a symbol of and for Southern whites in the same way that Thurgood Marshall, appointed by Lyndon Johnson, had been a symbol for African-Americans. Nixon wanted to reward the South for its political support, and one way to do it was to nominate justices who would lead the Court away from the liberal leanings of the Warren Court.

Nixon said he would use four main criteria in selecting his nominees for the Court: they should have a great legal mind, they should be young enough to serve at least ten years, they should have experience both as a practicing attorney and as a judge, and they must share his view that the Court should merely interpret the Constitution rather than amend it by judicial policymaking. His first nominee, federal appeals court Judge Warren Burger, a Northerner, won quick Senate confirmation. Nixon's second and third nominees were Southern judges, both of whom were rejected by the U.S. Senate. Nixon fumed. Democrats controlled the Senate, and leading civil rights and union leaders marshalled evidence that Nixon's nominees were either prejudiced or had various conflicts of interest that made them unfit for confirmation. Nixon then nominated a Midwesterner, Harry Blackmun of the federal appeals court based in Minnesota. One of Nixon's later nominees was Lewis Powell of Richmond, Virginia. Powell, a conservative Democrat, was a distinguished lawyer and past president of the American Bar Association. Although he had no experience as a judge, he was easily confirmed. Note, however, that to make his "Southern strategy" work, Nixon had to sacrifice his criteria of experience as a judge, and he nominated a person who at least nominally was a member of the opposition party.

In recent decades, presidents have been slightly more partisan in judicial selection. Jimmy Carter did not get the opportunity to appoint any justices to the Supreme Court, but Ronald Reagan was able to appoint three: Sandra Day O'Connor, Antonin Scalia, and Anthony Kennedy.

Reagan also elevated William Rehnquist to chief justice in 1986. George Bush appointed David Souter and Clarence Thomas. And Bill Clinton appointed two justices in his first term: Ruth Bader Ginsburg and Stephen G. Breyer. All these justices came from the appointing president's party. (See Table 8.1.)

The appointment of federal judges also reflects the partisan interests of the president and his party. Only 13 of 108 members of the Supreme Court came from a party other than the president's. And, as Table 8.2 reveals, Republican presidents tend to appoint more white men to the bench than do Democrats, who appoint a higher percentage of women and minorities.

Presidents have been influenced by a variety of factors when searching for potential Supreme Court nominees. First, as noted above, is finding someone who shares their ideological and philosophical views. Second, presidents are politicians and they naturally want praise and political credit for their appointments. Reagan's first nominee, Sandra Day O'Connor,

TABLE 8.1
Supreme Court Appointments, 1972–1996

Name	Year of appointment	Law school	Appointing president	Prior judicial experience	Prior government experience
William H. Rehnquist	1972	Stanford	Nixon	—	Assistant U.S. Attorney General
John Paul Stevens	1975	Chicago	Ford	U.S. Court of Appeals	—
Sandra Day O'Connor	1981	Stanford	Reagan	Arizona Court of Appeals	State Legislator
Antonin Scalia	1986	Harvard	Reagan	U.S. Court of Appeals	—
Anthony M. Kennedy	1988	Harvard	Reagan	U.S. Court of Appeals	—
David H. Souter	1990	Harvard	Bush	U.S. Court of Appeals	New Hampshire Attorney General
Clarence Thomas	1991	Yale	Bush	U.S. Court of Appeals	Chair, Equal Employment Opportunity Commission
Ruth Bader Ginsburg	1994	Columbia	Clinton	U.S. Court of Appeals	—
Stephen G. Breyer	1995	Harvard	Clinton	U.S. Court of Appeals	U.S. Senate Aide

TABLE 8.2
A Profile of Presidential Appointees to the Lower Federal Courts

	Number of appointees						
	Johnson	Nixon	Ford	Carter	Reagan	Bush	Clinton[a]
Gender							
Male	159	226	63	217	340	149	123
Female	3	1	1	41	28	36	57
Ethnicity or race							
White	152	218	58	202	344	165	128
Black	7	6	3	37	7	12	36
Hispanic	3	2	1	16	15	8	13
Asian		1	2	2	2		2
Native American				1			1
ABA ratings							
Exceptionally/well qualified	89	117	31	145	203	109	120
Qualified	68	110	32	110	165	76	60
Not qualified	4		1	3			
Total number of appointees	162	227	64	258	368	185	180

[a]Appointments made through 1995.
Note: One Johnson appointee did not receive an ABA rating.
Sources: Sheldon Goldman, *"Bush's Judicial Legacy: The Final Imprint," Judicature* 282 (1993); *Alliance for Justice,* Judicial Selection Project Annual Report, 1993; and Alliance for Justice Judicial Selection Project Annual Report, 1995 (Washington, D.C.: Alliance for Justice, January 1996). Reprinted by permission of Alliance for Justice.

conveniently satisfied both these standards. She was a strong Reagan supporter and her appointment won considerable praise for the president, especially from women's groups.

Still other factors are at work. Nominees have to be confirmed by a majority of the U.S. Senate. Presidents and their aides consult leading senators, especially those from a prospective nominee's home state. They also often consult with members of the Senate Judiciary Committee, the committee responsible for confirmation hearings and initial screening. Presidents also generally take into account the wishes of public and private sector leaders and interest groups who have a known interest in the nomination.

Presidents sometimes confer with, or at least obtain the views of, some of the sitting members of the Supreme Court. The justices, and especially some of the chief justices, have an obviously keen interest in any nomination. Chief Justices William Howard Taft and Warren E. Burger and Associate Justice Felix Frankfurter often expressed themselves and even lobbied for preferred candidates.

Since 1946 the American Bar Association has also played a role in presidential nominations to the federal judiciary. The ABA's Standing Committee on the Federal Judiciary evaluates the legal competence of presidential nominees, and most presidents and many members of the U.S. Senate take their "ratings" into account.

Who gets nominated and confirmed? They have all been lawyers, and in recent years most have been graduates of distinguished law schools such as Harvard, Yale, Stanford, Chicago and Northwestern. Almost all have been active in politics in some way or another. Some have served in local or state elective office. Some have been party leaders. Some were actively involved in presidential campaigns. A few even ran for the presidency themselves (Taft, Hughes, and Earl Warren). Approximately 30 percent served as executive branch officials in the national government before their appointment to the Court. Nine were attorneys general, four were solicitors general, at least eighteen held cabinet-level posts, and over a dozen held various other Department of Justice positions. Four justices, White, Rehnquist, Stevens, and Breyer, were law clerks to Supreme Court justices.

Perhaps the most common background in the recent past has been some experience serving as a state or federal judge. Justices Burger, Blackmun, Stevens, Brennan, Marshall, O'Connor, Scalia, Souter, Thomas, Ginsberg, and Breyer all fall into this category. Dwight Eisenhower, Richard Nixon, Ronald Reagan, and Bill Clinton all wanted candidates who had judicial experience. Why? In part because they believed this was ideal training for service on the nation's highest court. They also recognized it was easier to discern the predictable future responses of a potential Court member if that person's decisions could be studied to see if they conformed to a president's policy views.

If one were to prepare a profile of typical future court nominees (assuming the past is a reasonable prologue to the future), it might read like this: top graduate of a distinguished law school, law clerk to a federal judge, legal practice, involvement in party or presidential campaigns, Justice Department experience, and earlier appointment to federal district or federal appeals court. Nominees are typically also in their early or mid-50s, to ensure at least a twelve- to fifteen-year service on the Court.

The Constitution nowhere stipulates, however, that Court members be lawyers. Nor is there any congressional statute insisting on judicial experience. Yet no one today disputes the need for *legal* experience. The desirability of *judicial* experience, however, is less clear. Several distinguished justices did not have prior judicial experience, including John Marshall, Earl Warren, Louis Brandeis, Robert Jackson, and Lewis F. Powell. Intelligence and temperament, as well as being well read, would seem to be just as important as lower court experience. However, most future court nominees are likely to come from the lower federal courts.

Future presidents will probably consider these political factors:

(1) whether his choice will render him more popular among influential interest groups; (2) whether the nominee has been a loyal member of the President's party; (3) whether the nominee favors presidential programs and policies; (4) whether the nominee is acceptable (or at least not "personally obnoxious") to the home-state Senators; (5) whether the nominee's judicial record, if any, meets the Presidential criteria of constitutional construction; (6) whether the President is indebted to the nomi-

nee for past political services; and (7) whether he feels "good" or "comfortable" about his choice.[4]

Issues of race and gender will also doubtless play a role in presidential Court nomination decisions.

CONFIRMATION BATTLES

As noted, our Constitution does not vest the process of Supreme Court appointments solely in the president. Article II, Section 2 provides that a president shall have power, "by and with the advice and consent of the Senate," to appoint Supreme Court justices. If a president decides simply to assert his power to pack the court with good friends or unqualified or extremist persons, senators from the opposition party, or even from the president's own party, often try to use their constitutional power to defeat him. Over the history of the Republic the senate has rejected twenty-six presidential nominees to the Court (see Table 8.3). This amounts to about 15 percent of those nominated. This rejection rate increases about 25 percent for presidents in their last year in office and when the Senate is controlled by the opposition party. More important, this check on presidential power has forced presidents to not appoint persons who would fail to win Senate approval. In yet other instances some potential nominees may have taken themselves out of consideration precisely because they feared grueling confirmation hearings.[5]

Most of these rejections occurred before 1900. This century has witnessed only a handful of rejections, and few of these were based on a candidate's political views.

Why are some nominees rejected? Sometimes it is because a president is highly unpopular in the country and in the Senate. President Andrew Johnson's stock in the Senate was so low he probably could not have had any nominee of his confirmed; his one nominee fell victim to Senate rejection.

On occasion a practice known as "senatorial courtesy" contributes to the rejection of a nominee. This occurs when one or both senators from a nominee's home state indicate to the White House that they strongly oppose the proposed Court appointment. This practice is more common in cases of federal district judge appointments, yet it has in times past undercut a presidential Supreme Court appointment.

The political and philosophical views of a nominee can sometimes stir up the opposition of major interest groups and the opposition party, as was the case in Reagan's 1987 nomination of Robert Bork. This can especially harm a candidacy when the opposition party controls the U.S. Senate, as it did when Nixon, for example, tried to appoint Clement Haynsworth in 1969. Many senators, especially Democrats, viewed Haynsworth as too conservative. The senate rejected another of Nixon's nominees,

TABLE 8.3
Supreme Court Nominations Rejected, Postponed, or Withdrawn
Due to Senate Opposition

Nominee	Year nominated	Nominated by	Actions
William Paterson[a]	1773	Washington	Withdrawn
John Rutledge[b]	1795	Washington	Rejected
Alexander Wolcott	1811	Madison	Rejected
John J. Crittenden	1828	J. Q. Adams	Postponed
Roger B. Taney[c]	1835	Jackson	Postponed
John C. Spencer	1844	Tyler	Rejected
Reuben H. Walworth	1844	Tyler	Withdrawn
Edward King	1844	Tyler	Postponed
Edward King[d]	1844	Tyler	Withdrawn
John M. Read	1845	Tyler	No action
George W. Woodward	1845	Polk	Rejected
Edward A. Bradford	1852	Fillmore	No action
George E. Badger	1853	Fillmore	Postponed
William C. Micou	1853	Fillmore	No action
Jeremiah S. Black	1861	Buchanan	Rejected
Henry Stanbery	1866	Johnson	No action
Ebenezer R. Hoar	1869	Grant	Rejected
George H. Williams[b]	1873	Grant	Withdrawn
Caleb Cushing[b]	1874	Grant	Withdrawn
Stanley Matthews[a]	1881	Hayes	No action
William B. Hornblower	1893	Cleveland	Rejected
Wheeler H. Beckham	1894	Cleveland	Rejected
John J. Parker	1930	Hoover	Rejected
Abe Fortas[e]	1968	Johnson	Withdrawn
Homer Thornberry	1968	Johnson	No action
Clement F. Haynsworth, Jr.	1969	Nixon	Rejected
G. Harrold Carswell	1970	Nixon	Rejected
Robert H. Bork	1987	Reagan	Rejected
Douglas H. Ginsburg	1987	Reagan	Withdrawn

a = Reappointed and confirmed.
b = Nominated for chief justice.
c = Taney was reappointed and confirmed as chief justice.
d = Second appointment.
e = Associate justice nominated for chief justice.
Source: Adapted from *Storm Center: The Supreme Court in Amercian Politics,* 3rd ed. by David M. O'Brien.
Copyright © 1993, 1990, 1986 by David M. O'Brien. Reprinted by permission of Addison-Wesley Educational Publishers, Inc.

Florida Judge G. Harrold Carswell, mainly on the basis of inadequate professional competence. Opposition to Carswell came fast and strong. One respected law school dean suggested this nominee presented more slender credentials than any other nominee for the Supreme Court put forth this century, and after the confirmation hearings were completed, most senators agreed with this verdict.

Sometimes the Senate rejects a nominee to signal its opposition to the policy or recent record of the incumbent Supreme Court. Sometimes senators are unsure of the political reliability of a nominee. And sometimes the Senate merely makes a mistake, as some observers believe it did when it rejected President Herbert Hoover's 1930 nomination of Judge John J. Parker of North Carolina.

The Senate usually takes its constitutional responsibility seriously. Senators view an appointee to the Court as more consequential than most members of a president's cabinet. Court nominees are seen as more important because of their longer tenure, their independence from the White House, and their impact over a whole range of important public policy issues. Usually the Senate Judiciary Committee and then the Senate itself scrutinize these nominees closely. Poor selections such as Madison's choice of Alexander Wolcott, Grant's choice of George Williams, and Nixon's choice of Harrold Carswell have been rejected often enough to set reasonable standards. Although the Senate rejected only four nominees in the twentieth century, presidents and White House advisers with any sense of history are aware that the Senate jealously prizes this shared power of appointment.

In late 1987, Reagan's nomination of Robert H. Bork incited sharp differences of opinion about how much the Senate should take into account the philosophical views of a nominee. Some Democrats in the Senate proclaimed that the framers had intended the broadest role for the Senate in choosing members of the Court and hence they had wide authority in checking into a nominee's constitutional views and values. At least one Republican senator disagreed, saying that for the Senate to require that judicial candidates pledge allegiance to the political and ideological views of particular senators or interest groups could lead to paralyzing the Senate in a gridlock of competing interest groups, each hawking its own agenda. However, a New York Times/CBS News public opinion survey at the time found that Americans believed senators should attach importance to nominee's positions on constitutional issues when judging the nominee's fitness in the confirmation process. (See Table 8.4 for the poll data.)

TABLE 8.4
Judging Supreme Court Nominees: Should Their Views and Values Be Considered by the Senate?

Q. "When senators decide how to vote on confirming a president's nominee for the Supreme Court, after satisfying themselves about the nominee's legal experience and background, how much importance should a senator attach to the nominee's positions on major constitutional issues?"

A lot	62%
A little	25
None	6
No answer	7

Source: Based on a New York Times CBS News poll of 745 people interviewed by telephone, July 21–22, 1987, reported in *The New York Times,* July 24, 1987, p. 10. Copyright © 1987 by the New York Times Co. Reprinted by permission.

President Bush's first nominee for the Supreme Court was David H. Souter. Though Souter was eventually confirmed by the Senate in a 90–9 vote, the somewhat eccentric and shy bachelor had to endure an unusually intrusive scrutiny of his life style. His chief sponsor and champion in the Senate, Warren Rudman, was especially angered by how Souter was treated.[6]

The Clarence Thomas, Bush nomination fight in 1991 was one of the ugliest and most personal in U.S. history. Thomas barely won confirmation by a 52 to 48 vote, but both his nomination and especially his confirmation battle triggered a major political backlash for Bush and to some extent for the U.S. Congress.[7]

WAYWARD JUSTICES

Once confirmed, new members of the Court are in no way obliged to the president who appointed them. There is always an element of unpredictability in how justices will vote in future years. Loyalty to the president who appointed them is not considered a proper reason for judgment on the Court. Even justices who have "pleased" their sponsoring president acknowledge that they think and act differently once they join the Court.

Justice William H. Rehnquist said that any president who attempts to leave a lasting ideological stamp in the Court will typically fail. Unexpected legal developments, the Court's time-honored tradition of independence, the role of precedent, the influence of eight colleagues already there, and other factors undermine a president's efforts to leave a mark on the Court. Rehnquist added:

> The degree to which a new Justice should change his way of looking at things when he "puts on the robe" is emphasized by the fact that Supreme Court appointments almost invariably come "one at a time," and each new appointee goes alone to take his place with eight colleagues who are already there. Unlike his freshmen counterparts in the House of Representatives, where if there has been a strong political tide running at the time of a particular election there may be as many as 40 or 50 or 80 new members who form a bloc and cooperate with one another, the new judicial appointee brings no cohorts with him.
>
> A second set of centrifugal forces is at work within the Court itself, pushing each member of the Court to be thoroughly independent of his colleagues. Tenure is assured no matter how one votes in any given case; one is independent not only of public opinion, of the President, and of Congress, but of one's eight colleagues as well. When one puts on the robe, one enters a world of public scrutiny and professional criticism which sets great store by individual performance, and much less store upon the virtue of being a "team player."[8]

Dwight Eisenhower regretted his appointments of Chief Justice Earl Warren and Justice William Brennan. Both provided indispensable judicial leadership in civil rights, reapportionment, and criminal procedure issues. Cordial relations between Warren and Eisenhower diminished perceptively soon after the landmark *Brown v. Board of Education* (1954)[9] decision. Years later, Ike told friends "the appointment of that dumb son of a bitch Earl Warren" was his biggest mistake as president.[10] Earl Warren had no regrets. He loved his years on the court and was extremely proud of his votes and opinions. He obviously believed a justice owed no special loyalties to the person who appointed him to the Court. Moreover, a justice, Warren observed, had an obligation to change and adapt to the needs of the times as well as to new developments in legal and constitutional research. Warren said he did not "see how a man could be on the Court and not change his views substantially over a period of years . . . for change you must if you are to do your duty on the Supreme Court."[11]

How often do presidents fail to get what they want in the people they appoint to the Supreme Court? One student of the judiciary concludes about 25 percent of the justices have "deviated" from the expectations presidents held for them. In some instances, such as the Earl Warren case, we can relate judicial performance to presidential expectations with considerable precision. More often assumptions have to be made based on the general political views of the presidents and on the situations in which they operated. In still other cases of short tenure, it is near impossible to discern the "loyalty" or "deviation" of a justice. Political scientist Robert Scigliano examined the fit or lack of fit between justices and presidents and concluded:

> . . . that about three-fourths of those justices for whom an evaluation could be made conformed to the expectations of the Presidents who appointed them to the Supreme Court . . .
>
> Our conclusion is an important one in that it indicates limitations upon the ability of Presidents to influence the policies of the court through appointments and assures us, retrospectively at least, of a certain, but crucial, measure of judicial independence from Presidential attempts to bring the Court closely into line with the executive branch of government.[12]

Who are some of the prime examples of justices who deviated from their sponsors? John McLean, a Jackson appointee, became a noted apostate on the Court. Salmon Chase and Stephen Field, two Lincoln appointees, turned against Lincoln in significant Court rulings in 1866 and 1870. Theodore Roosevelt considered one of his appointees, Oliver Wendell Holmes, to be a "bitter disappointment." Woodrow Wilson believed he made a mistake in his selection of Justice James C. Reynolds. And Wilson's dismay would have grown had he lived to witness McReynolds' decidedly conservative record in later years. FDR apparently believed Felix Frankfurter was going to be more of a liberal than the cautious conservative he later became. Harry Truman expressed regrets that he appointed Tom Clark to the Court. When Byron White was nominated for the Court in 1962 he was expected to be a moderate to liberal justice, yet like so many others assumed to be "predictable" he charted his own course, after leaving the impression that he had reversed fields or, at the very least, defied traditional labels. Richard Nixon regretted his selection of Harry Blackmun, who became the most progressive of his four appointees, and one of the "liberals" in the 1980s.

Justice Blackmun, in a 1985 interview, said he did not believe he had changed much during his years on the Court. Yet he did say one obviously "grows" when one comes to Court. "You have to develop your own constitutional philosophy and look at issues and the Constitution differently than you did in earlier years." Blackmun added:

> I believe I've been pretty consistent over the years. I think some people who may be a little surprised didn't know me or my earlier lower court decisions very well. I hope I've become a force for or an influence for moderation on the Court. I am opposed to seeing the Court swing sharply to the right or sharply to the left.
>
> You grow when you win an appointment like this. I hope we all do. I really didn't have a well developed constitutional philosophy and this is what has made so much of a difference in my development over the years.
>
> One other factor. People wrongly thought Burger was the main reason I was appointed. He was not that important a factor. Nixon needed someone who could get confirmed. He wanted someone with federal court experience and several other people were backing me. And he had gone on record as saying he was not going to appoint any close friend or associate of his so this excluded people like Herb Brownell, John Mitchell and others, at least for this vacancy. So I came with a certain amount of independence.[13]

Blackmun readily agreed that a justice has several obligations, such as to the Constitution as well as to his or her own judgment, that far outweigh any loyalty to the president. And indeed, Blackmun's agreement with the Nixon tapes decision, his leading role in *Roe v. Wade*[14] (and other abortion decisions), and his opinions on certain affirmative action and church and state matters indicated both his independence from Nixon and perhaps his own personal "growth," to borrow from his own appropriate term.

Why have so many justices disappointed the presidents who appointed them? Sometimes it is because the presidents and their advisers failed to examine closely the already known views of the prospective nominee. Woodrow Wilson apparently overlooked some of the doctrinal conservatism of McReynolds. Eisenhower engaged in too much wishful thinking in his selection of Earl Warren. Ike was also in debt to Warren politically for his help at the last stages of the 1952 presidential nomination battle. Warren had not done a lot, but what he did do, Eisenhower greatly appreciated. It is also probably true that Ike didn't anticipate the kinds of new issues that were to come before the Court in the Warren era. Nor did Eisenhower delve into Warren's attitudes and try to fathom some idea of how Warren was likely to make decisions once he became chief justice. On occasion, such as with Hoover's selection of Benjamin Cardozo, a president knew in advance there would be differences. More important, however, are the institutional reasons and professional incentives that shape a justice's performance. The Constitution, precedent, the influence of legal research, the influence of rulings from the lower courts, life tenure, and the compelling facts of the case before them all influence justices.

The constitutional obligations of the court are different from those of being a friend or adviser to a president. The Court is not supposed to be a rubber stamp for anyone, and most of the justices come to view themselves as virtually sovereign in their own right. The Court, moreover, decides when violations of the law, including violations by the executive branch, occur. The Court further is charged with keeping the president and others in the executive branch in their constitutional place. The *Marybury v. Madison* (1803), *Ex parte Milligan* (1866), *Youngstown Sheet and Tube Co. v. Sawyer* (1952), and *U.S. v. Nixon* (1974) tradition is important to the Court and its new justices become defenders of this tradition.[15]

Justices are clearly affected by the Court's long-standing inclination to honor precedent except in unusual situations. The philosophy, interest, and training of the vast majority in the legal profession tend toward conservatism. Concern with precedent and respect for the past and for authorities, existing customs, vested rights and usages, and established relations characterizes the method of thinking in the legal profession. "No lawyer sufficiently devoted to the law to know our existing rules, the history of them, and the justification for them, will depart from them lightly," wrote Robert H. Jackson, shortly before FDR appointed him to the Court. "The contribution of legal philosophy to the balance of social forces will always be on the conservative side."[16] The job of the justices, most would agree, is not to promote the president's powers or policy initiatives, but rather to ensure that the political trends of the day don't overwhelm our enduring values of due process, liberty, and equality. As one legal scholar put it, the Supreme Court's "most important function is to provide an independent review of important governmental actions that affect citizens' constitutional liberties." If the justices "become political clones of the incumbent executive and legislative majorities, they will provide a less-effective check on the excesses of the current majority."[17]

Life tenure adds an incentive to think, act, and decide independently. The framers knew that what Alexander Hamilton called "the least dangerous branch" would need to be at least somewhat free from the influences of ambition and interest if they were to perform their responsibility. Life tenure gives justices "high honor, high responsibility, and guaranteed tenure and salary, so that they need neither seek higher office nor worry about retaining the one they have. These conditions, at once emancipating and greatly demanding, result in judicial behavior which may not conform either to presidential expectations or to the views that the justices expressed before joining the Court."[18]

Once a justice is on the Court, a president cannot effectively threaten, intimidate, or retaliate against a wayward or hostile justice. There may be times and circumstances when presidential "jawboning" would sway a decision but, as the following examples suggest, they would be the exception to the rule. Jefferson tried in vain to have a justice impeached. FDR tried to reorganize or pack the Court. Nixon, in a crucial case involving him, hinted he might not comply with the Court's decision. None of these presidential intrigues worked. At worst, a justice might be excluded from social functions at the White House.

Justices are also influenced by their colleagues. John Marshall unquestionably influenced some of the Republican appointees of Jefferson and Madison. Earl Warren was influenced by William O. Douglas and Hugo Black, who had been on the Court for over fifteen years. Frankfurter also tried to influence Warren, but with less success. Two of Nixon's appointees, Blackmun and Powell, changed or "developed" their view under the influence of the holdovers from the Warren-era Court. Students of national politics also find that justices deviate more from the positions favored by the appointing president in their later years on the Court. As issues change, so also does the president's influence. "While both presidents and judicial nominees may know the current constitutional issues of importance," Justice Rehnquist notes, "neither of them are usually vouchsafed with the foresight to see what the great issues of 10 or 15 years hence are to be."[19]

In short, there is, as there should be, a certain amount of unpredictability in the judicial appointment process. Presidents can never be certain their nominees will be approved by the Senate. As noted, the Senate has rejected almost one of every five nominees. Still others have declined the honor when presidents have offered the position. Nor can a president ever be sure a justice will not change when he or she dons the judicial robes. A Yale law professor put it well: "You shoot an arrow into a far distant future when you appoint a Justice and not the man himself can tell you what he will think about some of the problems that he will face."[20] It should be noted, too, that in perhaps a majority of the cases that reach the Supreme Court partisan ideology has little effect on the Court's ultimate decision.

PRESIDENT LINCOLN AND THE COURT

Lincoln did not believe in a dominant role for the American presidency, at least not in normal times. As a Whig Party member in earlier years, he believed Congress should make the law, and he believed in a strong cabinet. He also thought presidents should rarely use the veto power.

Lincoln, however, came to office in extraordinary times. In one sense, he owed his election in 1860 to Chief Justice Roger B. Taney. For the Taney Court's controversial 1857 *Dred Scott* decision split the Democratic

Party into northern and southern wings over the issue of slavery. Lincoln rose to prominence because he opposed the Taney Court decision, and he won election because of the Democratic schism.[21]

Lincoln's expansion of presidential powers came because of the wartime emergencies he faced soon after he came to office. He would defer to Congress and his cabinet on domestic matters but would use his powers as commander-in-chief to the fullest. Indeed, he would push them beyond the boundaries the country had hitherto known.

Fellow members of Lincoln's Republican party were determined to reform the Supreme Court as Lincoln came to office. Some wanted to call a national convention to modify the power of the Court through a constitutional amendment. Others pushed a "court-packing" scheme that would have enlarged the Court to thirteen members. Still others proposed even more radical remedies.

Lincoln chose a more elusive and generally a politically successful approach. He would do what he deemed necessary to fight and win the Civil War and he would, whenever possible, try to prevent the Court's interference by denying it an opportunity to become involved in controversial cases. Lincoln exercised broad constitutional power, and in nearly every case he sought congressional approval only after he acted unilaterally. With Congress and public opinion on his side, Lincoln was usually not challenged by the Court.

Yet Lincoln clashed with the Court on some occasions, and the federal judiciary ruled against him in several cases (though many of these came after his assassination). His wartime leadership expanded the scope of executive power and tested the Court's ability to restrain a president in times of war or crisis.

Faced with the secession crisis in 1861, Lincoln, with Congress out of session, proclaimed a blockade of southern ports and ordered the navy to enforce the blockade. The blockade-running shipowners whose ships were seized went to court and challenged the legality of Lincoln's blockade. The owners claimed that under international law a blockade was legal only in a declared state of war. Yet no such declaration existed. Hence, they claimed, Lincoln's blockade was illegal and unconstitutional until July 1861, when Congress sanctioned the war.

Crucial constitutional questions were raised or raised anew as this case came before the Court. Was the blockade valid without a congressional declaration of war? Can a president commit the nation to military action without congressional approval? Are undeclared wars sanctioned by the U.S. Constitution?

The Court ruled (5–4) that a state of war had existed at the time of the blockade and its institution, thus justifying a resort to these means of subduing a "hostile force." Lincoln "won," yet only by a narrow margin. Still, the case established significant constitutional precedent. A president may determine the existence of an emergency and he can take whatever measures are necessary to meet it. A president cannot declare war, yet as commander-in-chief he may commit the nation to military action, especially in an ongoing domestic crisis. The president, at least in certain cases, can wage a "defensive war."

Had this case been brought up soon after the event itself, Lincoln might not have won. The Court delayed hearing the blockade cases for a year. By the time of the decision Lincoln had made three appointments to the Court; they constituted three of the five pro-Lincoln or pro-presidential power votes. Chief Justice Taney voted against Lincoln's position. Lincoln's initial war-making acts were thus held valid, the blockade was legitimated, the condemnation of the seized ships was sustained, and a convenient dual theory of the Civil War was outlined: it was both an insurrection, a hostility within a nation, and a hostility between independent nations.

During this time Lincoln created a national army out of state militias, calling up 75,000 volunteers for three months of military service, which the law authorized him to do. But he went beyond the law a short time later when he called up 42,000 volunteers to serve for three years in the army, thereby doubling the army's size and necessitating major spending for unauthorized projects. Congress later gave him retroactive approval, yet he was exercising his powers to their fullest. He also suspended the right of habeas corpus (a court order stipulated in the Constitution directing an official having a person in custody to produce the person in court and explain to the judge why he or she is being held).

The central issue in another clash that Lincoln had with Chief Justice Taney rested on whether the Constitution expanded in wartime to provide enlarged presidential discretion and authority. Lincoln insisted necessity forced him to take whatever measures were needed to subdue the enemy. On this and similar occasions, Lincoln believed measures otherwise unconstitutional might become lawful by being indispensable for the preservation of the Constitution through the preservation of the nation. Usually, however, Lincoln insisted the Constitution did provide the necessary authority for him to act as he had acted.

The Constitution, Lincoln believed, was nothing without the nation. National survival was his prime duty. Preserving all the rights and liberties provided for in the Constitution was a secondary duty, at least for the time being.

Chief Justice Taney disagreed. He wrote the president saying he had no constitutional power to suspend the writ of habeas corpus. A president certainly "does not faithfully execute the laws if he takes upon himself legislative power, by suspending the writ of habeas corpus, and the judicial power also by arresting and imprisoning a person (Merryman) without due process of law."[22] Taney lectured Lincoln that Merryman should be set free and that the President should reconsider his misuse of powers belonging to Congress.

Lincoln ignored Taney's judgment, although later in a special message to Congress Lincoln asked: "Are all the laws but one to go unexecuted and the government itself go to pieces lest that one be violated?"—which is a classic validation of executive power beyond the law for the alleged needs of national security. Taney's verdict wasn't a Supreme Court decision, it came in a directive from Taney acting as a circuit court judge and thus had less binding force than had it come as a Court decision. More important, in this case Lincoln enjoyed public opinion support and popular acquiescence for his extension of executive power. Even Taney conceded he could not match the president's power.

Lincoln, through his secretary of state, instituted censorship and military arrest and trials, and continued to ignore or even suspend habeas corpus requirements in cases of persons accused of aiding the Confederacy. Throughout this period Lincoln resisted the federal courts, and Taney in particular, on the ground a president's duty to obey judicial rulings may be suspended, in special circumstances, by the higher responsibility to "preserve, protect, and defend the Constitution." One part of the Constitution may be disregarded temporarily, he implied, in order to save the government as a whole.

What is the lesson to be drawn from this clash between Taney and Lincoln? At first glance it was simply a question of Taney as Supreme Court justice striving to preserve respect for the law. Lincoln as president was determined to preserve the Union. Constitutional historians draw additional verdicts. Clinton Rossiter comes close to saying that if a president has public support, he can virtually disregard both the Constitution and federal courts.

> The one great precedent [here] is what Lincoln did, not what Taney said. Future Presidents will know where to look for historical support. So long as public opinion sustains the President, as a sufficient amount of it sustained Lincoln in his shadowy tilt with Taney and throughout the rest of the war, he has nothing to fear from the displeasure of the courts. . . .[23]

Others concluded that Lincoln's sweeping assertions of authority revealed an inability of Congress and the Court to curb a dynamic president determined to act decisively in times of emergency. His leadership stunned even some of those who sympathized with the northern cause. Thus, the abolitionist Wendell Phillips called Lincoln an "unlimited despot." Lincoln may have lacked mass backing for certain of his war policies and civil liberties violations, yet he could count on the popular acquiescence that proved to sustain his leadership.

Taney died in October of 1864, just before Lincoln won reelection to the White House. Two years later, Taney, who while in office could never curb Lincoln's sweeping definition of emergency powers, won at least a symbolic victory in the Supreme Court in the case of *Ex-parte Milligan* (1866). It vindicated the Taney judgment in *Ex-parte Merryman* (1861). The Court, with Lincoln dead and at least a couple of Lincoln appointees now voting with the majority, ruled Lincoln's use of military courts outside the war zone unconstitutional.

As long as civil courts are open and operating, the Court said, an accused is entitled to a civil trial by jury. Congress had not clearly authorized the imposition of martial law. Thus Lincoln and the executive branch had acted unconstitutionally.

Note, however, that this after the fact "slap-on-the-wrist" decision came a full year or so after the Civil War was over. It had not restrained or affected Lincoln or his presidential performance. Many analysts point to this decision as typical of the timidity of the Court or its inability to act as a reliable check on a president during wartime. If the Court was unable or unwilling to restrain a prudent leader like Lincoln, will it be able to restrain a determined, guileful, and ambitious future president? Will future presidents learn from the Lincoln lesson that emergency actions in the gray area of the Constitution are unlikely to be met with Court resistance, at least not during the emergency or during their lifetime?

Lincoln scholar Richard N. Current disputes the suggestion that Lincoln was a despot or an American Cromwell. Lincoln, he writes, continued to be sensitive to the constitutional limits on executive action. Not only did he retain his Whiggish habits of deferring rather than dictating to Congress, he almost never used the veto and he continued to consult widely. Lincoln would sometimes even acknowledge he had little legal basis for what he had done, and he'd ask Congress for appropriate authority.

> Never did Lincoln show any hankering for the perquisites or trappings of exalted power. If absolute power corrupts absolutely, he was absolutely safe from corruption. As a young man, he warned against future politicians of a dictatorial bent. . . .
>
> Such, in actual life, were the concerns of the Lincoln whose image, more or less distorted, later presidents were to exploit. When, regardless of their party, they represented themselves as true followers, they were indeed doing as he had done, for he had said again and again that he was the true follower of the honored dead of both parties of the past. But when his successors asserted the broadest presidential power and took him as their prototype, they were making a very dubious use of history and biography. . . . Neither by word nor by deed did he give any real justification for the idea of "executive privilege" or of an "imperial president."[24]

FRANKLIN D. ROOSEVELT AND THE COURT

Like most strong presidents, Franklin D. Roosevelt had a series of clashes with the Court. Like Lincoln, FDR came to office during an emergency, and the times demanded extraordinary leadership. Roosevelt was more than happy to provide it. The Supreme Court, on the other hand, would not always approve his interpretation of what needed to be done.

Roosevelt's New Deal emergency relief and recovery programs had their presumed constitutional basis in emergency executive powers or in the power of Congress to provide for the general welfare and to regulate

interstate commerce. The combined efforts of FDR and a generally cooperative Congress became the most far-reaching assertion to date of national emergency leadership.

The Supreme Court of the early Roosevelt years was hardly a body
receptive to FDR's spacious uses of executive and federal power. Most of
its members were appointees from the Harding, Coolidge, and Hoover
years.

Roosevelt's first two years passed without notable clashes with the
Court. This changed in 1935. Between early 1935 and mid-1936 the Court
ruled against the New Deal in several decisions. Among other actions, it
held unconstitutional important parts of the National Industrial Recovery
Act, the Railroad Retirement Act, a federal farm mortgage relief act, the
Agricultural Adjustment Act, and the Bituminous Coal Conservation Act.
The Court also told Roosevelt he was without inherent power, which FDR
had claimed, to remove a Federal Trade commissioner.

The Supreme Court was not the only part of the federal judiciary hostile to Roosevelt. Hundreds of injunctions restraining officers of the federal
government from carrying out various congressional acts were granted by
federal district judges.

To Roosevelt and his advisers the Court's actions amounted to a reckless, partisan, and irresponsible claim of judicial supremacy. The Court
had virtually nullified vital, core chunks of the New Deal recovery effort.
Robert H. Jackson, a top FDR Justice Department official, who would later
become a Supreme Court member, describes the Court's attack on the
New Deal:

> In striking at New Deal laws, the Court allowed its language to run riot.
> It attempted to engraft its own nineteenth century laissez faire philosophy
> upon a Constitution intended by its founders to endure for ages. . . .
> The Court not merely challenged the policies of the New Deal but
> erected judicial barriers to the reasonable exercise of legislative powers,
> both state and national, to meet the urgent needs of a twentieth-century
> community.[25]

Then came 1936. Roosevelt won by a landslide. It was a defeat for the
Court almost as much as it was a Roosevelt victory. Tensions were high.
Senator Burton K. Wheeler proposed a constitutional amendment that
would allow Congress to overrule a Supreme Court decision merely by
reenacting it by a two-thirds vote in both houses. Neither FDR nor his
advisers wanted to go that far, yet they warned that some means had to be
found to adapt the legal system and judicial interpretation to contemporary
national needs.

On February 5, 1937, a frustrated Roosevelt, emboldened no doubt by
his impressive reelection victory, introduced his famous reorganization or
"court-packing" plan to Congress. Jefferson, Jackson, and Theodore Roosevelt all altered or threatened to alter the size of the Court, thus Roosevelt's actions were not entirely unprecedented. Yet the scope of his pro-

posal (from nine to fifteen) was an almost revolutionary court-packing effort.

FDR's plan recommended that if a federal judge (Supreme Court or lower courts) who had served at least ten years waited more than six months after his seventieth birthday to resign or retire, a president could add a new judge to the bench. A president, in fact, could appoint as many as six new justices to the Supreme Court and forty-four new judges to the lower federal courts.

Roosevelt's objective was obvious. Although claiming he sought efficiency and elimination of crowded backlogs facing the courts, everyone realized he was seeking to gain quick control of the Supreme Court. He told friends that what the Court needed was some Roosevelt appointments. "Then we might get some good decisions out of them." He had had no opportunity to appoint a justice in his first term. He believed he had no choice but to act now, and to act boldly.

The plan stirred angry protests. Even Roosevelt Democrats worried it might destroy the independence of the judiciary. Several Roosevelt confidants such as James Farley, Harry Hopkins, and Rexford Tugwell disagreed with FDR's strategy. The less than candid reasons he gave for the plan damaged his credibility. Chief Justice Charles Evans Hughes blasted the Roosevelt scheme, saying that adding members to the Court would actually hinder rather than promote efficiency in Supreme Court operations. Most all of the members of the then sitting Court resented FDR's attempt to make the Court more submissive to his political will. Justice Louis Brandeis, the Court's eldest member, was irked by the arbitrary and indiscriminate attack on age. Still the plan was given reasonably serious consideration in Congress—a Congress controlled by Democrats—until the Senate Judiciary Committee wrote a scathing report on the measure, which helped defeat it.[26]

Other reasons for the plan's defeat came as a result of changes by and on the Court itself. The Court began upholding as constitutional key New Deal measures that came before it. Two of the justices (Hughes and Roberts) shifted sides. Then in May of 1937 one of the conservatives announced his retirement. These developments gave FDR pretty much what he wanted, a likely 6–3 working majority. Other developments in Congress, also were at work, but the end result was a defeat for FDR's devious plan.

Scholars often say Roosevelt lost the battle yet won the war. Plainly his 1936 election victory combined with his subsequent campaign to alter the Court's size had an impact on the Court. "A switch in time," it was said "saved nine." The presidential power of appointment combined with congressional authority to determine the Court's size proved to be useful guns behind the door. Still this was one of the worst political defeats of the Roosevelt presidency.

In the end, however, FDR got what he wanted and won significant changes.

Roosevelt, denied the opportunity to make any appointments to the Supreme Court in his first term, was able before he was through to name eight justices and elevate Harlan Fiske Stone to the chief justiceship. The new Court, the Roosevelt Court as it was called, took so broad a view of the commerce and taxing powers that scholars speak of the Constitutional Revolution of 1937, for since that time the Court has not struck down a single significant piece of social legislation. Of all the many consequences of FDR's Court-packing endeavor, by far the most significant was the legitimization of the twentieth-century state.[27]

As with Lincoln, Franklin Roosevelt's most important ally in his conflicts with the Court was public opinion. Ultimately, the Court came to legitimize Roosevelt's more sweeping use of national power because he enjoyed the support of the public. Whether public opinion should sway the judgments of the judiciary is of course another matter, and one of the most profoundly debated aspects of the role of the Court and constitutional review in our democratic republic.

Roosevelt enjoyed even stronger public support for the foreign policy and wartime measures he initiated. Conservative Republican George Sutherland, a Harding appointee, shared Roosevelt's view of a strong president for foreign policy. Justice Sutherland said in the *U.S. v. Curtiss-Wright Export Corporation* case in 1936 that in foreign policy, "with its important, complicated, delicate and manifold problems, the president alone has the power to speak or listen as a representative of the nation." Congress had delegated broad powers in foreign affairs to the president and even extended the normal pattern. Sutherland called the president the sole organ of the nation, and came close to calling the president "the sovereign." It appeared from this that virtually any delegation of foreign policy power to a president would be upheld.

Political scientist David Gray Adler summarizes its impact:

> There can be little doubt that the opinion in *United States v. Curtiss-Wright Export Corp.* in 1936 has been the Court's principal contribution to the growth of executive power over foreign affairs. Its declaration that the president is the "sole organ of foreign affairs" is a powerful, albeit unfortunate, legacy of the case. Even when the sole-organ doctrine has not been invoked by name, its spirit, indeed its talismanic aura, has provided a common thread in a pattern of cases that has exalted presidential power above constitutional norms.[28]

Once again in 1942, in *Ex-parte Quirin,* the Supreme Court expanded presidential power in a period of wartime. The case involved the manner in which eight German saboteurs who had landed on American shores with explosives should be tried. Franklin Roosevelt ordered a secret military trial and closed the civil courts to them. Judicial scholar David J. Danelski explains this complex, fascinating, and tortured case well, but his conclusion is of primary concern here:

For the executive branch, the Saboteurs' Case was a constitutional and propaganda victory; it expanded Executive power, and it allayed public fears of subversion. For the Supreme Court, it was an institutional defeat. If there is any lesson to be learned from the case, it is that the Court should be wary of departing from its established rules and practices, even in times of national crisis, for at such times the Court is especially susceptible to co-optation by the executive.[29]

In 1944, another significant delegation of power was added to the list of precedents for presidential emergency leadership. The Court, in the *Korematsu v. U.S.* case (in which the Court approved the actions of the executive branch when in the early 1940s it forced West Coast Japanese-Americans to relocate to the Rocky Mountain region) gave the president foreign policy or national security powers even when these powers extended into the domestic arena.[30] A president's war powers were found to take precedence even over the constitutional rights of U.S. citizens on American soil. The Court based, as it did in the *Curtiss-Wright* decision, much of its decision on the reality of congressional approval of the president's action. Also, Roosevelt still had public opinion on his side. When a president has both Congress and the public supporting an initiative or a policy, nine out of ten times the Court is likely to legitimize it as constitutional.

Two later Presidents, Truman and Nixon, sometimes found themselves in the situation where neither Congress nor the public stood by them. The Court is more tempted to strike presidential actions down when a president stands alone. Such was Truman's fate in the 1952 *Youngstown Sheet and Tube Co. v. Sawyer* case.[31] Nixon suffered a series of defeats before the Court, including his impoundment policy, wiretapping, and the famous tapes case, *U.S. v. Nixon* (1974).

On balance, it is clear the Court will limit presidents only when they ignore congressional mandates, attempt to alter the structure of government, or seriously tread on individual liberties. (Even then the Court may find rationales for supporting extension of presidential power.) The Court becomes more bold than it otherwise might be when presidents lose the backing of the American public. Still, the judicial history of noninterference with the growth of presidential power contains just enough exceptions to make it unpredictable.

MOVING BEYOND THE LAW

Given the many roadblocks and veto points that clutter the president's path, it is not surprising that the more ambitious and goal-oriented presidents will get extremely frustrated as other actors block their way. An obstreperous Congress, demanding special interests, an uncooperative business community, an adversarial press, and others can at times seem to gang up on a president, preventing him from achieving his policy goals.

When faced with these multiple opposing forces, most presidents feel trapped. The choice may appear to be either to accept defeat or take bold action (always, the president believes, in the national interest). Making the complex separation of powers work is difficult under the best of circumstances, in normal times it may seem impossible to get the system moving. Thus, rather than accept defeat some presidents are tempted to cut corners, move beyond the law, stretch the constitutional limits a bit.

When all else fails, and it so often does, some presidents, knowing their future political success, not to mention their historical reputation, is at stake, simply cannot resist going beyond the law. If the choice is deadlock or illegality, some presidents will choose the latter. After all, presidents are convinced that what they want is for the best interests of the nation, so why let a corrupt Congress, an uninformed public, or a hostile press stand in the way of progress (self-defined by the president)?

Presidents not well grounded in the virtues of the American system may see the system itself as the enemy and thus feel justified in going beyond the law. Richard Nixon with Watergate[32] and Ronald Reagan with the Iran-Contra scandal are but the two most pronounced examples.[33]

Some presidents get away with it (e.g., Reagan); some do not (Nixon). But when the choice seems to be to stay within the law and fail (Ford, Carter) or go beyond the law and maybe you will succeed and maybe you won't get caught, the temptation is too great for some leaders to resist.

An attitude of arrogance may overtake the president and his top staff. "We know best" and "they are blocking progress" lead to the belief that the "slight" abuse of power is being done for the greater good. But such an attitude leads to the imperial presidency, and to further abuses of power.[34]

President Reagan was convinced that communism was an evil that had to be fought at all costs, that the Marxist government in Nicaragua was a serious and direct threat to the United States, that the Congress was soft on communism, and that public opinion, which opposed U.S. intervention, was uninformed in spite of herculean efforts by the administration. Therefore, Reagan was faced with the difficult choice of either accepting the will of Congress, the voice of the people, and the law or acting on what he believed to be in the national interest. He acted.

Putting aside for a moment the question of whether Reagan was correct about the threat Nicaragua posed, one thing is clear: the Reagan administration decided that the law was wrong and that they would not be bound by law. This was one of the chief reasons the framers insisted on checks and balances: to control abuses of power.

Is the president above the law?[35] No. Such a notion violates all precepts of the rule of law.[36] Yet are there certain prescribed circumstances when a president can go beyond the law? Are there times when the president may exceed his constitutional powers?

THE PRESIDENT'S EMERGENCY POWER

Is a president ever justified in stretching the Constitution? While the word *emergency* does not appear in the Constitution, there is some evidence to suggest that the founders may have envisioned the possibility of a president exercising "supraconstitutional powers" in time of national emergencies.[37]

During a crisis, the president often assumes extraconstitutional powers.[38] The branches that, under normal circumstances, are designed to check and balance a president will usually defer to the president in times of national emergency. The president's institutional position offers a vantage point from which he can more easily exert crisis leadership, and the Congress, Court, and public usually accept the president's judgments.

The idea that there is a different set of legal and constitutional standards for normal conditions than for emergency conditions raises some unsettling questions regarding democratic governments and constitutional systems. Can democratic regimes function in any but prosperous, peaceful circumstances? Or must the United States constantly rely on the strength of a despot or "constitutional dictatorship" to save it from disaster? Are constitutional governments incapable of meeting the demands of crisis?

In most instances, democratic political theorists have seen a need to accept a certain amount of authoritarian leadership in times of crisis. Locke called this executive "prerogative", Rossiter referred to a "constitutional dictatorship." To Rousseau it was an application of the "general will." In cases of emergency, many theorists suggest that democratic systems, to save themselves from destruction, must defer to the ways of totalitarian regimes. In other words, to preserve democracy one must abandon democracy.

For British theorist John Locke, in emergency situations the Crown retains the prerogative "power to act according to discretion for the public good, without the prescription of the law and sometimes even against it."[39] While this prerogative could properly be exercised only for the "public good," one cannot escape the conclusion that this is shaky ground on which democratic governments and democratic theory must stand. And what if an executive acts wrongly? Here Locke was forced to abandon secular concerns: ". . . the people have no other remedy in this, as in all other cases where they have to judge on earth, but to appeal to Heaven."[40]

Rousseau, writing in the *Social Contract,* noted:

> The inflexibility of the laws, which prevents them from adapting themselves to circumstances, may, in certain cases, render them disastrous, and make them bring about a time of crisis, the ruin of the State.
> . . . If . . . the peril is of such a kind that the paraphernalia of the laws are an obstacle to their preservation, the method is to nominate a supreme ruler, who shall silence all the laws and suspend for a moment the sovereign authority. In such a case, there is no doubt about the gen-

eral will, and it is clear that the people's first intention is that the State shall not perish.[41]

The chief difference between Locke and Rousseau on this matter rests in Rousseau's refusal to rely on an "appeal to heaven" in cases of possible abuse of the power of the "supreme magistracy." Rousseau said this power must have strictly enforced time limitations as a protection against a complete takeover of the system. Yet once extraordinary power is assumed by an individual ruler, who can be sure that such time limits can and will be adhered to?

John Stuart Mill, a defender of representative government, made this comment: "I am far from condemning, in cases of extreme necessity, the assumption of absolute power in the form of a temporary dictatorship."[42]

Political scientists J. Malcolm Smith and Cornelius P. Cotter summed up this problem in the writings of democratic theorists when they noted that ". . . democratic political theorists tacitly admit the existence of a fatal defect in any system of constitutional democracy: Its processes are inadequate to confront and overcome emergency."[43]

Niccolo Machiavelli also addressed the problem when he wrote:

> . . . in a well-ordered republic it should never be necessary to resort to extra-constitutional measures; for although they may for the time be beneficial, yet the precedent is pernicious, for if the practice is once established of disregarding the laws for good objects, they will in a little while be disregarded under that pretext for evil purposes. Thus no republic will ever be perfect if she has not by law provided for everything, having a remedy for every emergency, and fixed rules for applying it.[44]

Political historian Clinton Rossiter's *Constitutional Dictatorship* is a twentieth century discourse on the same problem.[45] The "constitutional dictatorship" is an admission of the weakness of democratic theory, or of its failure to cover the full range of governing requirements. To save democracy, we escape from it. To protect democracy, we reject it. Democratic theory, then, according to the democratic theorists, is incomplete; in cases of emergency, one is called on to reject democracy for the more expedient ways of the dictator. In this manner, democratic theory may open the door to a strong, power-aggrandizing executive.

Nowhere in the Constitution is it specified that a president will have additional powers in times of crisis or emergency. History, however, has shown us that in times of national emergency the powers of a president have greatly expanded, and while former Justice Abe Fortas writes that, "Under the Constitution, the president has no implied powers which enable him to make or disregard laws,"[46] under the microscope of political reality we can see that this is precisely what American presidents have done.

The consequence of this view of an enlarged reservoir of presidential power in emergencies was characterized by constitutional scholar Edward S. Corwin as "constitutional relativity."[47] By this Corwin envisioned a constitution broad and flexible enough to meet the needs of an emergency situation as defined and measured by the Constitution. The Constitution can be adapted, in short, to meet the needs of the times. If the times call for quasi-dictatorial action by the executive, the Court could find this acceptable.

The dilemma of emergency situations in democratic systems is not easily answered. If the potential power of the state is used too little or too late, the democratic state faces the possibility of destruction. If used arbitrarily and capriciously, the system could accept a form of permanent dictatorship. In a modern sense, the constant reliance on the executive to solve the many "emergencies" (self-defined by the executive) facing America could very well lead to the acceptance of the overly powerful executive and make the meaning of the term *emergency* shallow and susceptible to executive manipulation.

With each new "emergency" in American history, the public and our political institutions seem to have become more accustomed to accepting an enlarged definition of presidential power to meet it. In the twentieth century, it seems hardly a debatable point to say that a president is expected to assume responsibility and added powers to meet both domestic and foreign crises.

The Court under Rossiter's constitutional dictatorship will generally recognize the need for the president to have inflated powers with which to deal with the impending crisis, and will allow for a "flexible" interpretation of the powers of the president. For as Rossiter noted, "In the last resort, it is always the executive branch in the government which possesses and wields the extraordinary powers of self-preservation of any democratic, constitutional state."[48]

Under Rossiter's theory, the Court recognizes the emergency and allows the president to assume additional powers. But the constitutional dictator must recognize the limits of his responsibilities. Franklin D. Roosevelt, after claiming/requesting of Congress a grant of an unusually large amount of power in 1942, assured the legislature that: "When the war is won, the powers under which I act automatically revert to the people—to whom they belong."[49] The executive, in short, is to return the extraordinary powers he was granted during the crisis to their rightful place. But serious questions remain as to (1) whether presidents have in fact returned this power, and (2) whether, even if the president desired to do so, a complete or even reasonable return to normalcy is possible after the dictatorial or quasi-dictatorial power is placed in the hands of one person. In sum, can a democracy survive without a strong executive, and conversely, can a democracy exist with one?

PRESIDENTIAL ACTION IN TIMES OF EMERGENCY

In practice a president's emergency power has indeed been great in comparison to powers under normal circumstances. When faced with a crisis situation, presidents have made exaggerated claims of power, have acted on these claims, and generally have gotten away with these excessive, and often extralegal, uses of power. History provides us with clear examples of the enormous power of a president in an emergency situation.[50]

Presidents on occasion act with little regard for the wishes and dictates of the other branches. The necessity for quick, decisive, often extraconstitutional actions, which the crisis may demand, places a heavy burden on the president. Being the only leader able to move quickly, the president must shoulder the burden of meeting the crisis. According to Richard Longaker, "In time of crisis constitutional limitations bend to other needs."[51]

For the crisis presidency to be seen as legitimate, (1) the president must face a genuine and a widely recognized emergency; (2) Congress and the public must, more or less, accept that the president should exercise supraconstitutional powers; (3) Congress may, if it chooses, override presidential acts; (4) the president's acts must be public so as to allow Congress and the public to judge them; (5) there must be no suspension of the next election; and (6) the president should consult with Congress where possible. Lincoln and Roosevelt met (more or less) all of these requirements, Nixon and Reagan virtually none.

Even when the above requirements are met, however, one should not be sanguine about presidential usurpations of power. As the Supreme Court reminded us in *Ex-parte Milligan* (1866): "wicked men, ambitious of power, with a hatred of liberty and contempt of law, may fill the place once occupied by Washington and Lincoln."

COURT DECISIONS AND PRESIDENTIAL POWER

Five types of Supreme Court decisions are possible when presidential powers come into question (Table 8.5). Most of the time, as noted earlier, the Court approves or *expands* presidential authority. It often also *legitimizes* presidential power. The Court can also *avoid* or duck the question on the grounds it is a political matter to be settled by Congress and the president. On rare occasions a *two-sided decision* is possible, when the Court may restrict an individual president but add to the power of the office. The Nixon tapes decision in 1974 is illustrative of a two-sided ruling. The Court ordered Nixon to yield his tapes. The Court also, and for the first time, recognized executive privilege as having constitutional standing. Sometimes, of course, in a clash of views the Supreme Court *restricts* or curbs a president and presidential powers.

How often do we see the Supreme Court handing down these kinds

TABLE 8.5
Types of Court Decisions Regarding Presidential Power

Type of decision	Definition	Example
Expanding	Decision adding power to presidency	*U.S. v. Curtiss-Wright Export Corp.* (1936). Recognizes it is necessary for presidents to have more power in foreign than in domestic affairs.
Legitimizing	Decision giving Court approval for presidential activities that were questioned	*Korematsu v. U.S.* (1944). Approved FDR and executive powers to intern Japanese-American citizens in World War II.
Avoiding	Decisions the Court decided "not to decide"; avoids getting involved	*Massachusetts v. Laird* (1970). Denied to hear a case that questioned the president's broad power in the Vietnam War, thus avoiding a decision on the war.
Two-Sided	Decisions going against a president yet adding power to institution of the presidency	*U.S. v. Nixon* (1974). Nixon told to yield tapes, yet court recognizes "executive privilege" as valid in serious national security situations.
Restricting	Decisions curbing or even diminishing presidential power	*Youngstown Sheet and Tube Co. v. Sawyer* (1952). Truman and his Secretary of Commerce told they had exceeded their power in seizing the nation's steel mills to prevent a strike. Truman based his action on general powers of his office. Court held he could take no such action without express authorization from Congress.

of decisions? The two-sided rulings are rare. The Court, over its history, has often avoided questions affecting presidential power. Justices, for example, rarely want to deal with questions about the legality of a war such as in Vietnam.

When sensitive issues of national security arise the Court often invokes what is commonly called the doctrine of "political questions" to avoid head-on collisions with the president. The doctrine rests on the separation of powers theory, which holds that the Supreme Court exercises judicial power and leaves policy or political questions to the president and Congress. Chief Justice John Marshall invented the "political question" doctrine as early as 1803 when he said that matters in their nature political are not for this Court to delve into. He affirmed this later, as in 1829 when he refused to rule on a boundary dispute between the United States and Spain. The judiciary shall not, said Marshall, decide foreign policy. Questions such as foreign boundaries, he added, are more political than legal.

Expanding and legitimizing decisions are common. They are several times more likely than restricting decisions. Nearly every analyst of Court-presidency relations emphasizes the Court's role in the expansion of presidential powers. Most of the time the Supreme Court has supported the vigorous actions of our strong presidents. One scholar emphasizing this view in 1957 concluded that "in every major constitutional crisis between the executive and the judiciary, the president has emerged the victor." He also said the judiciary can neither force the president to do anything, "nor prevent him from doing anything he may decide to do."[52] Although his sweeping verdict exaggerated the case, many students of presidential-judicial relations modify this judgment only slightly today.[53]

The Supreme Court has been and will very likely continue to be a supportive ally in most potential showdowns, especially in the case of emergency or national security contexts.

PRO-PRESIDENCY COURT RULINGS: EXPANDING OR LEGITIMIZING PRESIDENTIAL POWER

Many of the cases cited here are cases in which the highest court in our nation essentially gave or approved of power that had been previously undefined or undetermined. The issue at stake is whether presidents are able to cite a law or precise wording in the Constitution, or whether their broad and vague "executive power" allow them to conduct certain activities.

The Prize Cases (1863) were not the first dealing with presidential power to come before the Court, yet they are the first dealing with extraordinary, independent actions taken during a crisis. The Supreme Court ruled five to four that Lincoln could wage war with the South, under his authority as commander-in-chief, without congressional declaration. Presidential discretion was deemed sufficient for exercising this power. Indeed it was deemed to be within the broad grant of executive power found in Article II of the Constitution. The maxim that "There are two Constitutions, one for peace, the other for war" has its roots in these Civil War decisions.[54]

Before 1863, under international law ships could be legally taken as prizes only when a conflict had been recognized as a war between two belligerent powers. Thus, if the Supreme Court said the South did not have belligerent status, it appeared the justices would have had to rule the blockade as illegal. They decided instead to give primacy to executive discretion.

Another pro-presidency landmark case, *In re Neagle* (1890),[55] dealt not with foreign affairs but rather with the domestic task of effectively carrying out the functions of government. David Neagle, deputy U.S. marshal in the San Francisco region, was assigned to travel with Supreme Court Justice Stephen J. Field. Field's life had been threatened by a Californian, whom Field had sentenced to prison. The U.S. attorney general assigned Neagle to help protect Field. In implementing his assignment Neagle had to shoot, fatally it turned out, the man who in fact did try to attack Field.

The central question involved in this case was whether a president has either the constitutional or the prerogative sources of power to execute orders of the kind that assigned Neagle to his duties. Is a president limited to enforcing only acts of Congress?

Here again the Supreme Court ruled the president had been granted broad executive powers that he must be able to use at his discretion to administer properly the laws of the land. The Court openly acknowledged it was interpreting the word *law* in a liberal manner, the only conceivable manner that would serve the interests of justice in this instance. The president's executive order was clearly constitutional under the "faithfully execute" clause of the Constitution. The president's duty to fulfill this clause is consequently not limited to the enforcement of acts of Congress, according to their express terms, "but includes also the rights, duties and obligations growing out of the Constitution itself, our international relations, and all the protection implied by the nature of the government under the Constitution."

This decision strengthened and expanded the powers of the presidency. This broad interpretation was underlined a few years later in a subsequent Court case that legitimized President Grover Cleveland's sending troops into the Chicago area to deal with a railroad strike. Workers at the Pullman railroad company in 1894 had gone out on strike over certain wage reductions. The American Railway Union carried out a secondary boycott against the Pullman company, a boycott that eventually threatened violence. A federal court issued an order seeking to halt the boycott on the grounds the strike crippled interstate commerce. Union President Eugene V. Debs and his aides ignored the court order, were arrested, and were convicted of contempt and sentenced to prison. President Grover Cleveland had dispatched federal troops to Chicago and ordered the U.S. attorney in Chicago to halt the strike. Debs not only ignored these federal initiatives but petitioned the Supreme Court for a writ of habeas corpus, challenging his detention as illegal. Writing for the Court in this famous *in re Debs* (1895) case, Justice David J. Brewer affirmed sweeping executive emergency powers. His pro-presidency opinion:

> While it is not the province of the government to interfere in any matter
> of private controversy between individuals, or to use its great powers
> to enforce the rights of one against another, yet, whenever the wrongs

complained of are such as affect the public at large, and are in respect of matters which by the Constitution are entrusted to the care of the nation, and concerning which the nation owes the duty to all the citizens of securing to them their common rights, then the mere fact that the government has no pecuniary interest in the controversy is not sufficient to exclude it from the courts, or prevent it from taking measures therein to fully discharge those constitutional duties.[56]

In a 1926 landmark decision, *Myers v. United States,* the Court in a 6–3 decision granted a president, and presumably the presidency, with further administrative power, authorizing presidents to remove governmental officials without obtaining the consent of Congress. Frank Myers, postmaster for Portland, Oregon, was removed by President Woodrow Wilson without Wilson's securing Senate consent. Myers brought suit for salary for the remainder of his term. Back in 1876 Congress had passed a law providing for Senate participation in the removal of postmasters. Thus the question now, in the 1920s, was: May Congress limit a president's removal power?

Here is a rare instance in which the Supreme Court went against the wishes of Congress, at least as represented by the old statute under which Myers claimed to have his job protected. As we have seen, the Court usually defers to congressional judgment when Congress has specifically addressed the issue at hand, yet in this case the Court did not. The majority in this case believed a strict separation of powers is necessary if the executive is to function effectively. The 1876 law was simply overturned as unconstitutional, for as Chief Justice Taft put it in his majority opinion, it was found to violate Article II of the Constitution. Taft said a president's power of removal of executive officers appointed by the executive with the advice and consent of the Senate is full and complete without Senate consent. Taft went on to say the sole power of removal is essential to the president if he is to faithfully execute the laws.

Justices Louis Brandeis and Oliver Wendell Holmes, in dissent, made the same arguments the majority would make in a case nine years later that would greatly narrow the *Myers* decision. Brandeis said Congress was given the power to establish offices and tenure; it was not specifically denied the right to participate in removals, so it should be able to do so. Removal, to Brandeis, was a condition or qualification of tenure. He added an observation widely cited today: "The separation of the powers of government did not make each branch completely autonomous. It left each, in some measure, dependent upon the others, as it left to each power to exercise, in some respects, functions in their nature executive, legislative and judicial. . . . The doctrine of separation of powers was adopted by the Convention of 1787, not to promote efficiency but to preclude the exercise of arbitrary power. The purpose was not to avoid friction, but, by means of the inevitable friction incident to the distribution of the government powers among three departments, to save the people from autocracy. . . ."[57]

Perhaps the next major expansion of presidential power came in a case previously discussed in the section on Franklin Roosevelt. In the *United States v. Curtiss-Wright Export Corporation* (1936), the Court in a 7–1 decision chose to affirm and expound at length on the broad foreign powers held by the president. The executive, the Court said, was the sole spokesman of the nation, not of the Constitution, of the national government, or of Congress. To many observers this expansive decision came close to saying the president was "the sovereign." Congress could delegate, or so it seemed, virtually any foreign policy authority to the White House.[58]

A year later, in *United States v. Belmont* (1937), a near unanimous Court ruled the national executive has the sole right to enter into executive agreements with other nations. State laws or policies do not supersede presidentially arranged international agreements even if they are not precisely in treaty form requiring Senate approval. Implicitly, if not explicitly, the Court ruled executive agreements had the binding force of treaties. Although an international compact is not always a treaty, the federal government has sole claim to "external affairs" and the external powers of the United States are to be exercised without regard to state policies or laws.[59]

In 1944, an even more decisive delegation of authority was added to the long list of pro-presidency precedents. The Court, in the already noted *Korematsu* and related cases, gave the president national security powers even when those powers extended primarily to domestic public policy considerations. On February 19, 1942, President Roosevelt issued an executive order empowering the secretary of war to clear the three West Coast states and parts of Arizona of all persons of Japanese descent, 70,000 of whom were American citizens, and place them in detention centers.

Fred T. Korematsu, an American-born citizen of Japanese descent, violated the civilian exclusion order, a part of the implementation of FDR's general order, and was given a suspended sentence of five years of probation by the federal court. He appealed his case and the Court began to consider his and similar cases in May 1943.

Lawyers for Mr. Korematsu in 1943 (and in later years when his conviction was overturned) said Roosevelt in effect was using his commander-in-chief authority to condemn a race to imprisonment. No charge had been issued; no trial conducted. With his action, FDR called into question several constitutional rights for all American citizens: personal security, the right to move about freely, and the right to not be deprived of those rights except on an individual basis after trial by jury. The force of the Fifth Amendment that guaranteed equal treatment under the law and due process was weakened. The character of U.S. citizenship and the wartime powers of the military over citizenry were also called into question.

On December 18, 1944, the U.S. Supreme Court ruled against Korematsu, effectively legitimizing FDR's actions as constitutional. During the trial, no claim was made that Korematsu was a disloyal citizen. His "crime" consisted solely of refusing to leave the restricted West Coast region, in the state where he was a citizen near the house where he was born.

Justice Hugo Black, writing for the majority, said Korematsu was not excluded from the West Coast because of hostility to him or his race. He was excluded because we were at war with the Japanese Empire, and because the properly constituted military authorities feared an invasion of our West Coast and therefore felt constrained to take appropriate security measures. The presence of an "unascertained number of disloyal members" of the Japanese group deemed, he said, the exclusion necessary. Black noted that he and his colleagues were aware of the hardships imposed by these presidential orders, "But hardships are part of war, and war is an aggregation of hardships. Compulsory exclusion of large groups of citizens from their homes, except under circumstances of direct emergency and peril, is inconsistent with our basic governmental institutions. But when under conditions of modern warfare our shores are threatened by hostile forces, the power to protect must be commensurate with the threatened danger. . . ."[60]

Justices Frank Murphy, Robert Jackson, and Owen J. Roberts dissented. Justice Roberts said the indisputable facts in the case exhibited a clear violation of constitutional rights. Justice Murphy believed the decision placed the Court in "the ugly abyss of racism." Justice Jackson said the armed services must protect a society, not merely its Constitution. His dissent was particularly disturbing; in effect, he said the army actions were unconstitutional, yet he also appeared to acknowledge that the Court could not be expected to do much, if anything, about such actions in wartime.

Korematsu is an example of the Supreme Court's reluctance to question and its readiness to affirm a president's determination of necessity under the conditions of wartime. In dissent, Justice Murphy said the judicial test of whether the executive, on a plea of military necessity, can validly deprive an individual of his rights must be reasonably related to a public danger that is so immediate as not to admit of delay and not to permit the intervention of ordinary constitutional processes to alleviate the danger. He obviously thought the situation did not warrant the conviction.

Most scholars now regard the *Korematsu* decision as one of the Court's dismal blunders. Racism, hysteria, and misleading military judgments all were at work. One legal scholar concludes:

> Given the tensions of the period after Pearl Harbor, one might charitably advance the excuse of wartime hysteria for the harried members of Congress and the executive branch who made the initial decisions. No such excuse can be entertained for the justices of the Supreme Court who abandoned their most sacred principles at the first whiff of grapeshot.[61]

A federal study concluded in 1982 that the Roosevelt administration's military orders to remove, detain, and intern Japanese-Americans was unjustified by military necessity. Forty years later, Fred Korematsu's conviction was overturned by a federal district judge in San Francisco. Citing new evidence, the judge accused the government of knowingly withholding information from the courts when the judiciary was considering the critical question of military necessity.

Thus, the original decision in the Korematsu case, albeit long after the fact, now lies overruled in the court of history. Yet as Judge Marilyn Hall Patel warned in 1984, that case is a powerful reminder that "in times of distress, the shield of military necessity and national security must not be used to protect governmental actions from close scrutiny and accountability."[62] But don't count on it.

After a brief respite, including ducking nearly all the issues surrounding the Vietnam War, the Court in the recent generation has for the most part continued to add powers to the presidency.

One of the ironic expansions came about as a result of the *U.S. v. Nixon* case in 1974. The case involved an appeal made by the Nixon administration to vacate an order by federal district court Judge John Sirica requiring Nixon to release tapes and transcripts of sixty-four White House conversations that had been subpoenaed by Watergate Special Prosecutor Leon Jaworski. The executive branch claimed the president had the right under the doctrine or custom of "executive privilege" to withhold the tapes. It was essential, it claimed, for presidents to be able to speak freely with their advisers without fear that such conversations would be available for public consumption.

Although the Court, speaking through Chief Justice Warren Burger, held Nixon's claims erroneous, it also believed it was its duty to define executive privilege. The Court seemed to say national security considerations would weigh heavily in balancing executive privilege against a competing constitutional claim. In the case at hand, in the summer of 1974, national security was not being threatened, this was a criminal case, and hence Nixon had to yield the tapes. To be sure, then, the Court limited President Nixon's claims of executive privilege, yet only after acknowledging for the first time the constitutionality of executive privilege.

In *Goldwater v. Carter 444 U.S. 996 (1979),* the Supreme Court in effect approved President Jimmy Carter's termination of the 1954 Mutual Defense Treaty with Taiwan. U.S. Senator Barry Goldwater and certain others in Congress had challenged the president's authority to end such pacts, yet they never brought it to a vote. Most members of the Court considered this a matter that should be settled between the legislative and executive branches and that it thus was "not ripe for judicial review." In effect the Court said if Congress refrains from confronting the president on a matter such as this, it is not our task to do so. Only when there is an irreconcilable conflict between the two branches would the Court enter this kind of treaty dispute. The Supreme Court's order to a lower court to dismiss the challenge to Carter's actions was hailed by the Justice Department and presidentialists as a victory for the president's prerogatives in the foreign affairs field.

Doubtless Carter's action made sense in improving the evolving relations between the United States and the People's Republic of China. Indeed, the termination of the 1954 treaty was one of the conditions for the normalization of relations between the two nations. Yet the Rehnquist decision in this case is viewed by some analysts as another case of judicial obeisance to presidential dominance in foreign policymaking. Criticizing this Court decision as an act of timidity and a mistake, David G. Adler wrote:

> When a court invokes the political question principle in the face of very dubious presidential action, it causes tremors among those who favor government of limited powers. It is understandable that the Court, when it is confronted with important questions involving the separation of powers doctrine, will proceed cautiously, so as not to encroach upon the powers of one or the other departments. It is one thing for [the] Court to exercise judicious interpretation; it is quite another to shy behind the smokescreen of the political question doctrine when it prefers to evade a difficult problem.[63]

The Court deferred to the presidency again in mid-1981 when it upheld the executive agreements by Carter and Reagan that helped end the Iranian hostage crisis. Carter and later Reagan agreed to transfer large sums of Iranian funds and security out of the country in 1981 in exchange for release of the hostages. Yet an estimated 450 lawsuits were filed in U.S. courts by American companies laying claim to these assets as partial compensation for broken contracts and seized property. Justice William H. Rehnquist, speaking for a unanimous Court, emphasized the Court was making no sweeping decision in favor of the presidency in all such instances, only in the narrow set of circumstances involved in this case. "But where, as here, the settlement of claims has been determined to be a necessary incident to the resolution of a major foreign policy dispute between our country and another, and where, as here, we can conclude that Congress acquiesced with the President's action," Rehnquist wrote, "we are not prepared to say that the President lacks the power to settle such claims."[64]

A divided Supreme Court in June of 1982 ruled 5–4 that a president enjoys absolute immunity from civil damage suits for official actions exercised while on duty in the White House. Toward the end of 1995 President Clinton, however, found himself facing charges of sexual harassment. Paula Corbin Jones accused the president of improperly approaching her while Clinton was governor of Arkansas. She was a state employee at the time. In 1997, the Supreme Court ruled unanimously that a sitting president is not immune from civil suits of this type, and the Paula Jones case was allowed to proceed.

Back in 1968, a nettlesome "whistle-blowing" Pentagon employee, A. Ernest Fitzgerald, testified to a congressional committee about outrageous cost overruns on the C5A plane. Two years later he lost his job, ostensibly due to a reorganization to increase efficiency. Fitzgerald filed suit, claiming he had actually lost his job because of policy disagreements with senior political appointees in the Nixon administration. After lengthy litigation he regained his job in 1973. Later the Watergate tapes revealed President Nixon boasting having ordered "to get rid of that son of a bitch." Fitzgerald proceeded to sue Nixon, who was by now a former president.

The legal issue involved in *Nixon v. Fitzgerald*[65] was whether a president could be sued by individuals alleging violations of their legal or constitutional rights by the nation's chief executive in the performance of official presidential duties. The Court, in a close vote, decided presidents must be immune from civil suits. The general view of the majority was that although a president would of course remain subject to criminal prosecution for serious misdeeds or abuse of power, the threat of civil suits would or might impede a president's freedom of judgment or action. A president, their reasoning implies, is so important and in such sensitive positions he must be protected from frivolous and diversionary civil charges.

Speaking for the Court, Justice Lewis Powell said the president's unique status distinguished him from other executive officials. "We consider this immunity a functionally mandated incident of the President's unique office, rooted in the constitutional tradition of the separation of powers and supported by history." This pro-presidency decision, while confined to the case before the Court, stressed the need for presidents to enjoy wide latitude and freedom in the exercise of their responsibilities. Thus Powell said, "Because of the singular importance of the president's duties, diversion of his energies by concern with private lawsuits would raise unique risks to the effective functioning of government."

Powell was well aware of the charge the dissenters in this case were making, namely that this grant of power to the presidency virtually placed presidents above the law. Did a president need such sweeping protection from reasonable attempts to recover damages when wrongdoing was clear? Was the Supreme Court once again too friendly to the president and foolishly insulating the presidency from the normal pressures to curb abuses of executive power?

Apparently Powell did not agree, but Justice Byron White vigorously dissented, saying:

> I do not agree that if the office of the president is to operate effectively, the holder of that office must be permitted, without fear of liability and regardless of the function he is performing, deliberately to inflict injury on others by conduct that he knows violates the law. . . .
>
> The Court concludes that whatever the president does and however contrary to law he knows his conduct to be, he may, without fear of liability, injure Federal employees or any other person within or without the government.
>
> Attaching absolute immunity to the office of the president, rather than to the particular activities that the president might perform, places the president above the law. It is a reversion to the old notion that the king can do no wrong.[66]

In a sense, the *Fitzgerald* decision undercut the rule of law for the sake of executive power and greater capacity for bold presidential action. It was a Hamiltonian ruling and one that recalls Patrick Henry's fear stated in the Virginia ratifying convention, namely that the proposed constitution "squints in the direction of monarchy."

In 1983 the Supreme Court declared Congress could no longer exercise a legislative veto when it was displeased with the manner in which a president was carrying out a congressional mandate. In the 1960s and 1970s, Congress enacted scores of laws giving the executive branch qualified authority to act, subject to the later disapproval of one or both houses of Congress. These legislative vetoes (sometimes they were called congressional vetoes) were attached to measures on subjects ranging from presidential war powers, foreign aid, and arms sales to health and safety bills, energy regulation, the budget, and the economy. This was a definite departure even from the extensions of presidential power in the past, which were usually accompanied by congressional approval. *INS v. Chadha* (1983) is a decision based on an interpretation of the separation of powers doctrine and a literalist reading of the relevant provisions of the Constitution.[67]

In what started as a minor immigration case, the Court ruled, seven to two, that the House of Representatives exceeded its constitutional powers when, in exercising a legislative veto provision in the Immigration and Naturalization Act, it blocked the attorney general's decision to waive deportation for a Kenyan student who had overstayed his visa.

Chief Justice Burger, speaking for the Court, said these cumbersome legislative veto provisions violate principles of separation of powers and the system of checks and balances embodied in our Constitution. Legislative vetoes violate the constitutionally prescribed requirement that every piece of "legislation" must be approved by both chambers of Congress and then presented to the president for signature.

At the time of the decision most scholars of the Court viewed it as one more example in the Court's long tradition of increasing or at least protecting presidential powers. A disturbed Justice Byron White said this single decision struck down in one fell swoop more laws enacted by Congress than the Court had cumulatively invalidated in its history. White, in dissent, said the legislative veto was an important if not indispensable political invention allowing presidents and Congress to resolve their constitutional and policy differences. "The history of the legislative veto also makes clear that it has not been a sword with which Congress has struck out to aggrandize itself at the expense of the other branches. . . . Rather, the veto has been a means of defense, a reservation of ultimate authority necessary if Congress is to fulfill its designated role under Article I as the nation's lawmaker."[68] White concluded that the apparent sweep of the Court's decision was "regrettable."

Chief Justice Burger and a majority of his colleagues, however, expressed no such regrets. On the contrary they sensed Congress had gained or usurped constitutional powers.

Supporters of a strong presidency were delighted by the *Chadha* decision, saying it was about time something was done to curb this congressional encroachment on presidential power. They viewed the legislative veto, especially as it was used in the 1970s, as a naked grab for congressional supremacy that ultimately could transform the president into a weak prime minister. Supporters of Congress took a different view and faulted the Court for failing to understand the new realities and needs of modern day congressional-presidential relations. Congressionalists also predicted, perhaps correctly, that the Court's ruling in the *Chadha* case would encourage Congress to find new means to achieve the same general objectives served by the legislative veto. In fact, Congress has continued to use several variations of the legislative veto, and the verdict is by no means clear that the *Chadha* decision has increased the power of the executive. At least on paper, however, it was a traditional response in support of a strong presidency.[69]

In 1986, in *Bowsher v. Synar,* the Supreme Court struck down a provision of the Gramm-Rudman-Hollings law that would have allowed the comptroller general to determine budget cuts when the provisions of the law went into effect. Chief Justice Burger, writing for the Court, claimed such a provision gave Congress a power that was executive; and thereby gave to the president the power to determine from where the budget cuts would come.[70]

Despite the inconclusiveness of the *Chadha* outcome, the impression should by now be well established that the Supreme Court, in its decisions, typically favors the American president.

Although FDR suffered some initial setbacks, he eventually got most of what he wanted from the Court and indeed shaped, with his nine appointments, a decidedly pro-national government and pro-presidency Court. Nixon also suffered several setbacks. He lost a number of cases both in the Supreme Court and in lower federal courts. Since Watergate, however, Court rulings have gradually shifted back in the direction of supporting, legitimizing, or even expanding presidential powers. There are, to be sure, some exceptions. The next section reviews some notable Court decisions constraining presidential power.

PRESIDENTIAL LOSSES BEFORE THE SUPREME COURT

On this side of the ledger, the cases are less numerous. The Court has taken decisive stands against presidents only after the Civil War, during the first New Deal, briefly near the end of the Korean War, during the Watergate affair, and recently in the second Clinton term (the Paula Jones case and Whitewater executive privilege issues). Moreover, these anti-presidency stands were often tempered by the fact that they generally were not direct, independent, or confrontational challenges to a president. Thus the Court stood up to Lincoln only after his assassination. The Court backed off after its brief though significant foray against the early New Deal legislation, and reacted to Watergate only after the press, the public, and Congress were already "up in arms."[71] The Court, then, has not been a very secure protector of rights and freedoms during times of crisis.

Still, when a president went directly against provisions of the Constitution or an explicit congressional directive, or both, the Court was inclined to stop him from treading further. And the Court, understandably enough, is usually a bit bolder in these assertions when the president has suffered a notable decline in public approval.

Because we have treated some of these cases earlier, especially in the discussions of Lincoln and FDR, and because the major pattern in the Court-presidency relationship has been in expanding presidential power, we shall discuss these restraining decisions in briefer detail. Note, however, that these counterpoint decisions are no less important; they are merely the exceptions to the pattern.

The first significant case arose in Jefferson's term when Chief Justice John Marshall, Jefferson's antagonist, explicated the doctrine of judicial review. The *Marbury v. Madison* (1803)[72] decision, while cleverly avoiding a direct confrontation with the president, gave the Court the means for serving as the primary interpreter of what laws or executive action conform to the U.S. Constitution. In the process of handing down this decision John Marshall also said President Jefferson and his Secretary of State James Madison acted improperly. More important, however, was the establishment of the Court and its authority to serve as a vital branch if not exactly a co-equal branch with the other two branches of government.

A year later, the Court in a unanimous decision instructed executive branch officials to pay for damages in the seizure of a Danish vessel, a seizure the Court said took place because of improper or invalid presidential instructions. The *Little v. Bareme* case involved the following question: Whether a naval captain could be found civilly liable for following what the Court found to be an executive order that went beyond what Congress had authorized.[73]

Chief Justice John Marshall confessed that initially he believed that although President John Adams' instructions of 1799 could not give a right for the seizure, they might yet excuse the navy captain, Captain Little, from paying damages. He added, however, "I was mistaken." The instructions were invalid and therefore furnished no protection for the navy captain, who obeyed his president at his peril, because the congressional legislation in question authorized only the seizure of vessels proceeding to French ports. Executive instructions cannot, the Court ruled, ever change the nature of an administrative transaction or legalize an act that would have been a plain trespass. If the Congress had been silent on the matter, the president's general authority as commander-in-chief would probably have been sufficient, but once Congress had prescribed the manner in which the law was to be carried out the president and all other executive officials were obliged to respect the limitations imposed by congressional statute.

Another major case arose during a military emergency (just as nearly all limiting or presidency-constraining decisions arose during unusual circumstances), the Civil War, during which Lincoln had suspended the writ of habeas corpus. Chief Justice Taney, acting in his additional role as a circuit court judge, ruled Lincoln had no constitutional power to suspend the writ. Lincoln never complied with this order, and a major lesson was learned. Rarely again would the Court attempt to stop a presidential action while it was taking place. Presidents can seldom if ever be enjoined from taking an action, only reprimanded once they have taken it. Still, *Ex-parte Merryman* (1861) was important in that a Supreme Court justice recognized limits to presidential power, even in wartime.

In *Ex-parte Milligan* (1866), the Court did place clear limitations on the emergency powers of the president, limiting his military authority. Of course this happened after the fact, after the war was over and after Lincoln had died. And in the *Korematsu* case of 1944, when the danger to the Union was almost nonexistent, a different Court ignored this *Milligan* Civil War precedent and granted to the chief executive all the powers Lincoln had assumed and more. Thus, in many respects these two Civil War limitations can be said to be insignificant in the overall definition of presidential power, even as they are important because they were among the first weak attempts by the Supreme Court to recognize that individual rights deserve to be considered as important as a strong national executive.

Other clashes between the Court and the presidency arose again in emergency circumstances, during the height of the Depression. Once more, the Court believed the president lacked the constitutional authority to act alone in "saving the Union." Roosevelt's multiple plans to shore up the economy or regulate economic behavior met with Court disfavor. The Court's several rulings about the unconstitutionality of the early New Deal measures have already been discussed. Congress, in one sense, had erred by giving Roosevelt too general a mandate. Had FDR been carrying out specific congressional plans, he would have probably won the Court's approval in most instances (although it is doubtful Congress could have agreed on specific plans for him to administer). These anti–New Deal Court rulings represented a distinction the Supreme Court has always made between foreign and domestic affairs. In foreign affairs, a president hardly needs even a general mandate from Congress. But domestic and economic affairs have usually required more detailed statutes from Congress.

This maxim was proved true once again in the 1935 case of *Humphrey's Executor v. United States*.[74] In this case, a congressional prescription for removing a Federal Trade Commission commissioner was violated by Franklin Roosevelt. And the Court responded by narrowing and in some ways overturning the *Myers* decision it had made just nine years earlier. In this 1935 ruling, the Court acknowledged that Congress, in creating agencies to carry out judicial and legislative duties, could restrict a president's removal power in specified cases, and thus it overruled FDR. *Humphrey* became one of only a handful or so of the actual major precedents serving to limit presidential powers.[75]

Another case in which a president was limited came during what most people might consider less than a national crisis, although President Truman apparently thought otherwise. This was the *Youngstown Sheet and Tube Co. v. Sawyer* (1952) case. Sawyer was Truman's secretary of commerce who was instructed by Truman to seize and operate certain steel mills that otherwise would have been shut down by union strikes (or, depending on your outlook, by the failure of the steel mill companies to improve pay and benefits).

The Korean War was dragging on and on. The pending shutdown of the steel mills so concerned Truman that he appealed to both sides but to no avail. Steel supplies were low already and some Pentagon officials believed a major Chinese offensive was in the making. The United States Steel Corporation was willing to meet the United Steel Workers terms, but only if it could get some relief from the wartime price controls the Truman administration was exercising over the steel industry. Truman refused this. He also refused to invoke provisions of the Taft-Hartley Act, recently passed by Congress over his veto, that might have permitted yet another way out of the impasse.

With deadlock and a prolonged strike likely to occur, Truman faced a dilemma: shutdown of the mills for even a short time could seriously impair, he believed, the Korean supply effort and jeopardize the lives of American soldiers. On the other hand, this executive action was a bold initiative sure to arouse critics and possibly legal and constitutional objections.

Truman, however, was privately advised by his close friend and the man he had nominated as chief justice, Fred Vinson, to go ahead with the steel seizure. Vinson based his recommendations on legal grounds. Moreover, Truman placed considerable faith on a memorandum that Associate Justice Tom Clark, another of his appointees to the Court, had prepared for him while Clark was earlier serving as Truman's attorney general. Clark had informed the president he could use various inherent powers of the presidency to settle labor disputes of this type.

Truman had appointed four of the then sitting members of the Court. The rest were FDR appointments. Hence, Truman thought his steel seizure plan would win approval in the Court if it did come to a test.

In any event, Truman seized, through his secretary of commerce, all the steel mills affected by the strike. This accounted for eighty-six companies, over two hundred steel mills, 600,000 workers, and 95 percent of the nation's steel production.

The seizure was implemented without much administrative trouble. Executive order number 10340 was issued and all employees of these mills now worked for the U.S. government and did so at the same wages as before. Every company was ordered to fly the American flag above its mills. Everyone complied. Yet the steel companies didn't like it and they soon filed suit in federal district court. The district court agreed with the steel mills and issued a preliminary injunction against the seizure. Hours later, the court of appeals issued a stay of the preliminary injunction and the seizure was allowed to continue until the Supreme Court could be brought into the matter.

Meanwhile Truman, his popularity already low, was attacked for being a bully, a usurper, a lawbreaker, and an architect of a labor dictatorship. "Newspapers, magazines, steel executives, business organizations, and Republicans [exceeded] their own performances of the Roosevelt years. They attacked Truman as a Caesar, an American Hitler or Mussolini, an author of evil. . . ."[76] Within a few days the Court responded and heard the case. The steel companies hammered away at the unconstitutionality of Truman's actions. He had, they said, acted without congressional approval and without constitutional justification.

Truman's Justice Department countered with every plausible precedent, especially from the Lincoln and FDR eras. It didn't work. Justice Hugo Black, speaking for a 6–3 majority, agreed with the steel companies. No clause in the Constitution justified Truman's action, nor had it been authorized by Congress. Although Black commanded the majority in this ruling, every single justice, on both sides of the issue, wrote his own opinion. This resulted in an overall set of opinions consuming nearly 130 pages. More questions were probably raised than settled in this case, and even as Truman's initiative was being denied, several members of the Court implied he could probably have seized the mills if he had relied on emergency powers granted his office under several congressional acts rather than on the inherent powers of the office he claimed he possessed. Still, the reality was that Truman had to get the government out of the steel mills and thus suffered one of the worst setbacks of his presidency.

The Court in effect ruled that the Korean crisis was not a full-scale emergency justifying the full invocation and exercise of presidential war powers. This case obviously had serious political ramifications. Truman long remained furious about this decision and bitter toward his appointee and former Attorney General Tom Clark, who voted against him.

That damn fool from Texas that I first made Attorney General and then put on the Supreme Court! I don't know what got into me. He was no damn good as Attorney General, and on the Supreme Court, it doesn't seem possible but he's been even worse. He hasn't made one right decision that I can think of. And so when you ask me what my biggest mistake was, that's it. Putting Tom Clark on the Supreme Court of the United States.[77]

Other notable cases limiting executive power involve Nixon. Richard Nixon several times tested the limits of the office, sometimes succeeding in gaining power for the office and sometimes having his authority curbed. In the celebrated Pentagon Papers case, or *New York Times Co. v. United States* (1971),[78] the Supreme Court acted for the first time under wartime conditions since it had done so belatedly during and after the Civil War to protect institutional liberties against inroads from the executive branch. Vietnam was, however, an undeclared war, being waged at that time without much public or congressional support and by a somewhat unpopular president.

The Nixon administration attempted to halt publication in *The New York Times* and the *Washington Post* of a collection of classified but leaked essays and documents entitled "History of U.S. Decision-Making Process on Vietnam Policy." The real question, the only question on which, ultimately, this decision is based, is whether publication of these articles and documents would have yielded "direct, immediate, and irreparable damage to our nation and its people."

Nixon and the Justice Department said the publication of the so-called "Pentagon Papers" assaulted the principle of government control over classified documents. There are other parts of the Constitution that grant power and responsibilities to the president, they claimed, and the First Amendment was never intended to make it impossible for the president to function or to protect the security of the nation.

The Court was not persuaded. They ruled six to three in favor of *The New York Times* and permitted publication. The reasoning of the justices, however, is not simple to explain. There were six separate opinions for the majority and one for the three dissenters. The common bond linking the six majority judges was that the executive branch had simply failed to prove that the publication of these documents posed a major threat: "Any system of prior restraint of expression comes to this Court bearing a heavy presumption against its constitutional validity," the Court said. "The Government thus carries a heavy burden of showing justification for the enforcement of such a restraint."[79]

While the Supreme Court in this important case did not allow the Nixon administration to run roughshod over the First Amendment, it left the door open for executive discretion. Several justices acknowledged that had the publication of the Pentagon Papers truly threatened national security, they might have allowed for prior restraint. The Court still would not put an absolute limit on executive authority.

Justices Black and Douglas saw no need for this flagrant, indefensible suppression of information. A few others said it was the function of Congress, not the executive, to decide when an emergency might be great enough to justify this sweeping type of restraint.

A second decision that somewhat limited presidential power was a unanimous one in *United States v. U.S. District Court* in 1972.[80] Here the Supreme Court ruled that domestic surveillances required a warrant from the courts. And in fact, a lower court extended this decision to foreign wiretaps (made for national security purposes), and the Carter administration later pledged to do no wiretapping without prior judicial approval.

In this case the Court insisted there was a difference between domestic security and national security, and ruled that warrants would be required for domestic surveillance. The Court did recognize, it is important to note, that wiretaps for national security might sometimes be upheld, although they did not necessarily address that question in this case.

Critics of Nixon's and of the growing expansion of presidential power especially welcomed this decision. Clear limits, they believed, were placed on the executive through this opinion. Consequently the FBI had to disconnect several wiretaps.

There are indications, however, that the executive branch's compliance with this Court ruling has subverted its spirit. In practice, the executive branch pretty much wins regular approval through convenient procedures the judiciary and the Justice Department subsequently established.

Several questions remain unsettled in this area. Our political system still has a way to go to work out acceptable methods whereby both the rights of individual citizens to their privacy, and the responsibilities of the chief executive to ensure that national security interests, are reconciled in domestic wiretapping cases. Still, while some abuses may continue, the Supreme Court was effective in raising crucial issues and in halting a few of the obnoxious abuses of the Fourth Amendment that occurred in the Nixon years.

In response to Watergate and other charges of presidential abuses of power, the Congress, through the Ethics in Government Act of 1978, established an office of independent counsel. Designed so that the executive branch would not be allowed to investigate itself (via the Justice Department) when charges of wrongdoing are leveled, this law has been controversial from the beginning.

The Court rebuffed President Reagan's bid to have the independent counsel law declared unconstitutional.[81] Chief Justice Rehnquist, writing the majority opinion for the Court, ruled the law did not usurp the power of the president and thereby gave the Congress the power to appoint independent counsels to investigate allegations of wrongdoing in the executive branch, a power oft used in recent years.[82]

The intent of those who created the independent counsels was to insulate the administration of justice from partisanship and to ensure that the executive branch did not have a conflict of interest in investigating itself. But in recent years the job of independent counsel has been anything but nonpartisan. The journalistic scandal mill in Washington D.C. sometimes uses the independent counsel to go after political opponents and, in effect, criminalize partisan differences. This intermingling of law and politics creates a questionable precedent, and opens the door for abuses of power.

Since Watergate, almost twenty court-appointed independent counsels have been assigned to investigate executive branch officials. They have spent over $100 million, and rarely have convicted top officials. Some people now believe the post-Watergate efforts have not achieved their intended results. It is important, they say, that we hold executive branch officials to high standards, and history sadly reminds us that some presidents and their top appointees will, at times, abuse power. But the hyperpartisanship of some independent counsels may have weakened legitimate efforts to root out corruption in the executive branch of government.

The Court has rebuffed the Clinton White House on at least two notable decisions. As mentioned earlier, the Court saw no reason a sexual harassment suit against Clinton, a civil suit, could not proceed in the federal courts. Clinton's lawyers had said that such proceedings would interfere with the president's duties and might open up a president for multiple and distracting civil suits. The Court did not agree and in effect said no president was above the law. It plainly rejected the notion that a president could not take the time to answer questions about a matter that arose from civil legal proceedings.

The Clinton White House also lost a battle with the Court when the Court in 1997 left undisturbed a federal appeals court's ruling that notes from private conversations between Hillary Clinton and White House lawyers were not protected by the lawyer-client privilege.

This was a blow to the White House because it had frequently cited lawyer-client privilege as a justification for withholding various documents sought by congressional and federal investigators. As one White House counsel said in disappointment with this ruling:

> We [the Clinton White House] continue to believe that government lawyers must be allowed to have confidential discussions with their clients if they are to be able to provide candid and effective legal advice, and we regret that the Court has decided not to resolve the important issue.[83]

This decision had the effect of strengthening the hand of the independent counsel investigating Whitewater and related legal challenges that the Clintons faced.

CONCLUSION

Supreme Court justices tend to share with presidents a similar national perspective and a common outlook about the national interest. However, they have different obligations and responsibilities and operate in a different political forum with different roles and vantage points. More often than not, the Supreme Court (and the lower federal courts) has aided presidents as new demands and emergencies encouraged presidents to stretch the formal powers of the office. This trend or pattern is now well established and is rooted in Sutherland's famed 1936 *Curtiss-Wright* decision.

This judicial deference exists especially in wartime and when presidents enjoy high popularity.

The Court is well aware that only the president is elected by the nation. Still it is a paramount function of the Court to insist that presidents do not go beyond what is necessary or to try to singularly embody the nation's sovereignty. No president should be allowed to make the law or disregard for long the general commands of the U.S. Constitution.

Strong presidents invariably look for ways to expand their prerogatives and authority. Often, of course, this may be due to the emergencies confronting the nation. Their popularity or prestige does have an important bearing on whether they can win victories in their clashes with the Court.[84] Strong presidents try to nominate new justices who will advance their policies and values. For several generations now it has become common practice for presidents to appoint justices with a clear reference to their policy leanings, and this is likely to continue.

During the Reagan years, prospective appointees were queried about their views on a variety of issues, most notably abortion. This litmus test was an indication of how seriously Reagan took the federal court and how he hoped to use the judiciary to further his political agenda.

One of the strengths of the Supreme Court is its ability to function as an independent and at least a somewhat unpredictable institution. No president can shape it for long. As noted earlier, members of the Court develop their own independent constitutional philosophies, the more so the longer they remain on the Court. The Supreme Court remains, in Hamilton's phrase, "the least dangerous branch" and it is most assuredly vulnerable to the pressures placed on it by the nuclear age. Its power remains the ability of its members to persuade through reason.

A chief finding here is that we need an enlightened citizenry that will support the Court's occasional efforts to prevent the international requirements of the chief executive's job from leading to an undesirable and irreversible tip of the balance of powers in favor of the American presidency. We cannot expect the Supreme Court to be the sole or even main check on this tendency, yet we must encourage a vitality and independence and a climate of expectations in the country that will maintain a Supreme Court that is not intimidated by the other two branches. Our system functions best when Congress, the president, and the Court each energetically promote its own independent vision of good government. Preferring a strong presidency should not lead one to want a corresponding weakness in the other two branches, but rather a corresponding strength and assertiveness.

The Court and the other components of the federal judiciary can try to encourage the balance that is needed, yet in the larger scheme of things they are dependent on their ability to inspire others to play their proper role in the Republic. Justice Robert Jackson suggests this direction even though he speaks in a more narrow context of the *Youngstown Steel Seizure* case:

. . . I have no illusion that any decision of this Court can keep power in the hands of Congress if it is not wise and timely in meeting its problems. . . . We may say that power to legislate for emergencies belongs in the hands of Congress, but only Congress itself can prevent power from slipping through its fingers.[85]

The same can be said for citizens, state governments, the press, and every other vital sector of American society. The Court can try to lead and encourage, yet it is only a strong check on the abuses of power insofar as we, the people, are willing to support it, encourage it, and insist it carry out the functions that only it can carry out.

CHAPTER 9

The President and Cabinet

A good cabinet ought to be a place where the large outlines of policy can be hammered out in common, where the essential strategy is decided upon, where the president knows that he will hear, both in affirmation and in doubt, even in negation, most of what can be said about the direction he proposes to follow.

Harold J. Laski, *The American Presidency: An Interpretation* (Harper & Brothers, 1940), pp. 257–58

Cabinet meetings in the Bush White House were stilted, boring affairs. Instead of creative ferment and the clash of ideas, there were droning reports. . . . The truth is, Cabinet meetings are an anachronism.

Dan Quayle, *Standing Firm,* (Harper Paperback, 1995), p. 109

February 18, 1994—The White House. The first cabinet meeting in months. We sit stiffly while [President Clinton] talks about current events as if he were speaking to a group of visiting diplomats. I've been in many meetings with him, but few with the entire cabinet, and it suddenly strikes me that there's absolutely no reason for him—for any president—to meet with the entire cabinet. Cabinet officers have nothing in common except for the first word in our titles.

Secretary of Labor Robert B. Reich, *Locked in the Cabinet* (Knopf, 1997), p. 150.

The presidential cabinet in America is a misunderstood political institution. This is in part because the cabinet in several parliamentary systems has substantial influence as a policymaking group. This is not the case in contemporary America. Paradoxically, however, enormous media coverage is paid to a president's cabinet appointments and much fanfare is given to a new president's first cabinet meetings.

The American cabinet is too big and too divisive a group to function effectively in policymaking. As several scholars have noted, many cabinet members are appointed as much for their representativeness as for their policy or managerial expertise. Then, too, there is usually a lot of turnover in the cabinet, so much so that "cabinet secretaries are unlikely to form a policymaking unit."[1] They serve mainly as presidential emissaries to a department and as policy advisers to the president singly or occasionally as part of a selective team, for example as a member of the national security or economic policy team.

President Jimmy Carter, an engineer and more of a manager than most presidents, talked of his growing frustrations in trying to make the cabinet into a policymaking group. He acknowledged that he used his cabinet less and less as his four years wore on. "After a few months, the cabinet meetings became less necessary," Carter wrote. "As a result, in the first year we had thirty-six sessions with the full cabinet, then, during the three succeeding years, twenty-three, nine and six such meetings respectively."[2]

President Bill Clinton, in his first term, enjoyed more stability or continuity in his cabinet than most presidents. Still, he rarely felt the need to hold cabinet meetings. In his first two and one-half years in office, Clinton held twelve full cabinet meetings; in the following year and a half, he held only six.

A president may be but one person, yet the presidency is an institution—a collection of people and departments organized to, at least in theory, assist a president in making wise decisions and effectively implementing the laws of the land. But, "The plain fact is," writes Peri E. Arnold, "that no modern president has fully managed the executive branch."[3] We recruit politicians, not managerial leaders, to the White House, yet we evaluate them everyday on how well they are running the government, managing the bureaucracy, and leading their cabinet departments.

The way the cabinet is used and the value of the cabinet to a president obviously fluctuates according to the personality and the needs of the president and changing national conditions. It also changes over the course of an administration. After reviewing the origins of the cabinet and how cabinet members are selected, this chapter will assess how presidents, their top advisers, and cabinet members have interacted and managed, or mismanaged, the executive branch of government.

The framers discussed at length the possibility of creating some form of executive council that would comprise the president, heads of the departments, and the chief justice of the Supreme Court, yet they decided to leave things flexible. The Constitution makes no provision for a cabinet. It merely implies there are to be principal offices of the executive departments. The first Congress passed statutes that provided for the creation of three principal departments, State, War, and Treasury. An attorney general was also authorized, yet this would be a part-time adviser, not the head of a department.

From the outset President George Washington regarded the principal officers as assistants and advisers. "He began the practice of assembling his principal officers in council. And this practice became in the course of time a settled custom. The simple truth is that the cabinet is a customary, not a statutory body."[4] The term *cabinet* was probably first used in 1793, but mention of it in statutory language did not occur until 1907. Presidents over the course of our history have used the cabinet in greatly differing ways. Thus President Andrew Jackson did not even convene his cabinet in joint session during his first two years in office, whereas President James Polk held at least 350 cabinet sessions during his four years. Lyndon Johnson told his cabinet secretary that the thing that bothered him most about cabinet meetings was they were so dull; he said, "I just don't want them falling asleep at the damned [cabinet] table."[5] Reports had it that President Reagan did indeed doze off on a few occasions, prompting one wag to opine that the "cabinet room seats 12 and sleeps one."

As a general rule, recent presidents have often met with their full cabinet in formal cabinet meetings their first year or two in office and then, like Carter and Clinton, worked out other arrangements to get together with them singly or in small groups. A president who wants to dispense with cabinet meetings simply does so. Still, presidents and their staffs have to work out processes for regular communication with cabinet heads.

PRESIDENTIAL CABINET MAKING

Few acts of presidential leadership are as important as the appointment of a president's cabinet members and senior advisers. Presidents are thought to have a free hand in choosing their cabinet members, yet this is not exactly the way it works. In addition to administrative competence and experience, loyalty and congeniality are basic considerations in selections. Other factors are at work as well. Party rivals often have to be disarmed and placated either with an appointment to the cabinet or selection as the vice presidential running mate. Reagan's selection of Bush as vice president and Alexander Haig as his first secretary of state are examples. Regional, ethnic, gender, and geographical considerations are almost always important. Nowadays, for example, it is a custom to have at least one African-American, one Hispanic, one westerner, and one southerner in the cabinet, and several women. After capturing the presidency, a president usually goes about selecting the cabinet in such a way as to try to win the confidence of major sections and sectors of the nation. "I knew," wrote Richard Nixon, "that some of my choices for Cabinet posts would have to serve, even if only symbolically, to unite the country and 'bring us together.' "[6]

President Carter and his top campaign aides said he was going to appoint a crop of fresh faces from a new generation, not from the generation that had already served in high posts for previous administrations. Carter's attitude on this matter ran counter to the usual theory that to be effective in Washington you should have had extensive earlier experience learning the ropes there.

In theory a president can nominate anyone for a cabinet post. In practice the constraints are many.[7] First, each cabinet member has to be confirmed by a majority of the members in the U.S. Senate. Nowadays, nominees are being examined more carefully at both ends of Pennsylvania Avenue and by interest groups and the media. This automatically precludes those who will not submit to this process. Others are unwilling to give up their much higher salaries and fringe benefits. Some are unwilling to join the cabinet because it would mean having to give up a seat in the Senate or would hinder their chances of running for a national leadership position. Still others sometimes pose conditions under which they would accept nomination and the conditions are not acceptable to the president. Others are put off by the conflict-of-interest regulations and the Ethics in Government Act that recent presidents have imposed on all top appointees. Clinton had one top White House aide he might have considered for a cabinet post but for the fact that the individual was so blatantly partisan and politically controversial that the Republican-controlled Senate was unlikely to confirm him.

A president has to select certain cabinet officers because of the needed expertise they will bring to the administration and the policy needs of certain departments. Typically, for example, the secretary of defense is someone who has worked closely with that department in some previous capacity. The labor secretary will often be someone who has had extensive negotiating background and, if a Democrat, some advisory ties to the AFL-CIO leadership. A treasury secretary is traditionally viewed as the financial community's representative in the cabinet. In selecting a treasury secretary a president-elect usually wants someone who can simultaneously serve as a "spokesperson" to the financial world and as a "spokesperson" for those interests.

Generalists are often appointed to head up some of the domestic departments such as Commerce, Transportation, and HUD, while politicians especially close to clientele groups are often appointed head of Agriculture and Interior. President Bush appointed members of Congress to head Agriculture, Interior, and Veteran's Affairs departments. President Clinton appointed former governors to head up Interior and Education, and former mayors to head HUD and DOT. Appointing generalists who are unlikely to carry messages and advocate the agenda from interest groups and from partisan constituencies advantages the White House over the bureaucracy in the short run, but in the longer run the bureaucrats acting *independently* of their own cabinet secretaries often seek to form their own alliances with clientele groups and congressional committees.

President Clinton's first labor secretary, Robert B. Reich, pointed out the obvious:

> No other democracy does it this way. No private corporation would think of operating like this. Every time a new president is elected, America assembles a new government of 3,000 or so amateurs who only some-times know the policies they're about to administer, rarely have experi-ence managing large government bureaucracies, and almost never know the particular piece of it they're going to run. These people are appointed quickly by a president-elect who is thoroughly exhausted from a year and a half of campaigning. And they remain in office, on average, under two years—barely enough time to find the nearest bathroom. It's a miracle we don't screw it up worse than we do.[8]

THE JOB OF THE CABINET MEMBER

Defining the job of the cabinet member depends on one's vantage point. Members of Congress believe a cabinet officer should communicate often and well with Congress and be responsive to legislators' requests. Report-ers want a cabinet officer to be accessible, to make news; they applaud style and flair as well as substance. Interest groups want a cabinet mem-ber who can speak out for their interests and carry their messages to the White House. Civil servants in a department are generally looking for a cabinet leader who will boost departmental morale and appropriations. White House aides are as keenly concerned about a cabinet officer's loy-alty to the president as they are about ability. A president wants a cabinet member who will conserve his freedom of action and enhance his adminis-tration's reputation without embarrassing or overshadowing him.[9]

There is no effective training for new cabinet secretaries. They are thrown into the fire and expected to serve many masters. Robert Reich noted the feelings of isolation of a new secretary when he wrote, "As I walk out of her office it suddenly strikes me: I'm on my own from here on. There's no training manual, no course, no test drive for a cabinet sec-retary. I'll have to follow my instincts, and rely on whomever I can find to depend on along the way. I'll have to listen carefully and watch out for dangers. But mostly I'll have to stay honest with myself and keep perspec-tive. Avoid grandiosity. This is a glamorous temp job."[10]

What is a good cabinet officer? A Carter White House aide summa-rized how White House staff viewed the cabinet job:

> First, he should be clearly in charge of the department. . . . Second, he should be very sensitive to the department's interest groups and have

access to them and fully understand their views. But, third, he should also at the same time be able to distinguish the president's interests and political needs from the department's clientele interests. He should be able to say that this is what they want but it is or is not compatible with your interests. Fourth, the effective cabinet officer is one who can work on most of his congressional relations problems without running to the president for help. Finally, he should be able to follow-through on presidential policies, that is, to see that they don't get watered down, or lost in the shuffle.[11]

Presidents and their aides repeatedly say the last thing they want to see is a cabinet member who has become a special pleader at the White House for some of the special-interest groups that are strong in the particular department. "The Cabinet officer must certainly be attentive to his departmental business, and he should seek to ensure that the president has timely notice of the impact of other policies on his department's specific interests," a former White House aide writes. "But a Secretary should never choose his departmental interest as against the wider interest of the Presidency."[12]

The greatest test of cabinet members arises from the fact they are tied as closely to Congress as they are to a president. In the perpetual tug-of-war between these two branches, the cabinet officer is often like the knot in the rope. Bradley Patterson, a longtime civil servant and a staff aide in the Eisenhower White House, pointed this out: "His or her appointment is subject to Senate confirmation. Every power a Cabinet officer exercises is derived from some Act of Congress; every penny he or she expends must be appropriated by the Congress; every new statutory change the Cabinet officer desires must be submitted to the Congress and defended there. A Cabinet officer's every act is subject to oversight by one or more regular or special congressional committees, much of his or her time is accordingly spent at the Capitol and with few exceptions, most of the documents in his or her whole department are subject to being produced at congressional request."[13] A former secretary of education reflects that "all the while you must obey, simultaneously, the President, the Congress and the courts—even when they are moving in different directions." And if this isn't enough, you must also "explain to the nation, through the media, what it is that you are doing and why it is necessary."[14]

Presidents often believe cabinet members should serve as servants of the president and not make decisions independent of the president. When a president makes a decision the cabinet officer is expected to carry it out. In fact, it is inevitable that after a person has been in the cabinet for a time and has become enmeshed in the activities and interests of a department, he or she develops certain independent policy views. A certain hardening of view may set in as the cabinet secretary gets pushed by subordinates, interest-group leaders, or others in a direction that makes it likely he or she will come into conflict with the president. When this happens, the White House typically complains that the cabinet member has "gone native"—he or she has been captured by the interests native to that department. Often, the cabinet officer wants to extract more money out of the president's budget for the department. An old-time budget director once complained that "cabinet members are vice presidents in charge of spending, and as such they are the natural enemies of the presidents."[15] This is why Richard Nixon once bellyached that "rather than running the bureaucracy, the bureaucracy runs him."[16]

THE WHITE HOUSE WANTS LOYALTY ALONG WITH COMPETENCE

In reflecting on his old cabinet, former President Lyndon Johnson once blurted out that "I'll always love Dean Rusk, bless his heart. He stayed with me when nobody else did."[17] Though they might not always admit it, those at the White House judge cabinet officers by a formula that weighs ability as primary but ranks loyalty to the president and to the president's political future as a close second. For some White House aides, loyalty to the boss even outweighs competence. Definitions of loyalty not surprisingly vary depending on where you sit. Outsiders often complain that loyalty in the top reaches of the executive branch can too easily lead to mindless servility and a cabinet of "yes-people." Friends and supporters of a president insist a president needs and is entitled to loyalty from cabinet members and staff. A president's own authority and his very capacity for leadership can be jeopardized by disloyalty and independence of cabinet members.

Some presidencies, John Adams' for one, were undermined by cabinet disruptions. Woodrow Wilson had one secretary of state resign in protest and another who persisted in open dissent from the president's policies and even attempted to undermine Wilson's moderate war aims. Jealousies, political ambitions, rivalries, turf fights, and tactical differences can tear an administration apart and can leave considerable doubt in the public's mind as to who is really in charge. However, people with the president's point of view in mind often say loyalty can just as easily be undervalued as overvalued, adding that true loyalty should include the willingness to argue vigorously and speak one's mind—at least within reasonable bounds.

Taking both ability and loyalty into account, a simple fourfold matrix can illustrate how some recent cabinet members have been viewed by the White House (Figure 9.1). Keep in mind that these rankings are relative. Most of these cabinet officers were viewed as reasonably able and loyal. Yet interviews with White House aides and an examination of presidential memoirs and administration oral histories indicate some were viewed as abler than others, some more loyal and some less loyal. These rankings suggest the twofold standard used by most White House aides, and no doubt by presidents as well, as they pass judgment on cabinet member performance.

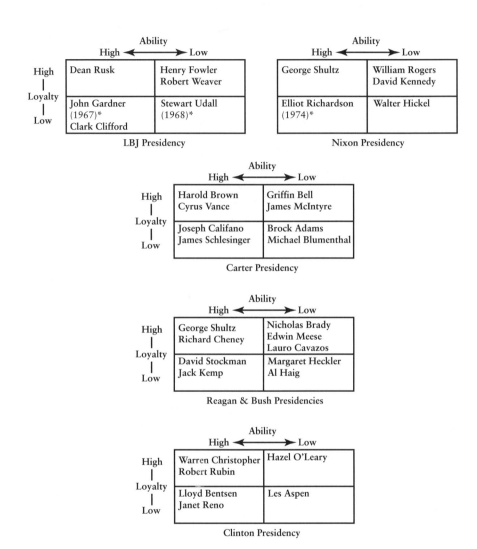

Figure 9.1: Illustrative White House Assessments of Cabinet Officers

*Note that these Low Loyalty rankings were sometimes temporary and that the years are indicated.

Firing cabinet members is the last thing presidents like to do. They do it, of course, but usually only after it is long "overdue." And they do it at the risk of considerable political backlash. President Andrew Johnson's removal of Secretary of War Stanton contributed to Johnson's impeachment. President Nixon's removal of special prosecutor Archibald Cox and the related resignations of Attorney General Elliot Richardson and Deputy Attorney General William Ruckelshaus badly damaged what remained of Nixon's credibility. Ronald Reagan fired Secretary of State Alexander Haig, White House Chief of Staff Donald Regan, and Secretary of Health and Human Services Margaret Heckler and encouraged some others, such as Raymond Donovan (labor secretary) to resign. To fire a cabinet member is to acknowledge you made a poor administrative decision in the first place. It is to acknowledge you were unable to create a balanced atmosphere in your cabinet of forthright disagreement and constructive congeniality. Sometimes the fired cabinet officer becomes a media celebrity, or joins the political opposition on either partisan or substantive issues, or both. Sometimes the fired cabinet member has inside information that would be politically damaging if revealed, or is sufficiently hurt by the removal to write a book critical about the inadequacies of the administration.

For example, Gerald Ford did not enjoy his relationship with Defense Secretary James Schlesinger. Ford considered removing Schlesinger soon after he became president. Later he regretted that he hadn't and he eventually did have to fire Schlesinger. Why did he wait? Part of the reason was that Ford wanted to have a smooth transition from the trauma of Watergate to his own administration. Legend has it that Ford finally fired Schlesinger because of Schlesinger's condescending rather than deferential manner. According to one report, Schlesinger seldom looked directly at Ford during cabinet meetings. Instead he looked at Secretary of State Henry Kissinger, as if to imply that his remarks would be lost on Ford.

Nixon fired Interior Secretary Walter Hickel after a celebrated "personal letter" from Hickel to the president was leaked to the press. Hickel was especially frustrated because he only saw the president once or twice for policy discussions. His letter included those pointed sentences: "Permit me to suggest that you consider meeting, on an individual and conversational basis, with members of your cabinet. Perhaps," continued Hickel, "through such conversations we can gain greater insight into the problems confronting us all, and most important, into solutions of these problems."[18] Hickel soon left the cabinet. They say he left in the same manner he had come to the cabinet—fired with enthusiasm.

Late in his presidency, President Johnson became particularly upset at several of his domestic cabinet members. Oral interviews at the Johnson presidential library in Austin, Texas, indicate several cabinet members were in his doghouse for crossing him on some program or another, or for failure to support the Vietnam War effort enough. More often than not, his chief domestic aide, Joseph Califano (who later, as secretary of HEW, would be fired by Jimmy Carter for lack of loyalty), was ordered to win compliance and greater cooperation from the particular cabinet member in question. Illustrative of Johnson's disappointment with some of his cabinet is his sharp denunciation of his former attorney general, Ramsey Clark: "You know ol' Harry Truman said his biggest mistake was appointing Tom Clark to the Supreme Court. Well, my biggest mistake was appointing Tom's son, Ramsey, as my Attorney General. He couldn't make up his mind about a fish fry." Johnson continued, "Ramsey wanted to go around preachin' bleeding-heart stuff, but he never did a damn thing. I heard Dick Nixon made a campaign speech against Ramsey Clark one night and I had to sit on my hands so I wouldn't cheer it."[19]

President Reagan, in one of his more shabby treatments of a cabinet member, fired Secretary of Health and Human Services Margaret M. Heckler in October 1985. A sixteen-year veteran of the House of Representatives, she lasted two and a half years in Reagan's cabinet. Heckler was a moderate Republican who failed to defer to conservative Reagan appointees at HHS, and more importantly, at the White House. Some of Reagan's White House aides ridiculed her managerial skills and saw her as ideologically "out of step" with the Reagan priorities. She had also reached a political impasse with White House officials over several appointees to top-level vacancies in her agency. More than anything else, however, Heckler clashed with the White House staff over policy direction. "The Department of Health and Human Services is where President Reagan most often collides with the modern social welfare state—where an Administration hostile to many human services must nonetheless administer them," editorialized *The New York Times* at the time. "As its head, Margaret Heckler tried to speak for fairness and compassion in a Government not renowned for either."[20] Heckler put up a fight, yet reluctantly yielded to Reagan when he "promoted her" to the post of ambassador to Ireland. Her departure was immediately hailed as another victory for both the White House staff and the conservative movement.

From the perspective of the White House, these are the questions asked about cabinet members. Have they managed the department well? Do they recruit talented officials to the department? Are they loyal to the president? Can they "handle" the interest groups associated with that department? Have they brought prestige to the department and to the administration? Does their presence in the administration boost confidence in it? Do they help politically as the next election nears? Have they handled the relationships with Congress and the press effectively? Have they come up with fresh and innovative ideas? Have they been able to implement the administration's programs in their department? Have they become an asset or liability? A cabinet level official sometimes can be both loyal and competent and still be viewed as a political liability.

THE ROLE OF THE CABINET IN POLICYMAKING

In practice, the cabinet is used as a forum for information exchange, as an occasion for a president to boost morale and highlight priorities, to review major issues that cut across all departments, to go over how members can help push for congressional approval of the president's program, and to discuss ways to improve implementation of major programs. It is, of course, also used as a feedback device through which presidents can obtain political and professional advice on how policies and programs are (or are not) working.[21]

A romantic view of the mythical American cabinet is in part encouraged by presidents themselves. Nixon, Carter, Reagan, and Clinton all emphasized the importance of the cabinet when they picked their original team. Reagan early in 1981 told his cabinet nominees he intended the cabinet to function much like a board of directors in a corporation, the president deliberating with them often and offering input prior to decisions. Truman said, "the cabinet is not merely a collection of executives administering different governmental functions but a body whose combined judgment the president uses to develop fundamental policies of the administration." But one of his biographers argues that "while President Truman used his cabinet to formulate decisions, in practice, important decisions were made by ad hoc groups consisting of cabinet officers and others. He did not wish to create a secretariat and to formalize the meetings."[22]

A consistent pattern seems to characterize president-cabinet relations over time. Just as a president enjoys a distinctive honeymoon with the press and partisan officeholders, White House–cabinet ties are usually the most conflictual or cooperative during the first year of an administration. A newly staffed executive branch, busily recasting the political agenda, seems to bubble over with new possibilities, proposals, ideas, and imminent breakthroughs. Ironically, White House staff, which soon will outstrip most of the cabinet in power and influence, receives less publicity at this time. In the immediate postinaugural months, the Washington political community, and the executive branch in particular, becomes a merry-go-round of cheerful open doors for the new team of cabinet leaders.

But this ends soon. Little over a year in office, Reagan's White House aides developed a distrust for many of their departmental secretaries, saying they had generally become advocates of their own constituencies. Department heads cannot be relied on to put the president's interests ahead of their departmental interests. "Cabinet government is a myth," said a Reagan staffer. "I'm not sure it has dawned on the [cabinet] members yet that they have been cut out of the decision-making process."[23]

Domestic crises and critical international developments begin to monopolize the presidential schedule. As a president has less time for personal contacts, cabinet members become disinclined to exhaust their personal political credit with him. A president's program becomes fixed as priorities are set, and budget ceilings produce some new rules of the game. Ambitious, expansionist cabinet officers become painfully familiar with various refrains from executive office staff, usually to the effect that there just isn't any more money available for programs of that magnitude; or that budget projections for the next two or three years just can't absorb that type of increment; and, perhaps harshest of all, that a proposal is excellent but will just have to wait until the next term.

A top White House staffer under Nixon captured the complex entanglements of time, presidential priorities, and interactions of people:

> Everything depends on what you do in program formulation during the first six or seven months. I have watched three presidencies and I am increasingly convinced of that. Time goes by so fast. During the first six months or so, the White House staff is not hated by the cabinet, there is a period of friendship and cooperation and excitement. There is some animal energy going for you in those first six to eight months, especially if people perceive things in the same light. If that exists and so long as that exists you can get a lot done. You only have a year at the most for new initiatives, a time when you can establish some programs as your own, in contrast to what has gone on before. After that, after priorities are set, and after a president finds he doesn't have time to talk with cabinet members, that's when the problems set in, and the White House aides close off access to cabinet members and others.[24]

After the first year or so, presidents become increasingly concerned with leaks from their administration. They worry about saying things or having key decisions made in large groups. They also, in President Carter's words, believe their time is "too precious to waste on many bull sessions among those who have little to contribute. . . ."[25] So what do they do? They begin to hold fewer cabinet meetings and hold smaller sessions with the inner cabinet. "We stopped meeting as often, substituting instead smaller sessions restricted to either domestic or foreign issues or to a single subject of interest to several departments."[26]

President Kennedy regarded the idea of the cabinet as a collective consultative body largely as an anachronism and often told close friends this in blunt terms. He once asked: "Just what the hell good does the cabinet do, anyway?" He felt that the nature of a problem should determine the group with which he met. Why should the postmaster general sit there and listen to discussions about Indochina, wondered Kennedy. He thought the historical custom of the cabinet should carry with it no special claim on his time. He believed there were few subjects that warranted bringing together, for example, the secretary of labor, the secretary of agriculture, and the secretary of defense. As Kennedy aide Theodore Sorensen recalled:

> [Kennedy] had appointed his cabinet members because he regarded them as individuals capable of holding down very difficult positions of responsibility. He did not want to have them sit through lengthy cabinet

sessions, listening to subjects which were not of interest to them, not of importance to them, at least not of an interest to their primary duties and their primary skills. So he called cabinet meetings as infrequently as possible. . . .[27]

Richard Nixon had sat through countless formal cabinet meetings during the Eisenhower years. In his memoirs he recalls that "most of them were unnecessary and boring. On the few issues that cut across all departments . . . group discussions would sometimes be informative. But the day had long since passed when it was useful to take an hour and a half to have the Secretary of Defense and Secretary of State discuss the Secretary of Transportation's new highway proposal."[28]

If Nixon found cabinet meetings boring in the Eisenhower period, Elliot Richardson found them the same way in the Nixon years. He wrote that cabinet meetings in both the Nixon and Ford administrations "ordinarily focused on bland common denominators like the economic outlook, displays of budgetary breakdowns, or reviews of the status of administrative proposals." Richardson added, sarcastically, that "in the Nixon Cabinet, as a special treat, Vice President Spiro T. Agnew would occasionally give us a travelogue."[29] A Reagan deputy chief of staff purposely avoided as many cabinet meetings as possible because he didn't want "to be bored unconscious."[30]

Bill Clinton, a "policy-wonk" who loved to talk policy in minute detail, rarely found the cabinet useful as a group. He was more comfortable with ad hoc or functionally based groups advising him on policy.

The point is that most of the recent presidents have found it neither comfortable nor efficient to meet frequently with their cabinets as a whole. When they have, it seems to have been as much for purposes of publicity, symbolic reassurance, or appearance of activity as for substantive debate or learning.

On the most important matters, the cabinet members as a whole are seldom called on for advice. The president, sometimes with a few senior advisers, makes up his own mind. What one FDR cabinet member said fifty years ago is still generally the case. "We never discuss exhaustively any policy of government. . . . Our cabinet meetings are pleasant affairs, but we only shine the surface of things on routine matters."[31]

Cabinet members come to be judged by the White House on their capacity to generate favorable publicity and to proclaim the virtues of the recent White House achievements. Strong-willed people of independence can hardly be expected to be enthusiastic about performing public relations tasks assigned to them by White House political counselors.

As time passes, cabinet members grow bitter about being left out of White House decisions, though they seldom make their opinions public. Some exceptions exist, of course, and many cabinet officers will talk about the problem privately. The case of Nixon's first secretary of the interior, Walter Hickel, who had only two or three private meetings with his president in the two years before his public protest and subsequent firing, is perhaps extreme. Most cabinet officers have more frequent contact with their president, but few of the domestic cabinet members have been wholly pleased by the quantity or quality of these meetings. A former secretary of commerce in the 1970s once joked of his infrequent ties to the White House by saying that his president "should have warned him that he was being appointed to a secret mission." Secretary of State Alexander Haig said he was "mortally handicapped by his lack of access to the president." And several others such as Cyrus Vance and George Shultz threatened to resign on several occasions because they felt bypassed or undercut by White House senior aides.

Nixon's increasing centralization of power in the White House was an apparent vote of no confidence in most of his cabinet members, although in naming his original cabinet Nixon claimed to have appointed men who had the potential for great leadership, men with "an extra dimension which is the difference between good leadership and superior or even great leadership." Five years later he had a totally new cabinet. Nixon ran through cabinet members faster than any president in recent history. During one period of about eighteen months he had five different attorneys general.

Nixon paid a price for keeping his cabinet officers at such a distance. Some of them were so often undercut that they sometimes purposely acted on their own just to show themselves, their departments, and the press that they were capable of independent action. Occasionally this made for a divided administration. At other times, as Nixon later admitted, it threatened to undercut the administration's credibility with foreign countries. More important, Nixon's disuse and downgrading of the cabinet increased his isolation from responsible sources of political and professional advice. An already insecure man—a man who needed friendship and peers—removed himself from a potential source of support. One result was the Watergate scandals.

Why are more spirited and substantive discussions absent from the modern-day cabinet? First, the number attending cabinet sessions is too large. President Ford sometimes had nearly thirty people in attendance. Most cabinet members are unlikely to talk about their troubles or highly sensitive topics in a group that large. The presence of twenty or more cabinet officers and senior aides such as the budget director and chief-of-staff and CIA director sitting around the cabinet table and along the walls permits fewer than three minutes of wisdom per elder per hour. Thus both presidents and cabinet members become disillusioned with government by cabinet meeting. But the alternatives cause problems too. "Of course, if you don't have cabinet government, you have government by a White House staff that has not been confirmed by the Senate," says former Attorney General Griffin Bell. "So you lose one of the checks and balances in our system. Because there is constant friction between the

White House staff and the cabinet secretaries, it is extremely difficult to find quality people to serve as cabinet heads."[32]

One of the president's dilemmas is that a fundamental separation of policy formulation and its implementation can develop. Whereas most major policy decisions are made by the president and a smaller number of his personal aides, the responsibility for enacting these programs rests, for the most part, with the cabinet officers and their departments. The gap between these two functions of the executive has been widening, a result of the transference of power from the more public institution of the cabinet to the relatively hidden offices of the White House staff. The result has been a divorce of the exercise of power from accountability for that exercise, a situation many people believe threatens the effectiveness of the presidency.

The very nature of the cabinet—a body with no constitutional standing, members with no independent political base of their own and no requirement that the president seek or follow their advice—helps contribute to its lack of influence as a collective body. Ultimately, the influence of the cabinet rests solely on the role that a president desires for it. Most presidents, particularly those since Eisenhower, have made that role an increasingly limited one.

To some extent, this downgrading of the cabinet was imminent from the late 1930s when Roosevelt first began to place top-level advisers in the White House itself. Roosevelt often relied on brain trusters scattered throughout the government in positions other than cabinet posts. Easy and frequent access to the president is an important determinant of an adviser's influence. As Daniel Moynihan aptly put it, "Never underestimate the power of proximity." But perhaps more important, presidents tend to view many of their cabinet officers as parochial and narrow in their outlook, playing only an advocacy role for their own departments.

Ironically, the presidential view of cabinet members as advocates for their departments often proves a self-fulfilling prophecy. Not only have presidents, as a result, tended to use their cabinets to appease interest groups with a spokesman at the head of their relevant department, the limitations presidents place on the time they spend with members forces them to use that precious time to push for their department's needs first. Paradoxically some presidents render the departments less effective in implementing their policies by denying them a say in program development. There is less incentive for cabinet officers or their departments to implement a program efficiently if they believe they have no real stake or say in it.

Experience suggests three basic ways presidents can use the cabinet as a group.

1. For discussing issues and problems in an informal and unfocused manner, primarily to exchange information back and forth

2. For focused, regular consideration of specific issues, set forth in papers authored by cabinet members and circulated ahead of time, aided by agendas, concise records of action, and a small secretariat

3. For summit discussions of issue papers, which have been through a searching interdepartmental review, with dissents identified and alternative language proposed.[33]

By general consensus the first format was used by Truman, Kennedy, Johnson, Nixon, Ford, Reagan, and Clinton, at least when they were willing to hold cabinet meetings. Eisenhower, however, followed the second alternative. Carter began with aspirations to the second format but his patience wore thin, as already noted. Probably no president has ever used his full cabinet according to format three, although Truman and Eisenhower sometimes did employ their National Security Council in approximately this manner.

THE REAGAN CABINET COUNCIL EXPERIMENT

More than other recent presidents, Ronald Reagan came to office with a comprehensive ideological program to implement. Although he too espoused the ideals of a strong cabinet, at the heart of his approach to governing was the centralization of policy decision-making in the White House and the Office of Management and Budget.[34] Reagan's advisers deliberately emphasized ideology and loyalty to the president in their appointments throughout the executive branch. Reagan's White House made explicit use of civil service reforms enacted in the Carter years to remove career officials from key posts, and filled them with Reagan supporters. They also used "reductions in force as a legal means of eliminating whole bureaucratic units staffed by careerists. This was done systematically, in the interests of presidential control and successful pursuit of the Reagan agenda."[35]

As governor of California Reagan had been a delegator to his staff and cabinet. He came to the White House prepared to continue this leadership style. He recognized, as had the presidents before him, that a president did not need to convene the entire cabinet to go over plans for defense or urban policy. And so, with Ed Meese's assistance, Reagan developed a highly compartmentalized structure of cabinet councils, grouped around specific subject areas: economic affairs, human resources, natural resources, food and agriculture, commerce and trade. They retained the statutory National Security Council and set up two additional cabinet councils that addressed items of general management cutting across the executive branch: legal policy, management and administration.

The intention behind these cabinet councils, or so the White House led people to believe, was to integrate both the cabinet officers and Executive Office staff into working bodies that could save the president from convening the whole cabinet and yet would still keep cabinet members involved. Edwin Meese, the principal architect of the Reagan cabinet council system, said these councils would enable cabinet members to have greater opportunities to advise Reagan and would ensure that important decisions would not be made without at least one member of the cabinet present. In theory, major policy matters would be considered by those cabinet councils together with their relevant White House counterparts. Then as the pros and cons were developed the cabinet council would meet with President Reagan serving as its chairman for fuller deliberation prior to a final presidential decision.

Here is how Ed Meese, at the time a senior White House aide, defended these councils:

> The value of such a system, augmented by senior members of the White House staff who occupy the back benches in the Cabinet room, lies in the cohesion developed between the White House and the heads of the various cabinet departments, and in a structure of decision-making that retains both flexibility and inclusiveness. When the President does make his decision, cabinet and staff alike know what the decision is. They also know, each of them having had their chance at input, that they are now accountable for carrying out that particular decision.[36]

Meese and Reagan emphasized they wanted to avoid a concentration of power in the White House, the same concentration that many observers believe isolated earlier presidents such as Johnson and Nixon. Yet a reading of the cabinet memoirs of Alexander Haig, David Stockman, and Terrel Bell suggest that Reagan's efforts, although they worked reasonably well for awhile, did not achieve the intended results.

In practice, what happened was that most cabinet members became members of subcabinet policy groups, chaired by a fellow cabinet member but usually directed or orchestrated by a senior White House official. Senior aides such as James Baker, Edwin Meese, or the national security assistant to the president loomed large. Meanwhile the president himself did not have to attend many meetings. Cabinet members, however, were forever going to the council meetings! They had a high degree of participation in policy discussion, yet with few exceptions they viewed themselves as "the grooms" who got the policy issues onto the track.[37] Senior White House aides, however, wielded the influence of "jockeys" who whipped the presidential decision into final shape.[38]

Participants complained too that the issues often overlapped into the jurisdiction of two or more cabinet councils, creating coordination problems. Cabinet members, many of whom served on at least three cabinet councils, soon complained about the number of meetings, and as their own schedules of congressional testimony and various out-of-town trips inevitably diverted them, they began to send their deputies or even lesser departmental aides to represent them.

After a while, the cabinet council system began to create more problems than it solved. When cabinet members stopped coming, White House aides tended to take control of these councils.

The cabinet council on economic affairs met with great frequency and was judged to be the most successful of these bodies. Midway through Reagan's first term, his cabinet councils on natural resources and food and agriculture seldom met. The councils on human resources and legal policy were almost as dormant.

By the beginning of Reagan's second term, half his cabinet had departed as had several of his senior White House team. All of the once highly acclaimed cabinet councils were eliminated except two general ones on domestic and economic policy. These would now be chaired by Ed Meese and Jim Baker, who had moved from the White House staff to the posts of attorney general and treasury secretary, respectively. And soon even these two cabinet councils fell into disuse. And, as the Tower Commission Report and Iran-Contra hearings made clear, President Reagan also failed to make sensible use of the far more institutionalized National Security Council as a means of involving his cabinet members charged with foreign policy responsibilities.

What had happened? The cabinet council system with its complicated support system collapsed of its own weight. Many of the most important policy decisions in both Reagan's terms were made neither through nor in the cabinet councils.

Another interpretation contends Reagan was simply allergic to public policy discussions save when it came to his deeply held views on slashing taxes, increasing defense spending, or matters such as aid to the Contras and for the Strategic Defense Initiative. Cabinet councils, then, were a way of keeping the cabinet busy and giving them a sense of participation without having to involve the president directly.

Reagan had centralized and politicized the upper echelons of government more than any president in memory. He had done so deliberately. The conservative agenda of the Reagan presidency had been forged before he took office. A close reading of the books written by Reagan staffers and cabinet officers suggests too that Reagan's top aides, Meese in particular, never really intended to allow people like Alexander Haig, Terrel Bell at Education, or Margaret Heckler at Human Services a major say in Reagan's policy decisions. To be sure, they had to be involved, but more to ensure they would be loyal cheerleaders and faithful implementers of the Reagan priorities. What he needed most was a means to enact, enforce, and implement his agenda. The cabinet members, through the cabinet councils, would play a role, but not nearly as influential a role as Reagan's White House staff, his Office of Management and Budget and also his Office of Policy Development. Political scientist Terry Moe described it well:

> There was no question that all the basic issues and problems related to
> policy had to be centrally processed and directed from the White House.

The only real question was one of organization: how to do it? Here, as usual, institutional memory played a guiding role. Every effort, in particular, was made to avoid the over-centralization and lack of meaningful cabinet participation that caused difficulties for Nixon, but also to avoid the administrative chaos and lack of direction that prevailed under Carter's more open arrangements. The heart of the new system was the Office of Policy Development, which was the functional equivalent of the domestic policy staffs that all presidents since Kennedy had relied upon.[39]

The OPD system did not become the central force behind policymaking. The agenda, after all, was already set, and much of the driving force was of necessity from the OMB, particularly during the first year. A few of the councils did play active, important roles, but the real contribution of the OPD system was its integration of diverse, potentially conflicting actors. It gave cabinet members and bureaucratic officials a sense of meaningful participation and a means of direct, continuing input, while it also gave the administration a regularized means of reemphasizing the president's agenda and the coordinated actions required through government for its pursuit.

In the end, the Reagan cabinet council experiment has to be judged a mixed success. For a time it gave cabinet members a sense of involvement. More important, however, it permitted Reagan's top staffers to control the cabinet departments and to protect the president. Policy debates did take place, but coordination of the president's program was at the heart of this process. It is doubtful that future presidents will adopt the Reagan system. It suffered from being too diffused and too complicated. It occasioned too many meetings. Cabinet members soon recognized the councils as another obstacle in the way of talking directly with the president. Reagan, never an avid consumer of advice and information, liked the system while it lasted, yet it also may have been the source of his greatest problems—failure to know what was going on in his own administration and failure to educate his senior people of the legal and constitutional as well as policy responsibilities that came with their jobs at the head of the executive branch.

A CABINET OF UNEQUALS

Cabinet roles and influence with the White House differ markedly according to personalities, the department, and the times. Each cabinet usually has one or two members who become the dominant personalities. Herbert Hoover's performance as secretary of commerce under Harding was of this type. George Marshall's performance in both State and Defense under Truman was similar. George Humphrey of the Treasury and John Foster Dulles of State clearly towered over others in the Eisenhower cabinet. Robert McNamara enjoyed especially close ties with both Kennedy and Johnson. And James Baker, George Shultz, and Edwin Meese all carried special weight with Ronald Reagan. Robert Rubin, Warren Christopher, and Madeleine Albright became major figures in the Clinton cabinet.

Certain departments and their secretaries have gained prominence in recent decades because every president has been deeply involved with their priorities and missions—Defense and State in the Cold War as well as in the so-called "detente" years, for example. The Acheson-Dulles-Rusk-Kissinger-Baker tradition of close and cordial ties with their presidents were founded in an era of continuous international tension during which diplomatic and alliance strategy loomed large. Other departments may become important temporarily in the president's eyes, sometimes because of a prominent cabinet secretary who is working in an area in which the president wants to effect breakthroughs: for example, John Kennedy's Justice Department headed by his brother Robert. HEW sometimes was thought to be developing into a presidential department during the mid-1960s when under John W. Gardner it grew rapidly in order to manage Johnson's major educational and health programs. The Department of Energy enjoyed special considerations in the early Carter years for similar reasons. Soon, however, HEW was divided into two departments and Energy was downgraded significantly by Reagan and later presidents.

Much White House–cabinet estrangement undoubtedly arises because presidents simply lack the time to spend with all cabinet officers, let alone the leaders of independent agencies or major bureau chiefs. Most of a president's schedule is consumed with national security and foreign policy matters.

Vast differences exist in the scope and importance of cabinet-level departments. The huge Defense Department and the smaller departments of Labor, Housing and Urban Development, and Education are not at all similar. Certain agencies not of cabinet rank—the Central Intelligence Agency, the National Space and Aeronautics Administration—may be more important, at least for certain periods of time, than certain cabinet-level departments. Conventional rankings of the departments are based on their longevity, annual expenditures, and number of personnel. Rankings according to these indicators can be seen in the first three columns of Table 9.1. Even a casual comparison of these columns reveals unexpected characteristics. Thus, although the State Department is about 190 years older than some of the newer departments, its expenditures are among the lowest. On the other hand, the Department of Human Services, formerly the Department of Health, Education, and Welfare, officially only about forty years old, ranks second only to the Treasury in expenditures.

The contemporary cabinet can be differentiated also into "inner" and "outer" cabinets, as shown in Table 9.1. This classification, derived from interviews, indicates how White House aides and cabinet officers view the departments and their access to the president. The occupants of the inner cabinet generally have maintained a role as counselor to the president; the departments all include broad-ranging, multiple interests. The explicitly

TABLE 9.1
Ways of Looking at the Executive Departments

Seniority	Expenditures	Personnel	Inner & outer Cabinet
1. State	1. Treasury	1. Defense	*Inner:*
2. Treasury	2. HHS	2. Veterans' Affairs	1. State
3. War/Defense	3. Defense	3. Treasury	2. Defense
4. Interior	4. Agriculture	4. Agriculture	3. Treasury
5. Justice	5. Transportation	5. Justice	4. Justice*
6. Agriculture	6. Veteran's Affairs	6. Interior	*Outer:*
7. Commerce	7. Labor	7. Transportation	5. Agriculture
8. Labor	8. Education	8. HHS	6. Interior
9. HHS†	9. HUD	9. Commerce	7. Transportation
10. HUD‡	10. Energy	10. State	8. HHS
11. Transportation	11. Justice	11. Energy	9. HUD
12. Energy	12. Interior	12. Labor	10. Labor
13. Education	13. State	13. HUD	11. Commerce
14. Veterans' Affairs	14. Commerce	14. Education	12. Energy
			13. Education
			14. Veterans' Affairs

*Sometimes inner, sometimes outer. "Inner" and "Outer" cabinet is our classification done according to the counseling-advocacy departments.
†HHS, Health and Human Services.
‡HUD, Housing and Urban Development.
Source: Expenditures Statistical Abstract of the U.S. 1996, p. 334. Personnel figures from ibid., p. 345.

domestic policy departments, with the exception of Justice, comprise the outer cabinet. By custom, if not by designation, these cabinet officers assume a relatively straightforward advocacy orientation that overshadows their counseling role.

THE INNER CABINET

A pattern in the past few administrations suggests strongly that the inner, or counseling, cabinet positions are vested with high-priority responsibilities that bring their occupants into close and collaborative relationships with presidents and their top staff. Certain White House staff counselors also have been included in the inner cabinet with increasing frequency. The secretary of defense was one of the most prominent cabinet officers during all recent administrations, for each president recognized the priority of national security issues. Then, too, the defense budget and the DOD personnel, the latter ranging over three million (including the military), makes it impossible for a president to ignore a secretary of defense for very long. Despite the inclination of recent presidents to serve as "their own secretary of state," the top people at the Department of State have nevertheless had a direct and continuous relationship with contemporary presidents. Recent treasury secretaries—George Humphrey, Robert Anderson, Douglas Dillon, John Connally, George Shultz, William Simon, James Baker, Robert Rubin—also have played impressive roles in presidential deliberations on financial, business, and economic policy. The position of attorney general often, though not always, has been one of the most influential in the cabinet.

The inner cabinet, as classified here, corresponds to George Washington's original foursome and to most memoirs of the Eisenhower through Clinton period. The status accorded these cabinet roles is, of course, subject to ebb and flow, for the status is rooted in a cabinet officer's performance as well as in the crises and the fashions of the day.[40]

A National Security Cabinet

The seemingly endless series of international crises—Indochina, Central America, South Africa, the Middle East, Bosnia, and Iraq, and the tensions with the former Soviet Union and North Korea—have made it mandatory for recent presidents to maintain close relations with the two national security cabinet heads. Just as George Washington met almost every day with his four cabinet members during the French crisis of 1793, so also all of our recent presidents have been likely to meet at least weekly and be in daily telephone communication with their inner cabinet of national security advisers. One Johnson aide said it was his belief that President Johnson personally trusted only two members of his cabinet, Secretary of State Rusk and Secretary of Defense McNamara. Kennedy aides say their president viewed his regional assistant secretaries of state as more important officers of the government than most of the cabinet. Harold Brown (Defense) and Cyrus Vance (State) were President Carter's best appointments as well as two of his closest counselors. Carter said the Vances were his and his wife's friends among cabinet family. Reagan appeared to be closest, although he kept some distance, to Shultz, Weinberger, James Baker, and Ed Meese—all usually involved in one way or another with national security. Bush was closest to Secretary of State James Baker and DOD Secretary Dick Cheney. The others were all more remote.

Throughout recent administrations, more than a little disquiet has been engendered in the White House by the operational lethargy of the State Department. Although the secretary of state is customarily considered by the White House to be a member of the president's inner cabinet, the department itself is often if not always regarded as one of the most difficult to deal with. White House staffers invariably cite the State Department to illustrate White House–department conflicts. They scorn the narrowness and timidity of the encrusted foreign service and complain of the custodial conservatism reflected in State Department working papers. In State, more than in the other departments, the method and style of personnel, the special selection and promotion processes, and the protocol consciousness all seem farther removed from the political thinking at the White House. Although they often may be gifted and cultured, State's personnel invariably become stereotyped by White House aides as overly cautious and tradition bound.

Another source of the department's problems is the way in which recent secretaries, especially John Foster Dulles, Dean Rusk, Henry Kissinger, George Shultz, James A. Baker, Warren Christopher, and Madeleine Albright have defined their jobs. The demands on Rusk personally were such that departmental management was hardly his major priority. John Leacacos surmised that the priorities appeared to be: "First, the President and his immediate desires; second, the top operations problems of the current crisis; third, public opinion as reflected in the press, radio and TV and in the vast inflow of letters from the public; fourth, Congressional opinion; fifth, Rusk's need to be aware, at least, of everything that was going on in the world; and only sixth and last, the routine of the State Department itself."[41] That the secretary of state so often serves as the president's representative abroad or before Congress is another reason so few secretaries have had the time or energy necessary for managing the department's widely scattered staff. Moreover, more than fifty federal departments, agencies, and committees are involved in some way in the administration or evaluation of U.S. foreign policy.

Attorneys General

The Justice Department is often identified as a counseling department, and its chiefs usually are associated with the inner circle of presidential advisers. That Kennedy appointed his brother, Nixon appointed his trusted campaign manager and law partner, Carter appointed a personal friend, and Reagan appointed his personal attorney and later one of his campaign managers to be attorneys general indicates the importance of this position, although extensive politicization of the department has a long history. The Justice Department traditionally serves as the president's attorney and law office, a special obligation that brings about continuous and close relations between White House domestic policy lawyers and Justice Department lawyers. The White House depends heavily and constantly on the department's lawyers for counsel on civil-rights developments, presidential veto procedures, tax prosecutions, antitrust controversies, routine presidential pardons, the overseeing of regulatory agencies, and for a continuous overview of the congressional judiciary committees and separation-of-power questions. That these exchanges involve lawyers working with lawyers may explain in some measure why White House aides generally are more satisfied with transactions with Justice than those with other departments.

Treasury Secretary

The secretary of the treasury continues to be a critical presidential adviser on both domestic and international fiscal and monetary policy, but this person also plays somewhat of an advocate's role as an interpreter of the nation's leading financial interests. At one time the Treasury Department included the Bureau of the Budget. With budget staff and numerous economists, particularly within the Council of Economic Advisers, now attached to the White House, however, Treasury has become a department with major institutional authority and responsibility for income and corporate tax administration, currency control, public borrowing, and counseling of the president on such questions as the price of gold and the balance of payments, the federal debt, and international trade, development, and monetary matters. In addition, Treasury's special clientele of major and central bankers has unusual influence. Although the Council of Economic Advisers and the Federal Reserve Board may enjoy greater prestige in certain economic deliberations, they are less effective as counterweights in international commerce and currency issues, in which Treasury participation is most important. The latter connection helps to draw the department's secretary into the inner circle of foreign policy counselors to the president. Also, the treasury secretary is almost always a pivotal figure on key cabinet-level committees and in crucial negotiations with Congress on tax and trade matters.

The importance of the treasury secretary as a presidential counselor derives in part from the intelligence and personality of the incumbent. Dillon and Connally, for example, were influential in great part because of their self-assuredness and personal magnetism. President Eisenhower found himself responding in a similar manner to George M. Humphrey: "In Cabinet meetings, I always wait for George Humphrey to speak. I sit back and listen to the others talk while he doesn't say anything. But I know that when he speaks, he will say just what I was thinking."[42] Leonard Silk has suggested, however: "Formally, the Treasury Secretary had a mystique and power potential fully comparable to those of the chancellor of the Exchequer in Britain or the Minister of Finance in France. The mystique may not be all that mysterious to explain, it derives from money. Power over money, in the hands of the right man, can enable a Secretary of the Treasury to move into every definite action of government—in military and foreign affairs as well as in domestic economic and social affairs."[43]

Reagan's second-term treasury secretary, James Baker, was widely regarded as one of the most important individuals in government for several reasons. He had the confidence of the President, he enjoyed a reputation for competence, and he was nearly always a central player in policy decisions concerning the dollar, trade and tariff legislation, tax reform, and our economic relations with key allies. Clinton's Robert Rubin was given considerable credit for the successful economic strategies that helped win Clinton his 1996 reelection. Rubin was widely regarded as one of the most influential, if not the most influential, individuals in Clinton's cabinet.

INNER CIRCLES

The inner-circle cabinet members have been noticeably more interchangeable than those of the outer cabinet. Henry Stimson, for example, alternated from being Taft's secretary of war to Hoover's secretary of state and then back once more to war under FDR. Dean Acheson was undersecretary of the treasury for FDR and later secretary of state for Truman. Dillon reversed this pattern by being an Eisenhower undersecretary of state and later a Kennedy secretary of the treasury. When Kennedy was trying to lure McNamara to his new cabinet, he offered him his choice between Defense and Treasury. Attorney General Nicholas Katzenbach left the Justice Department to become an undersecretary of state; Eisenhower's attorney general William Rogers became Nixon's first secretary of state; and John Connally, once a secretary of the navy under Kennedy, became Nixon's secretary of the treasury. Within a mere four and a half years, Elliot Richardson, sometimes called "our only professional cabinet member," moved from undersecretary of state to HEW secretary to defense secretary to attorney general. He became, his unexpectedly short tenure notwithstanding, the fourteenth head of the Justice Department to have served also in another inner-cabinet position. Cyrus Vance, who had served as deputy secretary of defense for LBJ, became Carter's secretary of state. Harold Brown, LBJ's secretary of the air force, became Carter's secretary of defense. George Shultz, treasury secretary and OMB director under Nixon, served for seven years in the Reagan cabinet as secretary of state. Treasury Secretary Donald Reagan moved from treasury secretary to White House chief-of-staff under Reagan, and top White House aide Ed Meese served Reagan later as attorney general. Jim Baker went from White House chief-of-staff to treasury and then under Bush to state and finally back to chief-of-staff. Occasionally shifts have occurred from inner to outer cabinet, but these have been exceptions to the general pattern.

This interchangeability, or musical chair rotations, may result from the broad-ranging interests of the inner-cabinet positions, from the counseling style and relationships that develop in the course of an inner-cabinet secretary's tenure, or from the already close personal friendship that has often existed with the president. It may be easier for inner-cabinet than for outer-cabinet secretaries to maintain the presidential perspective; presidents certainly try to choose people they know and respect for these intimate positions.

In recent years members of the White House staff have performed cabinet-level counselor roles or even assumed cabinet-level roles and prerogatives. Eisenhower, for example, explicitly designated Sherman Adams to be a member of his cabinet ex officio. Kennedy looked upon Theodore Sorensen, McGeorge Bundy, and some of his economic advisers as perhaps even more vital to his decision-making than most of his cabinet members. Johnson and Nixon also assigned many of their staff to cabinet-type counseling responsibilities. Carter adviser Robert Strauss performed several cabinet-level responsibilities in the later 1970s, ranging from anti-inflation matters to diplomatic negotiations. Carter also relied heavily on National Security Adviser Zbigniew Brzezinski. Reagan plainly relied on James Baker and Edwin Meese to supervise or even control cabinet officials.

The people to whom presidents turn for overview presentations to members of Congress and cabinet gatherings are another indicator of inner-counselor status. When Kennedy wanted to brief his cabinet on his major priorities, typically he would ask Secretary Rusk to review foreign affairs, chairman of the Council of Economic Advisers Walter Heller to review questions about the economy, and Sorensen to give a status report on the domestic legislative program. When Lyndon Johnson held special seminars for gatherings of members of the Congress, he would invariably call upon the secretaries of state and defense to explain national security matters and then ask his budget director and his chairman of the Council of Economic Advisers to comment on economic, budgetary, and domestic program matters. Nixon and Ford usually called on Henry Kissinger, the director of the Office of Management and Budget, and one of their chief White House domestic policy counselors. Similarly, Carter and Reagan increasingly turned inward for coordination and crucial political strategy. Clinton relied heavily on OMB Director, then Chief-of-Staff Leon Panetta, as well as White House economic adviser Robert Rubin, who later became secretary of the treasury.

THE OUTER CABINET

The outer-cabinet positions deal with strongly organized and more particularistic clientele, an involvement that helps to produce an advocate relationship with the White House. These departments—Health and Human Services, HUD, Labor, Commerce, Interior, Agriculture, Transportation, Energy, Education, and Veterans' Affairs—are considered the outer-cabinet departments. Because most of the president's controllable expenditures, with the exception of defense, lie in their jurisdictions, they take part in the most intensive and competitive exchanges with the White House and the Office of Management and Budget. These departments experience heavy and often conflicting pressures from clientele groups, from congressional interests, and from state and local governments. These pressures often run counter to presidential priorities. Whereas three of the four inner-cabinet departments preside over policies that usually, though often imprudently, are perceived to be largely nonpartisan or bipartisan—national security, foreign policy, and the economy—the domestic departments almost always are subject to intense crossfire between partisan and domestic interest groups.

White House aides and inner-cabinet members may be selected primarily on the basis of personal loyalty to the president; outer-cabinet members often are selected, as already mentioned, to achieve a better political, geographical, ethnic, or racial balance. In addition to owing loyalty to their president, these people must develop loyalties to the congressional committees that approved them or to those that finance their programs, to the laws and programs they administer, and to the clientele and career civil servants who serve as their most immediate jury.

President Reagan and majorities in Congress pressed for the creation of a cabinet-level Department of Veterans Affairs in 1987 and 1988. The Veteran's Administration and its supporters wanted more recognition and status for their $30 billion program and nearly 250,000 employees. Critics viewed this move as just one more effort to create a "special interest" advocacy cabinet slot. Other campaigns have been mounted in recent years to add a cabinet-level Department of Trade.

THE PRESIDENT'S SPOUSE

Presidential spouses have a long history of behind the scenes influence. They certainly enjoy great proximity.[44]

The notable paradox about First Ladies is that the greater their perceived political and policy influence, the more they are criticized because of their unelected and unconfirmed (by Congress) status. Edith Wilson, Eleanor Roosevelt, and Hillary Clinton, and to a lesser extent Nancy Reagan, became feared or resented, and sometimes both, because they were thought to wield power and influence that had no basis in constitutional provisions.

While the Constitution may be somewhat vague concerning the precise powers of the presidency, it is absolutely silent on the role of the president's spouse. The First Lady or First Spouse's public position has evolved from a ceremonial hostess role to that of close and sometimes influential policy adviser.[45]

Initially there was confusion over how to address the president's wife. A nation accustomed to royalty and unschooled in the language of democratic politics understandably had a difficult time coming to grips with this new position. Martha Washington was often referred to as Lady Washington, and some even called the president's wife "Presidentress" or "Mrs. President." Over time the lack of (and some would say the desire for) royalty led to a reemergence of regal address, and it wasn't long before the president's wife was regularly referred to as "first lady."

Anthropologist Margaret Mead once said that, "Kings and queens have always focused on people's feelings and since we're not very far from a monarchy, the President's wife, whoever she is, has little choice but to serve as our queen."[46] But almost every president's wife has hated the title First Lady. Jackie Kennedy so loathed it that for a time she forbade her staff to use the term.

At first the role of First Lady was publicly downplayed. This reflected early confusion over the proper place of the president's wife in the public world, and the restricted role of women generally in early American society. Women were relegated to the home, certainly not to the public arena of politics. If First Ladies could not publicly display their political skills, they often exerted significant influence behind the scenes.

The gradual evolution of the First Lady's role from ceremonial/hostess to symbol surrogate, issue highlighter, campaign worker, and political adviser reflects the impact of certain activist First Ladies, the changing role of women in American society, increased power in the executive branch, and the increased public and press attention focused on the presidency's politics and personalities.

The first prominent First Lady was Dolley Madison, who remained a visible social figure for decades. But Dolley Madison was one of the early exceptions—the truly public First Lady. It was not until the time of the Civil War when Mary Lincoln and shortly thereafter Julia Grant basked in the public spotlight.

The first influential First Lady was probably Edith Wilson. When in 1919 her husband had a stroke, Mrs. Wilson became de facto president, prompting Massachusetts Senator Lodge to comment, "a regency was not contemplated in the Constitution."[47]

In later years the outspoken Eleanor Roosevelt would stretch the limits of her role as First Lady. The active and independent Mrs. Roosevelt became publicly identified with a number of progressive causes, and often served as the eyes and ears of FDR.

Edith Mayo, creator of the exhibit at the Smithsonian Institute entitled, "First Ladies: Political Role and Public Image," has written "[i]t is sad and telling that the press and public alike are unaware that Presidential wives since Abigail Adams have been wielding political influence. Further, it is disheartening to find that virtually every First Lady who has used her influence has been either ridiculed or vilified as deviating from women's proper role or has been feared as emasculating."[48]

In fact, the "office" of First Lady has been institutionalized, and is served by a sizable staff and a significant budget. Often the president's spouse is a key player in the internal affairs of the White House.

The president's spouse is nonelected and nonappointed, and no formal job description exists for the office.

More recent First Ladies have played an even greater public and political role than their predecessors. Jackie Kennedy was a valuable partner with her husband along the campaign trail. Her style and beauty contributed to the Camelot mystique that surrounded the Kennedys. Betty Ford, outspoken and direct, said she often used "pillow talk" to influence her husband.[49] Rosalyn Carter attended some cabinet meetings, used her position as a platform from which to advance women's issues, and was active in the field of mental health education.

Nancy Reagan was also influential. When her husband went through cancer surgery, Mrs. Reagan somewhat downplayed the vice president and announced herself as "the President's stand-in."[50] Mrs. Reagan also held veto power over her husband's schedule, sometimes consulted an astrologer as a guide to her husband's actions, influenced the hiring and firing of key personnel, and along with Michael Deaver, worried a lot about her husband's image and place in history. It is believed that, in the aftermath of the Iran-Contra scandal, Mrs. Reagan encouraged her husband to negotiate an effective arms deal with the Soviet Union in order to advance his reputation.

Barbara Bush was a more "conventional" First Lady. Grandmotherly, warm, and devoted to her husband, she preferred the role of "First Homemaker" to that of political adviser.

Hillary Rodham Clinton became one of the most controversial First Ladies in history. Mrs. Clinton was a respected attorney before her husband decided to seek the presidency, and was active in the legal rights for children movement. Independent and outspoken, Mrs. Clinton rubbed many conservatives the wrong way. Radio talk shows hosts blasted her as a radical liberal and criticized her influence.

Mrs. Clinton plainly did not fit the role of subservient trophy wife. She was the administration's unofficial leader on health care reform in 1993 and 1994, regularly attended policy strategy meetings, was an influential adviser to husband, and was a prominent spokesperson for the administration. Her political influence was not, by historical standards, unusual, but her openness about it was. In recognition of the political role of First Ladies, a 1993 U.S. Court of Appeals decision, *Association of American Physicians and Surgeons v. Hillary Rodham Clinton,* concluded that the First Lady was "a full-time employee of the government." Hillary Clinton emerged to political consciousness in an age when the women's movement opened doors for women. But the nation had not caught up, and Mrs. Clinton received more than her share of criticism for pushing the boundaries of the First Lady's position to new heights of public power.

In effect, the president's spouse is in a no-win situation. Expected to pursue a serious issue agenda, she is also held up to ridicule and suspicion if she has too much influence; asked to serve as first hostess and presidential partner, she cannot make the president appear weak or "unmasculine"; expected to support traditional values, she is also accomplished and increasingly has a career outside the home and independent of her spouse. Tensions are inevitable.

Overall, the public seems more comfortable with the Barbara Bush model than the Hillary Clinton model. A survey (January 1993) found that 70 percent of those polled preferred a traditional role for the First Lady. Apparently the public prefers its First Ladies to exercise influence behind the scenes, rather than in public.

THE WHITE HOUSE STAFF

The post–World War II era has witnessed a rise in the size, importance, and power of the president's staff (Table 9.2) and a corresponding decline in the importance and power of the cabinet. Many of the functions once performed by the cabinet are now the responsibility of the staff.

Part of the reason for this can be traced to the failure of the cabinet and bureaucracy to supply presidents with the help they feel they need. As stated earlier, presidents come to question the loyalty of cabinet officials, and the bureaucracy is often a creature of habit where the president may want creatures of politics. The staff has emerged to fill this vacuum.

The way a president organizes his staff depends on *personality, experience,* and *circumstances.* Bill Clinton likes to engage with his staff. This leads to an open, unstructured staff arrangement. In Clinton's case it was too unstructured, however, and in 1993 the president replaced Chief-of-Staff Mack McClarty with former Congressman and former OMB Director Leon Panetta. A large part of Panetta's job was to organize and discipline President Clinton.

Experience also plays a big role in the choice of staffing structure. Eisenhower, accustomed to a hierarchical and formal command structure from his career in the military, chose a formal staffing system for his presidency. There is not one perfect staffing structure for all seasons. In a crisis, for example, presidents often abandon the more formal or customary staff system and rely on a select few close advisers, as Kennedy did during the Cuban missile crisis, and as Bush did during the Gulf War.

One of the first questions a president must decide when putting to-

TABLE 9.2
Growth of the White House Staff

Year	President	Full-time employees
1937	Franklin D. Roosevelt	45
1947	Harry S Truman	190
1957	Dwight D. Eisenhower	364
1967	Lyndon B. Johnson	251
1972	Richard M. Nixon	550
1975	Gerald R. Ford	533
1980	Jimmy Carter	488
1984	Ronald Reagan	575
1992	George Bush	605
1993	Bill Clinton	543

Source: Adapted and updated from Thomas E. Cronin, "The Swelling of the Presidency: Can Anyone Reverse the Tide?", in Peter Woll, ed., *American Government: Readings and Cases,* 8th ed. Copyright © 1984. Reprinted by permission of Addison-Wesley Educational Publishers, Inc.

gether his administration is how strong a chief-of-staff he will have. James P. Pfiffner, a leading student of presidential management, says there are "two firm lessons of White House organization that can be ignored by Presidents only at their own peril: No. 1, A chief of staff is essential in the modern White House; No. 2, a domineering chief-of-staff will almost certainly lead to trouble" (e.g., Sherman Adams under Eisenhower, H. R. Haldeman for Nixon, Donald Reagan for Reagan, and John Sununu for Bush). Sununu, instead of catching lightning for President Bush, created it, and was eventually forced out. Pfiffner concludes that "the preferred role for a chief of staff is that of a facilitator, coordinator, and a neutral broker."[51]

Apart from the chief-of-staff decision, presidents must also decide how to organize their staff. There are a variety of different possible models: FDR's competitive approach, which set some staff members against other staff members in a dynamic tension; JFK's collegial style, which sought a cooperative, bonding approach; Nixon's hierarchical model, with a closed, rigid pyramid of access and line of authority; and Reagan's delegating style, which transferred power and authority to underlings.[52]

Each style of staff organization has costs and benefits. The key is for presidents to know themselves, their strengths and weaknesses, and to model the staff in such a way as to take advantage of their strengths while ensuring that their weaknesses do not lead to serious mistakes.

Presidents must be aware of several things lest they risk managerial (which translates into political) failure: (1) they need to spend time on managing; (2) they need to spend political capital on managing; (3) they need a mix of old experienced White House staffers *and* some new fresh thinkers from the outside; (4) they need a devil's advocate close to the center of power whose job is to poke holes in the ideas and programs that are proposed (all presidents need someone who can tell them to their face that they are screwing up); and (5) they need to be somewhat flexible in style.

The increased importance of the president's staff has spawned tension between aides and department heads. Rather than a creative tension, this conflict often degenerates into deep-seated mistrust.[53]

Can the White House and the departments arrive at functional working arrangements?

PROPOSALS FOR IMPROVED WHITE HOUSE DEPARTMENT RELATIONS

White House aides cite several sources of conflict between the White House and the executive departments; they also suggest a number of different remedies. The appropriateness of any suggested reform depends not only on the type of problem but also on which staff functions at the White House and which departments are involved. Rather than calling uniformly for a strengthened presidency, as might be expected, White House staff members often support a middle way, that is, an enhanced cabinet role as well as a strong, assertive White House role.

Most White House aides acknowledge that numerous remedial efforts are needed within the White House as well as between the White House and the departments. Many former presidential aides begin their discussion of reforms by pointing out that presidential styles, as well as policy preferences, differ, so "each president should organize his office more or less as he sees fit."

Even those White House aides who acknowledge the arrogance of the White House staff often conclude that presidents have to have tough and aggressive staff help. The following responses from White House staffers, suggested as remedies for reducing problems with the departments, provide some flavor of the strong "presidentialist" beliefs of many of these aides.

> The presidency has to be the activist within the very conservative federal bureaucracy. The bureaucracy is the . . . custodian of old laws and old policies. They fight against anything new suggested by the White House, ... The inability of department institutions to be creative or to take on new responsibilities is fantastic! In my view, the most important thing for a president is to know how to shake up the bureaucracy!
>
> I think it is impossible to run the White House staff without having tough men to do the work of the president. [Kennedy aides] were of this type. They could be very tough, abrasive, and uncompromising. But they had to be tough because if they were not the people in the agencies and departments just wouldn't respect the communications that came from the White House. I think it is a fundamental dilemma that people working for a president have to be arrogant, and almost be bastards, in order to get White House work done with the departments.[54]

Although considerable overlap exists between those aides who support a presidential and those who support an integrative perspective, the latter approach receives relatively more support from among the administrative, public relations, and national security policy aides than from among the domestic and budgetary policy advisers. Integrative recommendations seemingly are based on the assumption that the White House is unlikely to have much effect on the implementation of federal programs unless it can win support and cooperation from the middle and upper echelons of the executive branch departments.

White House staff aides note that even a strong presidency cannot succeed without an executive branch characterized by strong cabinet and departmental leadership. Many of these aides said the president and senior staff had underestimated the importance of these factors in making the government work. In interviews, one aide insisted that allowing the domestic cabinet departments to become so divorced from the White House was a major mistake: "One way to improve things is to have the president and the cabinet members, particularly in domestic areas, meet at least six or seven times a year and talk in great detail and in highly substantive terms about the major priorities of the administration. You have to have better communication. Basically you have to make the cabinet less insecure." Other aides criticized certain of their colleagues for having usurped operational responsibilities of the regular agencies, adding that if they accomplished anything, these aides enlarged their own importance more often than they expedited programs.

The American system is not a parliamentary one, and the cabinet does not, should not, and cannot exist as a collegial body in any formal sense. We elect the president, not the cabinet. We do not want a system in which the president is merely the first among equals. The essential question of the role of the cabinet in presidential government is an attitudinal, not a structural, one. A strong and "healthy" cabinet can exist today only because a president wants it. Multiple-advocacy systems that would enhance the constructive advice he might get cannot be imposed on a president. A president must find it compatible with his style. Ultimately, operating with a strong cabinet requires a president who is at ease with different and new opinions, who has high confidence in both his own personal and political position and the people he appointed. One lesson is clear: presidents will always have strained relationships with several members of their cabinets.

ADVOCACY CONFLICTS

White House aides generally view outer-cabinet executives as special pleaders for vested clientele interests over the priorities of the president. One of Franklin Roosevelt's commerce secretaries frankly acknowledged his representational and advocacy obligations when he explained: "If the Department of Commerce means anything, it means as I understand it the representation of business in the councils of the administration, at the Cabinet table, and so forth."[55] Few White House aides speak of the virtue of having a cabinet member reflect his or her constituency and how this helps, or could help, educate a president. Most speak of the cabinet as a burden and an ordeal for presidents rather than as a chance to forge coalitions and exercise leadership.

Because presidents have less and less time to spend with the outer-cabinet members, the advocate role becomes less flexible and more narrowly defined. Outer-cabinet members find they have little chance to discuss new policy ideas or administrative problems with the president. They must make the most of their limited meetings.

The interpretation of the advocate role by both the outer-cabinet member and the president may vary. It is much easier to listen to an advocate whose point of view fits with the White House philosophy than to one who continually transmits substantively different arguments or who encroaches on other policy arenas. As departments have grown and their administration has become more exhausting, fewer of the department heads have had time to be well versed on problems beyond their domain.

OUTER-CABINET ISOLATION

As tensions build around whether, or to what extent, domestic policy leadership rests with the departments, with the Office of Management and Budget, or with the White House, and as staff and line distinctions become blurred, the estrangement between the domestic department heads and the White House staff deepens. White House aides believe they possess the more objective understanding of what the president wants to accomplish. At the same time the cabinet heads, day in and day out, must live with the responsibilities for managing their programs, with the carrying out of laws delegated to them by Congress, and with the multiple claims of interest groups. Outer-cabinet members often complain about the unmanageability of their departments, the many pressures on them, and the undermining of their efforts by White House aides.

Joseph Califano, HEW cabinet head under Carter, says White House aides were generally both incompetent and hostile. They were constantly leaking negative comments about him and other cabinet members. White House operatives pressured cabinet members to do political chores and make campaign stops, and sometimes they opposed vigorous enforcement of the laws. In Califano's case this meant they would have liked him to ease up on his antismoking campaign and university desegregation efforts. Califano contends he had good, and usually excellent, relations with President Carter. But White House staffers under Carter, with one or two exceptions, rarely answered his phone calls, seldom took an interest in his work, and hardly ever made his demanding job any easier. Mutual resentment was the result. In the end, these aides appeared to have succeeded so well at cabinet member bashing that even the president lost faith in many of his cabinet members. Carter, Califano reports, grew tired of acting like a referee between the cabinet and the White House staff.

Carter's National Security Adviser Zbigniew Brzezinski's attitude toward cabinet meetings may have reflected his attitude toward many of Carter's cabinet members as well as the futile efforts to make cabinet meetings work. Brzezinski remembers that these meetings "were almost useless. The discussions were desultory. There was no coherent theme to them and after a while they were held less and less frequently."[56]

Carter criticized his own White House staffers for failing to establish good relations with cabinet officers. Yet time and again he told his cabinet: "I need your absolute loyalty."[57] He told Califano, one of those he fired, "the problem is the friction with the White House staff. The same qualities and drive and managerial ability that make you such a superb Secretary create problems with the White House staff. No one on the staff questions your performance. . . . But you and some members of the staff . . . have not gotten along."[58] Califano, never a shrinking violet, was a high-profile entrepreneurial cabinet officer. In earlier years he had been a top White House aide to President Johnson. By now, however, he was a famous and wealthy lawyer and he was doubtless "a difficult customer"—at least from a White House staffer's point of view. Califano knew a cabinet officer had to put a little distance between himself and the president, especially on controversial matters. Califano, always a high achiever, was a threat to younger White House aides. They suspected, perhaps with some justification, that he harbored political ambitions of his own (such as running for governor or U.S. senator in New York) and thus sometimes put his own ambitions over the interests of the White House.

John Ehrlichman, one of Nixon's senior aides, wrote that Nixon viewed many of his domestic cabinet officers as "crybabies" and would often threaten to fire them or transfer them to some more remote position. He disliked meeting with them, disliked their demands on his time, and disliked their constant efforts to increase their budgets. Ehrlichman's depiction of the problems in Nixon's cabinet is as revealing about Nixon and Nixon's staff as it is about his cabinet members: "What went wrong with the Nixon cabinet? Surely something did. Most of the cabinet members were discontented most of the time, and many of them failed to manage their departments well."

At root were the President's own shifting and variable concepts of the cabinet—of what it should be and do—and what the president expected from it. At first, during its selection, he wanted the cabinet to symbolize that all elements of the body politic were to be represented in the Nixon administration. There would be liberals and conservatives, ethnics and even Democrats. At the same time there was some rather vague intention that the cabinet would perform a collegial advisory function.

But soon Nixon changed his view: he looked on the cabinet principally as managers of their respective bureaucracies. A good secretary keeps things under control. Cabinet members were to be spokespersons, too. They should be out in the country making the case for the president and his policies. A good secretary is a good PR operative.[59]

Every administration appears to reenact an identical script. Domestic advisers—or the outer cabinet—are led to believe they will be close intimates of the president, key policy advisers, and consulted on a range of matters. For awhile this happens. Inevitably, however, the president's time becomes scarce. Inevitably, the president's budget director wants to cut the cabinet member's departmental budget and the president usually sides with the budget director. Invariably, too, a president's political advisers grow impatient at a cabinet member's failure to "be more political" or failure to protect a president from politically damaging controversies. Thus Nixon, Ford, Carter, and Reagan all fired or accepted the contrived resignations of several of their cabinet officers.

The size of the bureaucracy, distrust, and a penchant for the convenience of secrecy lead presidents to rely heavily on White House staffers. The White House and executive office aides increasingly become involved not only in gathering legislative ideas but also in getting those ideas translated into laws and laws into programs. Program coordination and supervision, although often ill managed, also become primary White House interests. To an extent, these additional responsibilities transform the White House into an administrative rather than a staff agency. Outer-cabinet departments, understandably, begin to lose the capacity to sharpen up their programs, and the department heads feel uneasy about the lack of close working relations with the president.

One of the problems with this tendency to pull things into the White House and to exclude departmental officials is that excellent ideas or proposals that may exist lower down in the bureaus of the permanent government seldom get the attention they deserve because, according to people in the departments, nobody asks them or, at least, no one pays them serious attention. As social psychologists have suggested, one group tends to develop stereotyped perspectives that not only dehumanize the outsiders but also cut off the very communications channels that might provide valuable and even vital information.

Rather than try to strengthen departmental capacities to come up with broader, more innovative proposals, White House aides, impatient for action, instead have developed a wide array of advisory committees, commissions, and secret task forces or White House operational groups.

Clinton's first secretary of labor, Robert B. Reich, a prominent economist, devotes most of his well-written memoir about his four years in Clinton's cabinet bemoaning his fate in the outer-cabinet. He laughs at his situation, yet he clearly felt removed from access and he also felt his views were far less listened to than those of the president's inner circle of White House aides. Reich's memoir entry for June 6, 1994, captures both his sense of isolation and his frustration:

> Federico Pena, the secretary of transportation, phones me to ask me how I discover what's going on at the White House. I have no clear answer for him. The place is so disorganized that information is hard to come by. The decision-making "loop" depends on physical proximity to [Clinton]—who's whispering into his ear most regularly, whose office is closest to the Oval, who's standing or sitting near him when a key issue arises.
>
> [My aides] each have their sources over there. I also rely on Gene Sperling and a few other campaign veterans to keep me abreast.
>
> One of the best techniques is to linger in the corridors of the West Wing after a meeting, picking up gossip. Another good spot is the executive parking lot between the West Wing and the old Executive Office Building, where dozens of White House staffers tromp every few minutes.
>
> In this administration you're either in the loop or you're out of the loop, but more likely you don't know where the loop is, or you don't even know there *is* a loop.[60]

Ironically, or paradoxically, Clinton told people he admired his cabinet members but felt too constrained and often let down by his White House aides. Listen to what Clinton tells political strategist Dick Morris:

> I spent all my time before I took office choosing my cabinet. . . . It's a great cabinet. But I didn't spend the time I should have choosing my staff. I just reached out and took the people who had helped me get elected and put them on the staff. It was a mistake.[61]

DEALING WITH THE BUREAUCRACY

As the nation's chief executive officer, presidents are supposed to sit at the top of the bureaucracy and control its actions. But their control is incomplete at best. A president operates under a four-year time restraint; the bureaucracy has no such time constraint. The old saying, "Presidents come and go but bureaucrats stay and stay," speaks volumes. The president is the temporary occupant of the White House; the bureaucracy is the permanent government.

The federal bureaucracy, greatly swollen by the New Deal and the Cold War agencies, as well as by Great Society programs, is one of the most visible checks on a president. Indeed presidents are quick to fault the bureaucracy for the many problems that beset the implementation and evaluation of presidential programs. One is reminded of Harry Truman's observation about what it would be like for his successor to be a president rather than a general: "He will sit here and he'll say, 'Do this! Do that!' And nothing will happen. Poor Ike—it won't be a bit like the Army! He'll find it very frustrating."[62]

The problem of how to control the bureaucracy has become a major preoccupation for presidents. Even persons who championed the New Deal now recognize that the executive bureaucracy can be a presidential curse. Arthur Schlesinger, Jr., wrote that, "as any sensible person should have known, the permanent government has turned out to be, at least against innovating Presidents, a conservatizing rather than liberalizing force."[63] Concern about taming the bureaucracy comes from the right, from the revisionist left, and from moderates. Participants in the Nixon administration constantly embraced the same theme, even to the extent of claiming that their programs were being sabotaged from all directions. One of the key factors in former President Nixon's continuing attempts to centralize more authority either in the White House or in the hands of a few trusted and strong cabinet officials was this suspicion about the loyalties and parochialism of federal civil servants.

Gaining control over existing bureaucracies and making them work with and for the White House is an enormous burden on the president. They constantly delegate, they must be most precise about what they delegate, and they must know whether and for what reasons the agencies to which they are delegating share their general outlook. They must be sensitive to bureaucratic politics, to the incentives that motivate bureaucrats, and to the intricacies of their standard operating procedures. They must have some assurance (and hence an adequate intelligence system) that what they are delegating will be carried out properly.

Recent presidents have often misunderstood the workings of bureaucracy. They have little appreciation for bureaucrats' considerable concern about organizational essence, organizational morale, and organizational integrity. Presidents mistakenly yet invariably look upon the executive branch as a monolith, and they are especially offended when senior bureaucrats differ with them or otherwise refuse to cooperate. Presidents too often become defensive and critical of the bureaucracy. They fear, sometimes with reason, that their pet programs will get buried in inert custodial hands. Thus they have often sought shortcuts by setting up new agencies for each of their pet projects and relying increasingly on separate advisory and staff units within the executive office. However, the creation of a new agency does not guarantee presidential control. Indeed, some of the most independent, even maverick, federal agencies were originally set up to bypass the so-called old-line departments. Among such agencies are the Atomic Energy Commission, the Peace Corps, the Office of Economic Opportunity, and the Department of Energy.

Bureau chiefs and career civil servants often do avoid initiative, taking risks, and responsibility, opting instead for routine and security. The bureaucracy most assuredly has its own way of doing things, perhaps more conservative or more liberal than what the president wants. But the fact that bureaucratic interests and presidential interests often differ does not mean that the permanent employees of the federal executive branch constitute an active enemy force. Bureaucratic organizations act, rather, in rational ways to enhance their influence, budget, and autonomy. And they generally believe that in doing so they act in the nation's interest.

The bureaucracy often defines the national interest differently from the way it is defined in the White House. But a close examination of the two definitions may reveal that both are valid and representative views of what is desirable about which reasonable people can legitimately differ. The task for a president, then, is to understand the strategies and tactics of federal bureaus and appreciate the underlying motivations. Properly diagnosed, the bureaucratic instinct for competition, survival, and autonomy can be creatively harnessed by the White House both to educate itself and to develop cooperative alliances.

BUREAUCRATIC VETOES

One major cause of the distance and frequent distrust between the presidency and the rest of the executive branch lies in the nature of the bureaucracy and of bureaucrats.

> Consider the official who directs the day-to-day operations of even a broadly defined program; let us call him the bureau chief. He directs the work of large numbers of people, he disposes of large sums of money, he deals every day with weighty, intricate, and delicate problems. He has probably spent most of his adult years in the highly specialized activity over which he now presides. He lives at the center of a special world inhabited by persons and groups in the private sector who stand to gain or lose by what he does, certain members of Congress who have a special interest in his actions, and a specialized press to which he is a figure of central importance. The approbation which is most meaningful to him is

likely to be the approbation of the other inhabitants of this special world. The rest of the federal government may seem vague and remote, and the President will loom as a distant and shadowy figure who will, in any event, be succeeded by someone else in a few years. It would be unreasonable to expect this official to see his program in the Presidential perspective.[64]

Seldom is there a passion as keen as that of functionaries for their functions. The bureaucrats whose basic loyalty is to the established way of doing things and whose job survival and promotion are tied directly to program survival and expansion, instinctively but rationally resist threats to program control or coordination from the White House.

Consequently, although the bureaucracy was intended to be neutral and nonpolitical, the bureau chief will extend his allegiance to persons or groups who will aid and abet the enlargement of his program or, at the least, its stability. Knowing that presidential and congressional support are only won when the bureau can demonstrate widespread and intense public support, that a president wants to be associated with popular programs, many bureaus and even departments willingly invite capture by special interests to gain what they perceive as indispensable clientele and grassroots support. Where little special-interest support exists, shrewd bureau chiefs do everything possible to create it. Moreover, they spend considerable time forging alliances with well-situated members of Congress and their staffs.

How entrenched these relations may become can be inferred from the fact that, whereas tenure on the White House staff averages less than three years and even less than that among cabinet appointees, the bureau chiefs, senior members of Congress, senior staff in Congress, and veteran Washington lobbyists often endure in their posts for ten or fifteen years or more. Former Secretary of Health, Education, and Welfare John W. Gardner told the Senate Government Operations Committee:

> As everyone in this room knows but few people outside of Washington understand, questions of public policy nominally lodged with the Secretary are often decided far beyond the Secretary's reach by a trinity consisting of (1) representatives of an outside body, (2) middle level bureaucrats, and (3) selected members of congress, particularly those concerned with appropriations. In a given field these people may have collaborated for years. They have a durable alliance that cranks out legislation and appropriations in behalf of their special interest. Participants in such durable alliances do not want the Department Secretaries strengthened. The outside special interests are particularly resistant to such change. It took them years to dig their particular tunnel into the public vault, and they don't want the vault moved.[65]

Practically any new presidential initiative, therefore, faces a strategically placed potential veto group with major allies within the executive branch itself: a social services bureaucracy to resist a negative income tax, an Office of Education to resist measures that would undermine traditional school-of-education training programs, a defense establishment to resist disarmament agreements, and so forth.

The professionalization of many sections of the federal government has created another potentially powerful constraint on presidential action. Professionals in government—for example, senior engineers and physicists in the National Aeronautics and Space Administration or the Defense Department, physicians and biologists at the National Institutes of Health, and economists throughout the government—ordinarily are more committed to the values of their profession than to the political fortunes of presidents. Professionals are more likely to move in and out of government than are other career civil servants. Once outside, the professionals can use their insider knowledge to become effective critics of official decisions as well as lobbyists for interest groups. Those remaining inside can resist the administration by leaking documents or key findings or by threatening collective resignations.

For a president to move the bureaucracy he must invest time and political capital in doing so. Most presidents believe this to be an unwise use of their limited resources.

If presidents cannot and do not control the bureaucracy, they still need the bureaucracy to implement their policies. But they are not helpless. Different presidents use different methods to move the bureaucracy. Some presidents do it by placing loyal followers in key positions, others assign a top aide to ride herd, some try to circumvent an unresponsive agency, others create new agencies staffed by their hand-picked administrators. Still other presidents devote their own time and effort to either persuade, bully, or negotiate with an unresponsive bureaucrat.

Bureaucrats can frustrate a president's wishes in a variety of ways. Orders are never self-executing, and it may not be clear exactly what a president wants. Also, some members of the permanent government may disagree with the president's directives and may intentionally sabotage the order by ignoring it, altering it in its execution, or failing to pass it on to the operational level. Delay is another common form of bureaucratic resistance.

CONCLUSION

Regardless of how organization charts are drawn, future presidential use of the cabinet and White House staff probably will give greater weight to the realities of the differentiated roles and activities of the federal departments. The experience of numerous cabinet councils was an experiment that failed in Reagan's first term. But having working groups of cabinet members along the national security, economics, and domestic policy lines makes sense so long as the groups do not preclude occasional full meetings of the cabinet. A president and the White House top advisers need the advice of some persons who are not directly concerned with a problem.

A redesigned and strengthened outer cabinet might enable the White House staff to abstain more often from the temptation to gather administrative responsibilities to itself. But reorganizational developments will not lessen the need for skillful and decisive executive branch mediators who, with the full confidence of the president, can preside over thorny and complex claims and counterclaims by competing domestic departments, and who also can know when important elements of a debate are being seriously neglected or misrepresented within cabinet-level negotiations. White House aides need to be chosen with these abilities in mind.

In sum, the American cabinet has had a strange and anomalous history. It is paradoxically one of the best known and least understood aspects of our governmental system. It seldom operates as a policymaking body, yet every administration strives to reinvent this mythical wheel. No law or constitutional authority commands a president to consult with a collective cabinet or even to hold cabinet meetings. The job of the department head, especially in modern times, necessarily demands that department heads heed the concerns of Congress, bureaucracies, the Constitution, and various professional and ethical considerations in addition to the desires and demands of a president. Even the most carefully chosen cabinet member will occasionally differ, if not clash, with certain presidential wishes.

No president will be wholly satisfied with the loyalty of department heads. Cabinet members invariably aspire to a president's intimacy, and many in the inner cabinet will indeed become effective counselors and sometimes even friends, yet even some of these will break with a president, often without realizing how serious their differing perspectives have been. A president, however, must welcome advocacy from cabinet officers. And although open and frank discussions in cabinet meetings will sometimes limit a president's discretion, a president will generally profit from using the cabinet and smaller policy clusters of cabinet heads as regular consultative bodies.

The historical record of the cabinet is one of frustration, yet a more modified conception of the cabinet holds out promise for making a presidency more effective and responsible. Structural changes that must take place are minimal relative to the attitudinal changes that must take place on the part of the president, White House staff, and cabinet members. In the end, it is the president who determines whether that promise will remain unrealized because of ineffectual disuse or be translated into a means for furthering presidential leadership.

The American Vice Presidency

Once there were two brothers. One ran away to sea, the other became Vice President of the United States, and nothing was heard of either of them again.

> Attributed to Woodrow Wilson's Vice President
> Thomas Marshall

What the Vice President will do or is permitted to do . . . is determined by what the President assigns to him or permits him to do. . . . The President can bestow assignments and authority and can remove that authority at will.

> Hubert Humphrey, *Current History,* 67 (1974), p. 58

The job [of vice president] doesn't lend itself to high profile and decision making. It lends itself to loyally supporting the president . . ., giving him your best judgment, and then when the president reaches a decision, supporting it.

> George Bush, quoted in Michael Duffy and Dan
> Goodgame, *Marching In Place* (Simon & Schuster, 1992),
> p. 40

During our first 150 years, vice presidents served primarily as ceremonial ribbon-cutters. They also served as president and presiding officer in the U.S. Senate. The post was often seen as a semiretirement job for party stalwarts, as a resting place for mediocrities or a "runner-up." Its occupants typically found themselves in a frustrating spectator role. In most administrations a vice president was at best a kind of fifth wheel and at worst a political rival who sometimes connived against the president.

Until recent decades, Americans were more inclined to joke about the office than think about it. For a variety of sensible reasons, however, presidents now usually take the vice presidency more seriously, and its responsibilities, while still mainly ad hoc, have expanded. As Paul C. Light has noted, "After two hundred years as errand-boys, political hitmen, professional mourners, and incidental White House commissioners, Vice Presidents can now lay claim to regular access to the President and the opportunity to give advice on major decisions."[1]

Still, the vice presidency remains an ambiguous and paradoxical office.[2] To the founding fathers it was pretty much an afterthought. To some scholars it is one of our most conspicuous constitutional mistakes. To some modern-day presidents it still appears to be more of a headache or a threat than an asset. To vice presidents it is often a confusing and unhappy experience. "A damned peculiar situation to be in," said Vice President Spiro T. Agnew, "to have . . . a title and responsibility with no real power to do anything." Lyndon B. Johnson said that much of the unhappiness in the office stems from knowing you are on "a perpetual death watch." Vice President Hubert Humphrey once said it was "like being naked in the middle of a blizzard, with no one to even offer you a match to keep you warm. . . . You are trapped, vulnerable and alone. . . ."

Paradoxes abound. A vice president's chief importance still consists of the fact that he may cease to be vice president. We yearn for someone to fill the post who has the competence to be president, yet it is a kind of an "off-the-job" training and waiting post. Perhaps the prime paradox is how we select vice presidents. We select our presidential nominees by a process of intensive democratic exposure and deliberation that is long and grueling. But we continue to leave the designation of the vice presidential running mate almost entirely to the personal judgment of the presidential nominee—a judgment that is sometimes made hastily, possibly by an overextended and exhausted nominee.

The main purpose of having such a position is to have a stand-by competent leader available if accident befalls the president. Frequently, however, the main criterion used in selecting the vice presidential nominee is balancing the ticket to help win crucial electoral college votes. These two needs, *competence* and *electoral utility,* need not be incompatible, but they often are. The question also arises: How can we get vice presidents of presidential quality if once in office they may have little to do and doubtlessly will be maligned before we really need them, if we need them? Some presidents only share "dirty work," such as press bashing, attacking the opposition party, and funeral duty responsibilities, with their vice presidents. Some do not share much at all for fear of being upstaged and outshone. Others have understandably refused to delegate responsibilities because of sharp policy differences. One student of the office concludes there really can be no deputy or alternate president, because of the "indivisibility of presidential leadership and the lack of place for tandem governance for two."[3]

Because vice presidents are always "a heartbeat away" from becoming president, there is often a strain in relations between vice presidents and their boss. Henry Kissinger wrote that the "relationship between the president and any vice president is never easy; it is, after all, disconcerting to have at one's side a man whose life's ambition will be achieved by one's death."[4] And Lyndon Johnson wrote of his vice presidential days, "Every time I came into John Kennedy's presence, I felt like a goddam raven hovering over his shoulder."[5]

Another paradox is that technically a vice president is neither a part of the executive branch nor subject to the direction of the president. In practice, however, recent vice presidents have shunned their Senate duties and taken up office in the White House, and have served primarily as presidential advisers.

A further paradox is that what it takes to be an effective vice president is vastly different from what it takes to be president. Foremost among the attributes of an effective president are independence and strength of character and skills as a national agenda setter and national unifier. A vice president, on the other hand, even while acting as an "understudy" for the president, must always be loyal and self-effacing while trying to avoid being obsequious. A vice president is constantly aware of the fact that the office and its importance are at least 90 percent dependent on the preferences, and even the whims, of the incumbent president. When a president talks, a vice president listens. To paraphrase Cat Stevens and his famous "Father and Son" tune, a vice president, in effect, says, "From the moment I was elected, I was ordered to listen."[6]

Originally the person who received the second highest vote in the presidential election became vice president. This was changed, after a bitter dispute over the election in 1800, when Aaron Burr and Thomas Jefferson both received equal votes. With the Twelfth Amendment to the Constitution (1804), the president and vice president are elected on separate ballots. In recent times, however, more and more presidents essentially dictate the choice of their running mates and successors. Thus Roosevelt, Kennedy, Nixon, and Reagan have picked their own successors, a notion the framers assuredly did not intend. Some people believe the practice of leaving this vital choice of future national leadership to the discretion of a single individual, the presidential candidate or the president, is unacceptable in a nation that professes to be democratic.

Support exists for devising better ways to pick vice presidential nominees. The search is made more urgent by the fact that eight presidents have died in office, four by assassination, four by natural death, and one (Nixon) has resigned. One-third of our presidents were once vice presidents, including five of our last nine presidents.

The vice presidency has been significantly affected by two post–World War II constitutional amendments. The Twenty-second, ratified in 1951, imposes a two-term limit for the presidency, which means vice presidents have a somewhat better chance of moving up to the presidency. The Twenty-fifth, ratified in 1967, confirms prior practice that, on the death or resignation of a president, the vice president becomes not acting president but president.[7] Of perhaps greater significance, this amendment provides a procedure, albeit somewhat ambiguous, to determine whether an incumbent president is unable to discharge the powers and duties of the office. Thus the amendment allows an incapacitated president to lay aside temporarily the powers and duties of the office without forfeiting them permanently, as Reagan did for a few hours in 1985 during a cancer operation. The constitutionality of this procedure had previously been in doubt. Further, the amendment creates a mechanism through which a vice president together with a majority of the cabinet may declare the president incapacitated and thus serve as acting president until the president recovers. This procedure answers several problems but also, as shall be discussed, raises new questions.

In addition, the Twenty-fifth Amendment established procedures to fill a vacancy in the vice presidency (a procedure used when Nixon selected Ford and when Ford selected Rockefeller). In the event of such a vacancy, the president nominates a vice president, who takes office upon confirmation by a majority vote of both houses of Congress. This procedure should ensure the appointment of a vice president in whom the president has confidence. If the vice president, under these circumstances, has to take over the presidency, he or she can usually be expected to reflect most of the policies of the person the people had originally elected.

The reasonably positive Mondale, Bush, and Gore experiences prompted some modest reappraisal of the vice presidency. The way in which these men were selected and subsequently used has no doubt enhanced the importance of the position. Their experience somewhat increased its prestige as a position of influence as well as an apprenticeship. Still, each of them either lapsed into or were forced into a rather unbecoming "cheerleader-in-chief" role. Inextricably, even the best of vice presidents find themselves having to defend administrative policies in which they have had little or no say in shaping.

The Dan Quayle experience—both the way he was selected and his inability to earn the respect of the American public—was a temporary setback to the office. Al Gore, under Clinton, won both White House and public praise.

This chapter looks first at the traditional problems of the office. Then it considers recent vice presidencies. Finally, it treats questions of vice presidential selection and succession.

TRADITIONAL PROBLEMS

The office, as noted earlier, is often condemned as a superfluous non-job. In fact at the Constitutional Convention of 1787, Benjamin Franklin ridiculed the position by proposing the occupant be called "Your Superfluous

Highness." The position, along with any serious duties or prerogatives attached to it, comes only at the pleasure of the president. "I am not in a leadership position," said Vice President Nelson Rockefeller. "The President has the responsibility and the power. . . . The Vice President has no responsibility and no power."[8]

Few offices have been so disdained and lampooned. Our first vice president, John Adams, wrote to his wife in 1793 that: "My country has in its wisdom contrived for me the most insignificant office that ever the invention of man contrived or his imagination conceived." Daniel Webster remarked in 1848 when he turned down the vice presidential nomination: "I do not propose to be buried until I am dead." The great Henry Clay is said to have spurned the job twice. John Nance Garner, FDR's first vice president, said the job, "isn't worth a pitcher of warm spit." Gerald Ford could not get himself to accept Ronald Reagan's 1980 offer of running mate because he knew from experience that the job did not amount to much. Reportedly, Mario Cuomo twice turned down Walter Mondale's request in 1984 that he be Mondale's running mate, and may have done it again when Clinton aides approached him in 1992.

American voters send 537 elected officials to Washington. Of these, 536 have a reasonably clear idea of their role and functions. Vice presidents, on the other hand, are never exactly sure whether their functions will change from week to week. Over time, however, the jobs of the vice president have grown. On paper they look somewhat impressive:

1. President of the U.S. Senate
2. Member of the National Security Council
3. Chair of several national advisory councils
4. Diplomatic representative of the president and U.S. abroad
5. Senior presidential adviser
6. Liaison with Congress
7. Crisis coordinator
8. Overseer of temporary coordinating councils
9. Presider over cabinet meetings in absence of the president
10. Deputy leader of the party
11. Apprentice available to take over job of the president, either on acting or full-time basis
12. President-in-waiting and possible future presidential candidate

Vice presidents have done all of these things. Several vice presidents have served their president and their nation well. Often, however, considerable political, psychological, and structural obstacles stand in the way of the constructive use of vice presidents. As we shall discuss later, Mondale, Bush, and Gore were able to overcome some of these.

PRESIDENT OF THE SENATE

Until about 1940 most presidents and most Americans viewed the vice presidency almost exclusively as a legislative job. Even as late as 1954, the recently retired vice president, Alben W. Barkley, would write that, "The Vice President's principal duty is to preside over the Senate."[9] Vice presidents were to serve as president of the Senate, preside there, and cast an occasional tie-breaking vote.

Originally the Senate comprised a small number of elder statesmen; twenty-six men plus the vice president. Tie votes were frequent: John Adams cast twenty-nine tie-breaking votes. Nowadays, however, mainly because the Senate has grown to one hundred members, tie votes occur less than once a year. Dan Quayle didn't get to cast a single tie-breaking vote in four years. Al Gore has on only one occasion.

The Senate was designed in many ways as an ideal place for a vice president to learn the business of government and to serve along with the nation's leading thinkers. The Senate was intended to pass laws, shape national policy, confirm major presidential appointees, oversee the approval of treaties, and, more generally, advise and counsel presidents, particularly in foreign affairs. Over time, the Senate delegated more and more of its powers to the president, cabinet, and even to Executive Office advisers, so much so that the Senate is now almost indistinguishable from the House of Representatives—neither a council of state nor the prime presidential counseling body.

Technically vice presidents could become frequent presiders in the Senate. In practice they have not. The job of presiding over the Senate has become largely thankless and powerless, and today offers little training for the position of chief executive. One observer concluded that

> the single job conferred on a Vice President by the Constitution—presiding over the meetings of the United States Senate—can be performed by any six-year old. The presiding officer need do no more than sit there like an elegant dunce, and repeat aloud whatever the parliamentarian stage whispers to him. It is about as challenging as having your hair cut. . . ."[10]

Senators today view vice presidents as intruders, certainly not as providers of instruction or leadership. Members of Congress now look on a vice president as a member of the executive branch. Lyndon Johnson, upon election as vice president, for example, sought to retain a certain measure of the control over Democrats that he enjoyed as majority leader. He apparently talked Senator Mike Mansfield into inviting him to preside over Democratic caucus meetings, even though he was about to become vice president.

The proposal that LBJ serve as presiding officer of all the Senate Democrats whenever they met in a formal conference while he also served as vice president was opposed not only by several of his liberal antagonists but also by members of his own so-called "Johnson Network." The liberals wanted revenge for his having ridden roughshod over them too long. One argument they put forth was that it was an unnatural mixing of the separate branches: "We might as well ask Jack Kennedy to come back up to the Senate and take his turn at presiding," snapped Senator Albert Gore. More telling was the opposition of a long-time Johnson intimate:

> Senator Clinton Anderson specifically noted his support at Los Angeles and the debt all Democratic Senators owed Johnson for his leadership in the Senate the past eight years. But the office of the Vice President, said Anderson, was more a creature of the executive branch than the legislative branch. Therefore, quite apart from the fact that the Senate Democrats would look ridiculous electing a non-Senator to preside over them, to do so would violate the spirit of separation of powers.
>
> The debate continued in a mood of embarrassment. Johnson was present. . . .[11]

Johnson understood the message and seldom again did he try as vice president to influence Senate proceedings.

Vice presidents are rarely used effectively in congressional relations work. This is often because they are not properly prepared by their president. Harry Truman, in his brief tenure as vice president, wanted to become an effective link between Congress and the White House. But this depended on close contact with FDR, and contact was virtually nonexistent. Lyndon Johnson was similarly underused. Johnson was simultaneously rebuffed by both Congress and the White House. Caught both coming and going, he retreated into a nearly three-year sulk.

Franklin Roosevelt's troubles with John Nance Garner were legendary. Garner was an important figure in the Congress and he skillfully assisted the passage of a number of New Deal measures in FDR's first term. But he grew upset with the pace and sweeping scope of later New Deal measures and became outspoken in his opposition to several of them. He opposed Roosevelt's scheme for packing the Supreme Court. He opposed FDR's bid for a third term. Toward the end of his second term, President Roosevelt consciously avoided meeting with Vice President Garner and considered him an impossible person.

The limits of vice presidential influence in the Senate are illustrated by one of Spiro Agnew's transgressions of political convention. Early in his tenure, Agnew tried to interest himself in mastering Senate rules as well as learning the Senate's informal folkways: "But he violated protocol by lobbying on the Senate floor in behalf of the tax surcharge extension supported by the administration. 'Do we have your vote?,' he asked Senator Len Jordan of Idaho, a Republican. The senator replied, 'You did have until now." Thus was established Jordan's rule: "When the Vice President lobbies on the Senate floor for a bill, vote the other way."[12] Thereafter, Agnew seemed to lose interest in trying to become presidential emissary on Capitol Hill.

Vice President Hubert Humphrey had served in the Senate for sixteen years before becoming vice president. He was aware that as presiding officer he was to be seen but not heard, except for procedural matters. This was a difficult adjustment for the loquacious Humphrey.

Until recently, a vice president was a full member of neither executive nor legislative branch. The office, because of the duties as presiding officer of the Senate, is clearly a constitutional hybrid, limiting somewhat the use a president can make of a vice president. Thus, President Eisenhower asserted repeatedly in his memoirs that whatever executive assignments the vice president performed were voluntary and by his request.[13]

Some say a vice president should be relieved of traditional Senate functions and freed to participate more directly in executive branch responsibilities. The late James F. Byrnes of South Carolina, a former senator, associate justice of the Supreme Court, as well as secretary of state, offered this perspective:

> If a motion (in the Senate) does not receive a majority vote, it should be considered lost. It is not wise that the Vice President, a representative of the executive branch of government, should affect the will of the legislators by casting a decisive vote. In short, participation by the Vice President in Senate voting, either in support of his own views or the President's, constitutes a violation of the spirit of the fundamental provision of the Constitution that the three branches of our government shall forever be separated.[14]

Byrnes urged we adopt a constitutional amendment to effect this change.

Constitutional scholar Joseph Kallenbach also favored such a change, contending the vice president could then become a sort of "minister without portfolio," subject to executive assignments as a president may direct. "This would not only insure he would function in a subordinate administrative capacity to the President and not be tempted to become a rival," it would also provide assignments over a period of time that would enable him to acquire a wider knowledge of the operations of government as a whole. "In this way," adds Kallenbach, "he could be given a better opportunity than at present to prepare himself for the responsibility of serving as chief executive in case fate should thrust the role upon him."[15]

This suggestion has been largely ignored, just as most recent vice presidents have dismissed their Senate "job" as virtually irrelevant. Vice presidents do show up on certain ceremonial occasions, such as an opening day and when the president addresses a joint session of Congress, but the job of presiding over the Senate is low on the vice president's priority list.

VICE PRESIDENTS AS "ASSISTANT PRESIDENTS"

Corporate officials are often baffled with the way presidents underutilize their vice presidents. The trend in top management circles in the private sector, particularly in recent years, has been to form corporate managerial teams. The chief executive officer usually relies heavily on an executive vice president and a vice president for finance, a vice president for marketing, and often a series of group vice presidents heading up major divisions or subsidiary companies. Collegial leadership, although not always a success in practice, is a celebrated ideal in business communities. The talented chief executive in the private sector delegates well and delegates often. Nurturing up-and-coming executive talent is widely applauded.[16]

Presidents do delegate some of their responsibilities—but less to vice presidents, who are often considered outsiders, than to their own insiders, such as Harry Hopkins, Joseph Califano, H. R. Haldeman, and James A. Baker, III. Have presidents tried to nurture their vice presidents, assigning them important executive apprenticeship experiences? Have vice presidents been called on to serve as cabinet counsellors, policy advisers, and "assistant presidents?" They occasionally have, especially in recent years. But curiously, not as much as one should expect.

The cabinet is often as nebulous an entity as the vice presidency. But a brief examination of vice presidential involvement in the cabinet helps to illustrate the changing, if still limited, vice presidential executive portfolio.

George Washington conferred on occasion with Vice President John Adams, but Adams acted in his official executive capacity only once, when he attended a cabinet meeting in 1791. Washington was away and wanted his department heads to get together in his absence, and he requested Adams to join them. Thomas Jefferson, when he served as vice president, declined President John Adam's invitation to attend cabinet meetings. Other vice presidents for the next 120 years stayed away from the cabinet sessions—either by personal or by presidential desire, usually both. Then in 1918 Thomas R. Marshall substituted for Woodrow Wilson when Wilson was in Europe. President Warren Harding invited his vice president, Calvin Coolidge, to regularly attend cabinet meetings.

FDR's second vice president, Henry Wallace, an experienced administrator who had already served two terms as secretary of agriculture, was the first vice president to be assigned major administrative duties. Roosevelt made Wallace chairman, successively, of the Economic Defense Board, the Supply Priorities and Allocations Board, the War Productions Board, and the Board of Economic Warfare. In the latter post Wallace became an aggressive administrator and an outspoken advocate.

Henry Wallace was idealistic, stubborn, and outspoken. He cherished the idea of succeeding Roosevelt as president, but he was always loyal. Roosevelt delegated considerable authority to Wallace, authority to oversee much of the domestic economy, strategic imports, exports, shipping, foreign exchange, and related matters. But Wallace's political base as vice president and the authority delegated to him proved inadequate to his responsibilities. He fast became embroiled in a massive collision with the departments of State and Commerce, and with secretaries Cordell Hull and Jesse H. Jones in particular. Both men vigorously defended their departments from nearly every Wallace intrusion. Both men attacked Wallace viciously, Jones often in public.

To perform his executive assignments Wallace had to secure State's and Commerce's approval with some frequency, but his efforts were in vain. He sought Roosevelt's approval for enlarging his authority to correspond with his duties, but Roosevelt yielded only a little. For a year and a half—1942 to June of 1943—Wallace sought to make a go of it. But after several public skirmishes and open warfare between the vice president and the secretary of commerce, Roosevelt abolished Wallace's board. So ended a year and a half experiment of having a vice president serve as an administrator. By most standards it was an experiment that failed, although factors other than the office of the vice presidency were involved. Thus, some people would claim that this failure was primarily due to Wallace's stubborn and apolitical temperament. Others might contend, perhaps correctly, that Roosevelt never gave Wallace adequate authority with which to do the job assigned to him. In any event, the Wallace example stands out as almost the sole instance up to the present of serious administrative responsibilities being delegated to a vice president.

Since 1943 vice presidents have, however, been invited to cabinet meetings and related policy councils with some regularity. By a 1949 amendment to the National Security Act of 1947, the vice president now sits as a regular member of the National Security Council. Since the Eisenhower presidency, the vice president has been made the cabinet's acting chairman in the president's absence. Nixon chaired several cabinet sessions during Eisenhower's several illnesses. LBJ and Humphrey chaired several presidential-level councils and commissions, and, like Wallace and Nixon before them, they traveled widely as goodwill ambassadors as well as on some sensitive policy missions. FDR actually began the tradition of sending vice presidents abroad. John N. Garner was FDR's representative at the installation of a Philippine president in 1935, and Henry Wallace as vice president traveled to Latin America as well as to China and Siberia. Nixon traveled to fifty-four countries, LBJ to more than thirty. Humphrey and Agnew were often sent to visit Asia. Mondale conducted a wide range of diplomatic missions to twenty-six nations during his four years as vice president. George Bush continued this tradition and traveled to even more nations; he logged nearly a million miles during his eight years in the Veep post. Dan Quayle recognized that vice presidential trips were almost the only way he could win positive press coverage.[17]

Nearly every vice president in recent times has found it difficult to interfere with the work of cabinet members. The departmental secretaries preside over congressionally authorized departments; hence a vice president is very much an intruder unless a problem arises that is definitely interdepartmental. Even then, what is an interdepartmental matter to some may not be to others. Thus Secretary of State Alexander Haig was reportedly irate in 1981 when President Reagan officially designated George Bush to be the administration's chief "crisis manager."

Reformers sometime urge that vice presidents be assigned specific cabinet responsibilities, such as secretary of defense or attorney general. This might answer the criticism that the vice presidency is a training ground for nothing at all. Critics of this suggestion say that the existing presidentially approved cabinet members are already hard enough to "keep in line." As an elected department official, a vice president who served as a cabinet secretary/department head would be even tougher to "control" or to "fire." A vice president as department head would also surely be compromised by the narrow clientele constituencies of a single agency. Such a precedent, if established, might also affect the kinds of people a president would pick for vice president. Picking someone who might be a good department head might unwisely subordinate breadth, political talent, or general leadership ability to managerial experience.

Finally, it is not wholly evident that a departmental secretary position provides presidential training. Under some circumstances it might, under others it clearly would not. How many recent department heads come readily to mind as especially promising presidential candidates? To be sure, Jefferson and Madison served as cabinet members prior to their presidencies, but so too did James Buchanan and Herbert Hoover.

Both Presidents Hoover and Eisenhower recommended the creation of an additional appointed vice presidential post. Hoover advocated an administrative vice president; Eisenhower a "first secretary of government" for foreign affairs. Neither proposal received much attention. President Eisenhower's brother and confidant, Milton Eisenhower, urged the creation of two executive vice presidents, to be appointed by the president— one to deal with domestic policy, the other with international affairs, but both to work "in close collaboration with the President." Milton Eisenhower claimed the elected vice president cannot hope to do what appointed vice presidents can. First, he said, echoing his older brother, the elected vice president is not a member of the executive branch. Second,

> Even though he normally is suggested to a convention by the presidential nominee as a runningmate, there is no guarantee that he will be in agreement with the president; there is no legal bar to his openly disagreeing with the president. Vice presidents have often disagreed with chief executives. . . . Most important of all, the president cannot discharge the elected vice president. This point is critical. In any organization delegation of authority is workable only if the chief executive has absolute confidence in the individual to whom he delegates some of his own duties, and this confidence exists only if there is mutual trust, a shared philosophy, and a recognition by the subordinate that final authority always rests with the chief; should directives of the president be unacceptable to the appointed executive vice president, and he felt that he could not loyally do what was expected of him, he would have to resign or be removed by the president.[18]

Eisenhower's suggestion has obviously not been acted on. Instead presidents rely on key cabinet members or those who serve on their staff.

More radical is a proposal that what we really need to do is to separate the presidency into two functions: the head of state and the chief executive. "What we can do, perhaps, is establish a head of state to greet foreign dignitaries and to visit them abroad, to officiate on occasions when a personification of the nation is required, to become the central figure even at the inauguration of the chief executives, and to live at the White House. The chief executive who would be elected every four years as at present, would live as cabinet members live."[19] Beguiling as this proposal might be, it has generally been considered naive. Doubtless it smacks too much of importing the royalty feature from Great Britain. Because of this alone it will never gain popularity here. Most experts contend too that the chief executive office would be significantly weakened if the chief of state functions were taken away. Presidents such as the Roosevelts and Reagan clearly mixed the two and strengthened themselves by doing so.

PSYCHOLOGICAL PROBLEMS
OR THE "THROTTLEBOTTOM COMPLEX"

Tensions between a president and a vice president are natural; after all, the vice president is the only officer who works closely with a president who cannot be fired by him.

Alexander Throttlebottom was a character in the successful 1930 Broadway play "Of Thee I Sing" by George S. Kaufman and Morris Ryskind. Throttlebottom was a stumbling, mumbling caricature of an unemployed vice president. Neglected by his president, John P. Wintergreen, Throttlebottom is artfully drawn as an unknown, unwanted, and improbable national leader. Not only did he not want to be vice president or know what was expected of him, but he was even refused the right to resign. The unhappy "Throttlebottom" sat around in the parks and fed the pigeons, took walks, and went to the movies. The character of Throttlebottom provided for splendid satire and comic relief, but the mocking portrait lives on and the name has become a familiar term for the real plight of contemporary vice presidents.

The estrangement of vice presidents from presidents began with Adams and Washington. Relations between the first president and first vice president were civil but scarcely friendly. Then during his tenure as vice president under President Adams, Thomas Jefferson refused certain diplomatic assignments. Jefferson said of Aaron Burr, his first vice president, that he was a "crooked gun whose aim or shot you could never be sure of." Time and again relations between presidents and vice presidents have been strained, abrasive, and cold to the point of open hostility. A few other examples underscore the mutual frustration.

Roosevelt and John Nance Garner:

> The once cordial relations between the two men had long since turned sour. They had little contact except at cabinet meetings, where Garner, red and glowering, occasionally took issue with the President in a trucu-

lent manner. Roosevelt hinted he would desert the Democratic cause before he would vote for the Texan for President (in 1940). By early 1940 even official relations between the two men had almost ceased; Roosevelt was hoping that the Vice President would not show up for cabinet meetings. The President was gleeful about Garner's tribulations as a presidential candidate. . . .[20]

John F. Kennedy and Lyndon B. Johnson:

They [the White House] made his stay in the vice presidency the most miserable three years of his life. He wasn't the number two man in that administration; he was the lowest man on the totem pole. Though he has never said this to anyone (perhaps because his pride would never let him admit it), I know him well enough to know he felt humiliated time and time again, that he was openly snubbed by second-echelon White House staffers who snickered at him behind his back and called him "Uncle Cornpone."[21]

Richard M. Nixon and Spiro Agnew:

Nixon barely knew Agnew when he selected him in Miami Beach in 1968. Nixon, himself, had been humiliated in the vice presidential job, a job he called a hollow shell, the most ill conceived and poorly defined position in the American political system. He proceeded nonetheless to visit an even greater humiliation on Agnew. By most accounts Agnew was scorned by the White House staff and given little of importance to do.

Why does the "Throttlebottom complex" persist? Is there something congenitally or structurally deficient in the relationship between the presidency and the vice presidency? Does the relationship always have to be hollow, hostile, and counterproductive? "Mistrust is inherent in the relationship," wrote Arthur Schlesinger, Jr., because vice presidents are inevitable "reminders of their president's own mortality."[22] This mistrust is also not without historical foundation. Some previous vice presidents have turned against their presidents, mobilizing opposition in the Congress, proposing alternative legislative programs, and preparing to run for the presidency against the president they were then serving. Vice President John Calhoun became so disenchanted that he simply resigned from the office.

There are many reasons for these psychological tensions, but at least one factor is that both the president and the vice president are, or should be, astute politicians, and as Harry Truman once suggested, this often means neither will take the other completely into his confidence.

Although some progress has been made, a psychological barrier, an intangible distance, still seems to prevent a significant delegation of power to a vice president. Presidents do fear being upstaged. Their staffs fear it even more than they do. Presidents sometimes consciously assign jobs that are unpresidential in character. Thus, "The very fact that a problem is turned over to the vice president argues that it's not very important. . . ."

Some vice presidents refuse to accept demeaning jobs. Nelson Rockefeller did for a while until he lost interest. Lyndon Johnson just went into a retreat. Others willingly become partisan "hit men," straight-arming the press, slashing back at presidential opponents, and stumping endlessly for the reelection of friendly members of Congress. Thus, Spiro Agnew was used as a mouthpiece and as a surrogate presidential speaker at partisan events. Some of George Bush's supporters worried that he was hurt by being too loyal a Reagan cheerleader in the 1984 election campaign. Quayle's handlers may have felt the same way in 1992. Such experiences fit Arthur M. Schlesinger's generalization that the vice presidency may be less a *making* than a *maiming* experience for its occupants.

Often, it is as if a vice president is entirely unacceptable and unwanted unless he is willing to merge his identity completely with that of the president. Vice presidents, it turns out, often do lose their own identity. They become cheerleaders for their boss. They become expert at laying wreaths at remote ceremonial functions. They become, in a way, part of the furniture in another person's house.

The following passages from a televised "Conversation with Vice President Hubert H. Humphrey" illustrate how one vice president seemingly subordinated his own personal ambitions and vowed nearly feudal homage to his president:

> I did not become vice president with President Johnson to cause him trouble. I feel a deep sense of loyalty and fidelity. I believe that if you can't have that you have no right to accept the office. Because today it is so important that a president and his vice president be on the same wave length. . . . I'd hate to have the president be worried about me, that I may do something that would cause him embarrassment or that would injure his administration. . . . There are no Humphrey people, there are no Humphrey policies, there are no Humphrey programs. Whatever we have we should try to contribute, if it's wanted, to the president and his administration. You can't have two leaders of the executive branch at one time. . . .[23]

In spite of this subservient "leaning over backward," Humphrey was never really trusted by LBJ. Humphrey frequently sought more substantive assignments, only to be disregarded. Johnson would simply exclude Humphrey from his team of insiders whenever the vice president sought independence or tried to develop his own line of thinking. This was especially true in the case of occasional policy differences he had with Johnson over the Vietnam War.

When is loyalty to a president carried too far? Vice Presidents Humphrey, Agnew, Ford, and Bush sometimes became such apologists for their administrations that they could not help but diminish their credibility, perhaps undermining their own future capacity to provide serious leadership. This raises the question of whether being an apologist isn't inherent in the role of a vice president, at least the role of a trusted vice president. Some promising presidential hopefuls have had their political futures destroyed or at best greatly diminished as a result of their allegiance and their images of being "yes men" to less-than-perfect administrations.

Experience suggests, however, that not only will most presidents not trust their vice presidents but the relationship in terms of real assignments will be a limited one. Can a vice president, even an outstanding one, become a significant force in the presidential establishment? Or was John Nance Garner correct when he reasoned that a great person may be vice president but can't be a great vice president because the office in itself is unimportant. It is near impossible to assess a vice president's contribution to an administration's policy achievements. For instance, observers who say Mondale, Bush, and Gore, in contrast to Agnew and Quayle, were ideal vice presidents are hard pressed to describe in any rigorous way their share of the Carter-Mondale, Reagan-Bush, or Clinton-Gore records. This is in part due to the "low profile" nature of their duties. What were their own achievements? Can Mondale be relieved of Carter's failures? Can Bush point to his own contributions? On what issues did Gore lead the way?

Gerald Ford saw tensions and jealousies develop between his own five hundred-person White House staff and Nelson Rockefeller's vice presidential staff of seventy aides and assistants. Ford allows that he and Rockefeller got along famously. "Nelson was absolutely loyal to me, and he would do anything I asked him to do," writes Ford in his memoir.[24] But Ford, despite the fact he believed Rockefeller was eminently qualified to "step into my shoes" if tragedy struck, still went along with his own political advisers' campaign to dump Rockefeller. Ford was motivated almost exclusively in this instance by the political considerations. Later, he acknowledged, "I was angry with myself for showing cowardice in not saying to the ultra-conservatives, 'It's going to be Ford and Rockefeller, whatever the consequences.' "[25]

Despite all these traditional problems, the office has endured, if not prospered. Some of the problems have been diminished somewhat by reducing the physical distance between president and vice president and by providing more staff and perquisites.

Until 1961 vice presidents had to work out of their office in the Capitol. Lyndon Johnson was the first vice president to be given a suite of offices in the old Executive Office Building, across an alley from the west wing of the White House. Spiro Agnew was the first to be given an office in the White House, although he only stayed in it for a few months. In 1975, Nelson Rockefeller was the first modern vice president to be given an official residence. In 1977, Mondale was given his choice of White House offices (excepting only the oval office), and he stayed on the main floor of the west wing for his entire four years, as did George Bush for several years thereafter (although Bush actually preferred to work in the elegant suite also assigned vice presidents in the old Executive Office Building, next door to the White House).

By the time Mondale became vice president the trappings were considerable. He had offices in the White House, at the Capitol, in the Executive Office Building, in the U.S. Senate office buildings, at the vice president's official residence, another one back in Minnesota (for reasons that are not entirely clear), and yet another one in "Air Force Number Two," the jet made available for the vice president's never-ending journeys. Not bad for a "non-job." Today, "the vice president's office is . . . a replica of the president's office, with a national security adviser, press secretary, domestic issues staff, scheduling team, advance, appointments, administration, chief of staff, and counsel's office."[26]

THE MONDALE MODEL

Was Vice President Walter Mondale's influence the result of amplified importance of the vice presidency, or merely the result of Carter's willingness to be persuaded on a number of issues? Or, perhaps, was Mondale's alleged prominence in the Carter White House due mainly to Carter's own lack of Washington experience and thus serious need of advice from a veteran (twelve-year) former member of the U.S. Senate?

The vice presidency "has its frustrations . . .," said Mondale. "But I went into it with my eyes wide open. I know there is only one president; there is not an assistant president. I'm his adviser."[27] Most students of the vice presidency agree that Mondale enjoyed a closer relationship with his boss, President Jimmy Carter, than any previous vice president. Much of the credit has to go to Carter. Carter not only said he would give his vice president a lot to do; but he tried in earnest to make the vice president as close as one can get to a full working partner. Mondale enjoyed a standing invitation to attend and participate in any of the president's meetings. He had access to all reports and cables. Mondale also had a regularly scheduled weekly luncheon date with Carter.

Mondale came to the job with several helpful assets. He was well known and well liked on Capitol Hill. He knew how Congress and Washington operated. In many respects, he had more of a national base or national following than did Carter. A protégé of Hubert Humphrey's, he nonetheless had an identity of his own.

Perhaps his greatest strength was his ability to get along well with people—all kinds of people. He was a natural team player and not a showboating, flamboyant credit-seeker. General Eisenhower had once said "There's no limit to how much influence you can have if you are only willing to let others get the credit for it." Mondale was a shrewd politician who understood this proverb.

Mondale also understood the vulnerable "tender nature" of the vice presidency. He knew that nearly everything contributing to his effectiveness as vice president depended on a basic personal relationship of confidence and trust with the president. He knew also that presidential powers are not divisible. He knew he could not long be effective as chief lobbyist or as chief-of-staff at the White House—two jobs that Carter and his staff sometimes tried to give him. Mondale rejected the chief-of-staff post because he knew if he had taken that assignment "it would have consumed vast amounts of my time with staff work and distracted me from important work."

"I see my role," Mondale said in his second year in the vice presidency, "as a general adviser on almost any issue, as a troubleshooter, as a representative of the president in some foreign affairs matters and as a political advocate of the administration."[28]

Mondale and Carter overcame some of the traditional problems of the vice presidency by treating Mondale's staff as part of the White House staff. Some senior members of the Mondale staff were assigned high-level White House responsibilities, often with supervisory authority over some White House aides. Mondale's staff won high marks for professionalism, competence, and ability to work with and for Carter as well as with and for their boss. In doing this, Mondale and his staff were able to set a helpful precedent—one that has helped to remove some of the tension and barriers to success so frequently witnessed in the past.

Vice President Mondale is credited with being perhaps the first in that job who regularly exercised substantive policy influence rather than merely an occasional input of ideas. Aides say he moderated certain extreme policies that might otherwise have been announced by the White House. Although he doubtless succeeded in blocking some bad initiatives, he also failed on a number of occasions. The Mondale experience suggests too that even though his views were regularly heard, his role as a policy initiator was usually a limited one. Initiatives typically came from the cabinet departments, from staff agencies, and from Congress. With all the other responsibilities a vice president has, it is difficult for the vice president to take the initiative.

When Carter was asked about his conception of their relationship he put an especially positive gloss on it:

> I probably meet with the Vice President on a daily basis more than all the other staff members that I have combined. . . . there is no aspect of my own daily responsibilities as President that are not shared by the Vice President. . . . I would say without derogating the other members of my staff that there is no one who would approach him in his importance to me, his closeness to me, and also his ability to carry out a singular assignment with my complete trust.[29]

Carter wrote in his memoir that even when Mondale disagreed with him, he always gave him full support. "As we served together, our relationship constantly improved—a pleasant surprise to the two of us as well as to other observers of the White House."[30] He also said no president has ever been blessed as he was to have such a good vice president. We "are actually almost as close as brothers."

Carter's trust in Mondale was doubtless a significant factor in their relationship. Carter was not intimidated by Mondale. Further, Carter valued Mondale's frankness and confidential discussions. Mondale became someone with whom Carter could sit down and discuss political and policy problems. "What the President needs is not more information, although that is helpful," recalls Mondale, "he needs a few people who can honestly appraise and evaluate his performance. . . . He needs to hear voices that speak from a national perspective. He has no limit to the number of people who want to talk to him, but that does not assure him of the confidentiality he needs to speak freely."[31]

Is the vice presidency a useful understudy position, a place where one can serve a sensible apprenticeship for the presidency? Mondale insisted it was for him:

> I think it may be the best training of all (for the presidency). I don't know of any other office, outside of the presidency, that informs an officer more fully about the realities of presidential government, about the realities of federal government and the duties of the presidency that remotely compares to that of the Vice President as it is now being used.
>
> I'm privy to all the same secret information as the President. I have unlimited access to the President. I'm usually with him when all the central decisions are being made. I've been through several of these crises now that a President inevitably confronts, and I see how they work. I've been through the budget process, I've been through diplomatic ventures, I've been through a host of congressional fights as seen from the presidential perspective.
>
> I spent 12 years in the Senate. I learned a lot there, but I learned more here about the realities of presidential responsibilities. I learned more about our country and the world in the last three years than I could any other way.[32]

As vice president, Mondale breathed life into the possibilities of the office instead of viewing his position as an incurably frustrating hybrid, half-legislative and half-executive. He adopted the view that the vice president is the only office of the national government that breaches the separation of powers. He successfully played a broker role in helping to gain ratification of the Panama Canal treaty. He campaigned to help elect Democrats to Congress. He also did a lot of "creative listening."

In addition he tried to assist Carter's campaign to fend off Edward Kennedy's bid to snatch the Democratic Party's 1980 nomination. Some critics think Mondale may have gone too far, that he was forced into and accepted an uncomfortable and unnatural role of criticizing Kennedy's record and purposes when Kennedy's voting record was nearly identical to his own Senate record. Thus James Reston of *The New York Times* wrote "even good men like Fritz Mondale are being corrupted by the political struggle in the process." And, "It is almost startling to hear what Mondale says in this struggle for votes and support. He defends policies as vice president which, when he was representing Minnesota as a liberal Senator, he opposed."[33] Mondale said Reston was being unfair and overly selective. Mondale claims he was justified in asking Kennedy what he would do differently, what policies he would pursue that would be significantly different. "I have never been a 'hatchet man,'" said Mondale, "and I never would be, and the President would never ask me to be."[34] The line may have become thin at times, however.

When his term drew to a close in early 1981 Mondale told his successor, George Bush, that there were some institutional lessons he wanted to pass along. Mondale's chief recommendations were as follows:

1. Advise the president confidentially. The only reasons to state publicly what you have told the president is to take credit for his success and try to escape blame for failure. Either way there is no quicker way to undermine your relationship with the president and lose your effectiveness.

2. Don't wear a president down. . . . Give your advice once and give it well. You have a right to be heard, not obeyed. A president must decide when the debates must end. . . .

3. As a spokesman for the Administration, stay on the facts. A president does not want and the public does not respect a vice president who does nothing but deliver fulsome praise of a president. . . .

4. Avoid line authority assignments. If such an assignment is important, it will then cut across the responsibilities of one or two cabinet officers or others and embroil you in a bureaucratic fight that would be disastrous. If it is meaningless or trivial, it will undermine your reputation and squander your time.

5. The vice president should remember the importance of personal compatibility. He should try to complement the president's skills and, finally in a real sense the most important of all roles, be ready to assume the presidency. . . .[35]

Did the Mondale-Carter relationship establish a precedent for succeeding administrations? Mondale believed so.

Mondale may have succeeded because he was willing to hew the Carter line and he remained exceedingly loyal—even when close friends of his such as HEW Secretary Joseph Califano were being fired and even when some of his cherished social programs were being cut. One Carter White House aide said virtually no one around the White House challenged Carter seriously, and that Mondale concurred with Carter's preference for "harmony above all other virtues." Others say Mondale's influence stemmed from Carter's dependence on him to explain how Congress worked and to maintain close ties with the labor movement. Carter needed Mondale more than most presidents need their vice presidents. Mondale's patient personality and relative youth meant he (1) could wait for his own chance to become president; and (2) could avoid being a threat to Carter. Even Mondale would admit, and surely Bush would, that it helps to be seen as a trusted lieutenant but not be heard as an exponent of too many original views. "In effect the vice president's influence is inversely proportional to his perceived influence."[36]

Questions remain—perhaps they can be called the paradoxes of the modern vice presidency. Vice presidents have to give their advice in private. They may give candid advice to presidents but they have to keep that advice to themselves. And then, whatever the president decides, the vice president has to support. Must they become apologists and cheerleaders even when they disagree with presidents? Also, can vice presidents earn election in their own right when they have had to be that loyal and when their major contributions (if they have made any) have to remain mostly or even entirely off the record? Few vice presidents are likely to solve these puzzles.

THE BUSH EXPERIENCE

Most observers believe George Bush enjoyed almost as much influence and probably as good a relationship with Ronald Reagan as Walter Mondale enjoyed with Jimmy Carter. Bush started out in a more compromised position. He had campaigned vigorously against Reagan in the Republican primaries of 1980. His criticisms of Reagan and his proposals had hurt Reagan personally and it is known that for this and perhaps other reasons Bush was not Reagan's first choice for vice president. But Bush was selected and the Reagan-Bush ticket was victorious not only in 1980 but again in 1984. Reagan never wavered in his support for Bush.

Bush benefited from the model established by Carter-Mondale. He inherited the "Mondale office" in the White House both physically and spiritually. He was given access to the president's daily briefing papers and other important paper flow in the White House itself. Bush's stock rose when he acted prudently and professionally during the time of Reagan's assassination attempt and subsequent hospitalization early in the first term. It rose even more as he became an ardent supporter and campaigner for Reagan's economic programs. Bush also performed countless partisan chores with effectiveness.

Gradually the personal friendship and trust between Reagan and Bush developed and indeed often seemed to flourish. At the time of the second inaugural, Reagan enthusiastically toasted his vice president as "the best Vice President this Republic has ever had." Bush, for his part, became about as supportive and cheerful about his president as any vice president had ever become. "I can't think of anything that the president has left undone that he might have done to make me feel more comfortable in my job," he said. "And I guess the best example of that is that I am free to walk three doors down the hall to the Oval Office and walk in there. And I don't have any inhibitions doing that."[37]

Reagan has called Bush the most loyal team member that anybody could ever want. Bush even changed some of his views during his first four years of serving Reagan. He had opposed Reagan's supply side economics during the campaign of 1980, calling it "voodoo economics." He also supported the right of women to have an abortion and he endorsed the Equal Rights Amendment to the U.S. Constitution. By 1984, however, he had adopted the Reagan position on each of these issues and supported Reagan on just about everything else. He made it a policy not to discuss his private advice for the president. He rarely spoke at cabinet meetings or at similar forums when the president was present. "I don't ever discuss what I talk to the president about, because if I did, I would undermine the one thing that matters—the confidentiality and the trust that I think exist between us." He prized access and proximity over the ego needs and rewards for public credit. Further, as he put it, "I don't want to end up like Nelson Rockefeller, miserable in the job because the staff cut him off at the knees."[38]

Bush took on the chairmanships of several task forces, notably one on reducing federal regulations. He enlarged the staff. He also traveled more than any vice president had ever traveled. And although it was difficult to discern whether Bush regularly "influenced" Reagan on important policy matters, both the public and Washington insiders credited both Reagan and Bush for continuing to enhance the office and the evolving advisory role of the vice president. He became, in effect, a White House aide and political adviser to the president.

Like those before him, however, he was unable to escape some of the criticism and even ridicule that still seem to come with the job. Thus, during the campaign of 1984 Bush was sometimes portrayed as a man without any ideas of his own. He had become such a "cheerleader-in-chief" that he seemed to have sold out or compromised some of his own views. The cartoon strip "Doonesbury" portrayed him as having "put his political manhood in a blind trust." *Time* magazine wrote that "Bush's emphasis on loyalty and his failure to put his mark on policy cast him in the image of a solid junior executive. His conduct suggests that he has expedient political positions, not deeply held convictions."[39] Poking fun at Bush at a Washington dinner in 1985, presidential press secretary Larry Speakes noted that Bush spent a lot of time attending many funerals representing the United States: The Vice President's motto seems to be, Speakes said, "You die, we fly." Sticking it to Bush even deeper, he added that he knows President Reagan "relies on his vice president. Just the other day he asked George Bush to take his nap for him." Bush put up with considerable humor, often telling these stories on himself. In fact, he said he was willing to be the loyal subordinate and maintain his close access to Reagan even if it meant that he would be accused of not making major contributions, or worse yet, of forfeiting his independence. He said he had adopted the following rules to ensure that his special relationship with Reagan endured:

Do not participate in the leaking game; don't leak information. Do all interviews on the record. Do not play the "inside the Beltway" game by letting it be known that on this issue or that, you, in your infinite wisdom, differ with the president.

As Edmund Burke said, you owe your constituents your judgment. The vice president should frankly and openly give the president his judgment on matters, but this must be done in confidence.[40]

Despite Bush's valiant efforts at loyalty, President Reagan refused to endorse him as his successor:

Well now, you have me between a rock and a hard place, because I have to tell you, I have said that he's been the finest Vice President I ever had any recollection of. He's been an integral part of everything that we're doing. I've always had the feeling . . . I had it about a Lieutenant Governor in California when I was Governor there. The vice president shouldn't be just someone standing by waiting to be called off the bench. He should be like an executive vice president in a corporation or business. You use him. And he's been all those things. But, at the same time in this job, you are titular head of the party and as such you've kind of got a responsibility to let the party function and make its decisions. Now it's not an easy thing for me to think about, but I have to keep that in mind.[41]

Bush, as a member of the National Security Council, also felt the sting of guilt-by-association when the Iran-Contra hearings and the Tower Commission report on the national security policymaking process portrayed a White House that was poorly run and engaged in frequent lying to Congress, the secretary of state, and the American people. Where Bush had gained from Reagan's successes and popular approval through 1985, his effectiveness and leadership abilities were necessarily called into question in the more troubled years of the Reagan presidency.

A public opinion survey on Bush and his performance as vice president, seven years into his vice presidency, underscored the liabilities of service as vice president. Roughly half of those polled in a New York Times/CBS Survey said "they did not regard the vice presidency as a job that prepared someone to be president" (see Table 10.1). Bush's vice presidency, paradoxically, was both his greatest advantage and his greatest liability. Those surveyed obviously knew him because of his visibility and generally (59 percent) viewed him as honest. Yet only 35 percent of those registered voters held a favorable opinion of him. Why such a low approval rating? Only 42 percent said they had confidence in Bush's ability "to deal with a crisis" or at least would "feel uneasy about him." At least a third of those polled thought Bush lied when he said he did not know that money from the Iranian arms sales was going to help the Contras in Nicaragua. Finally, very few people thought Bush said what he really believed when in fact he did have differences with President Reagan.[42]

TABLE 10.1
The Vice Presidency: Good Training Post for President?

Q. Which comes closer to your opinion? The Vice President's job gives someone especially good experience that should train him to be President, *or* the Vice President's job is mostly for show and doesn't really prepare someone to lead the country.

Doesn't prepare someone for leading the country	47%
Good experience and training/preparation	44
Don't know/no answer	10

Source: New York Times/CBS News Poll, July 21–22, 1987, N-745. Copyright © 1987 by the New York Times Co. Reprinted by permission.

THE GORE EXPERIENCE

President Clinton has continued the recent trend in relying more heavily on his vice president. Clinton, a Washington D.C. outsider, has used the talents and experience of Al Gore, a D.C. insider, often and well. "He doesn't do funerals," said one of Gore's aides. He is prominent and powerful in the Clinton White House,[43] and as another aide remarked, "the President has become so dependent on his No. 2 that he does not make any decision of significance without him."[44] Overstated perhaps, yet close to the mark.

Foreign policy, one of President Clinton's early weak areas, was one of Gore's strengths because of his work in the Congress. As vice president, Gore has gone beyond the "wreath-laying" function at funerals of foreign dignitaries, and has engaged in some high-profile negotiations with officials abroad. He has also been a negotiator of U.S. policy toward Russia.[45]

Gore has emerged "as the president's most trusted official adviser and his ultimate troubleshooter, the one most likely to seize opportunities, put out fires and enforce decisions."[46] The vice president has been especially influential in such domestic policy areas as the environment, emerging technology, and the "reinventing government" campaign. In foreign affairs Gore has been influential on issues such as U.S. policy toward Russia, South Africa, the Middle East, and nuclear nonproliferation. Al Gore debated Ross Perot on national television over the North American Free Trade Agreement—with great success. And he debated rival vice presidential candidate Jack Kemp in October of 1996—again with considerable success.

Gore is one of the few people able and willing to stand up and challenge the president during staff policy meetings,[47] no easy task given Clinton's recognized mastery of policy.[48] Political journalist Elizabeth Drew called Gore "the most influential Vice President in history."[49] In some respects, the Clinton-Gore team has worked like a team, with the president relying on his vice president more than tradition would suggest.

Gore was seen by White House staffers, the cabinet, and Washington journalists as a force within the administration, a view President Clinton did not attempt to change. To Gore, the vice presidency became a job of substance, a preparation should the need arise for him to assume the office of president.

As Clinton's labor secretary Robert Reich wrote: "Al [Gore] is the perfect complement to B [Bill Clinton]: methodical where B is haphazard, linear where B is creative, cautious where B is impetuous, ponderous where B is playful, private where B shares his feelings with everyone. The two men need one another, and sense it. Above all, Gore is *patient,* where B wants it all now."[50]

President Clinton appears less threatened by Gore than some other president–vice president pairings such as Johnson and Humphrey. Clinton is thus able to see Gore as a resource, not a threat, and employ Gore's skills more fully than most vice presidents in history. His reputation was somewhat tarnished in Clinton's second term because of some fund-raising issues stemming from the 1996 election campaign and because some environmentalists and other policy activists chided the Clinton-Gore team for a lack of leadership on certain policy issues. Overall, however, Gore's influence on policy and process rivals the Mondale and Bush experiences.

THE QUAYLE PHENOMENON

In sum, the office of the vice presidency took on an improved image in the Carter, Reagan, and Clinton years. The greater visibility and alleged "importance" attached to the office was probably due to the quality of the vice presidents and the fact they worked for the presidents who were outsiders, who hadn't served in the Congress or the cabinet. In light of the different experience with Dan Quayle, it is a safe bet that the office will have its ups and downs, and its definition and effectiveness will continue to depend primarily on the interpretation the incumbent president chooses to give it and who are in the two positions. Much too, of course, will depend on why and how we select vice presidents.

Bush's vice president, J. Danforth Quayle, presented something of a problem.[51] A relatively inexperienced Senator from Indiana, Quayle was a surprise pick for vice president. Chosen for his youth and his devoted following among the conservatives in the Republican party (just as Reagan chose the more moderate Bush, when his turn came Bush chose the more conservative Quayle—an ideological balancing act in both cases). Quayle was widely seen by pundits and the general population as not up to the job—and he kept supplying evidence to prove the point.

While Quayle was generally "out of the loop," President Bush did use him to boost the president's standing with the self-proclaimed religious right and other conservative groups. In policy terms, the main area where Quayle had impact was in deregulation of business. Quayle headed the Council on Competitiveness, a pro-business, antiregulatory group that aided business in its efforts to cut government regulations.[52] And Quayle was also marginally involved in policy deliberations on the Gulf War.[53]

While Quayle was more qualified than his media image suggested, there was a widespread feeling among political heavyweights in both major parties that he was not the preferred choice to assume the responsibilities of the presidency should the need arise. Sandwiched among such well regarded and influential vice presidents as Mondale, Bush, and Gore, Dan Quayle suffered in comparison.

SELECTION

As noted, apart from the problem of what to do with vice presidents, there is the enduring problem of how best to recruit people to the office and to fill it when vacancies occur. For years the vice presidential nomination process has tempted the fates. It's a pig-in-a-poke system, critics say, that not only is often abused but also denigrates the office.[54]

The process invariably begins with a search for someone who will strengthen the presidential ticket but often ends with a search for someone who will not weaken the ticket. The search for someone who will "balance the ticket" causes countless problems. Sometimes it means choosing party workhorses whose chief virtue is that they have antagonized nobody. Sometimes it means choosing persons mainly because they come from a politically crucial region of the nation. Sometimes this practice means nominating someone from a rival political wing of the party. Sometimes it involves a "unite the party by picking the person who came in second" approach. And then there is always the gamble or "bold stroke" approach of selecting someone who will be a big surprise and put everyone off balance, such as Nixon's choice of Spiro Agnew in 1968, Mondale's choice of Geraldine Ferraro as a 1984 running mate, and, to some extent, Bush's selection of the much younger Quayle in 1988.

In recent years the identity of the presidential nominee has been determined well in advance of the party convention. Not so the vice president. Some are last-minute choices (McGovern and Shriver), some surprises (Bush and Quayle, or Dole and Kemp), some bizarre (Reagan offering former President Ford a "co-presidency" role if he accepted the vice presidential spot in 1980).

In general, the president is free to choose whomever he wishes as his vice presidential running mate. In practice however, presidential nominees choose someone they believe will help them get elected.

Pressure builds up to subordinate considerations of proven leadership ability to those of vote-getting power. (Ticket balancing has allowed the Andrew Johnsons, Arthurs, Henricks, Tompkins, Nixons, and Agnews to rise from relative political obscurity to become vice president.) A problem with nearly all the approaches mentioned above is that after all the excitement fades, the presidential candidate must live with the choice and the country sometimes gets the person as the next president.[55]

Is it right to entrust the selection of the vice president to the presidential nominee? How many vice presidents have been merely an afterthought at the national convention? The 1972 selection of Senator Thomas Eagleton of Missouri by George McGovern and the 1968 Governor Spiro Agnew of Maryland selection by Richard Nixon prompted various reappraisals, as did the Dan Quayle selection and election.

In 1973 the Democratic Party created a Commission on Vice Presidential Selection, chaired by former Vice President Hubert Humphrey. The Commission made several suggestions, including these:

1. Extending the convention for a day, permitting a forty-eight-hour interval between the nomination of the president and the nomination of vice president.

2. To ensure even more deliberation, the selection of the vice president could be put off for three weeks or more and the nominee could then be ratified by the party's national committee in a mini-convention (a procedure used in 1972 when McGovern's choice, Thomas Eagleton, withdrew and Sargent Shriver was approved as the vice presidential nominee.) [56]

Other reformers suggested that every aspirant for the presidential nomination might pick his vice presidential nominee about 60 days before his party's national convention. At the convention, the two would run as a ticket. This method might, its sponsors say, allow the vice presidential choice to emerge with more stature.

Another proposal suggests that to democratize the convention selection, the presidential nominee could be required to submit several names for vice president to the delegates. Others have suggested convening the party committee perhaps a month or so after the national convention to act on a recommendation by the presidential nominee. Still another variation would have the national convention delegates vote on the recommendation of their standard bearer by mail-in secret ballot a month or so after the convention.[57]

Yet another alternative is for the runner-up in the presidential balloting to be declared the automatic winner of the vice presidential nomination.

Still another proposal urged that national conventions nominate only the presidential nominee. Then the November election winner would announce a choice of a vice president in December, and the new Congress meeting in January would consider the choice before inauguration day. If the first name was not approved, the new president would submit another. This proposal might not require a constitutional amendment, since the president would in essence be following procedures provided in the Twenty-fifth Amendment.

The most radical proposal of all would require a constitutional amendment: abolishing the vice presidency altogether. The idea of abolishing the office has been occasionally proposed since the beginning of the Republic. This bold measure was embraced by a dozen writers in the mid-1970s in the wake of Watergate and two periods of vacancy in the vice presidency. Frustrated by the abuse and misuse of the office, critics said: let's get rid of the office before it sinks us. If a president dies or becomes disabled or resigns, let a designated cabinet member (secretary of state or defense or treasury) take over for one hundred days and meanwhile let's conduct a special election for a new democratically elected president. Arthur M. Schlesinger, Jr., a proponent of this plan, concluded that the multiple problems that now depreciate the vice presidency could simply be eliminated by eliminating the office itself:

> There is no escape . . . from the conclusion that the Vice Presidency is not only a pointless but even a dangerous office. A politician is nominated for Vice President for reasons unconnected with his presidential qualities and elected to the Vice Presidency as part of a tie-in sale. Once carried to the Vice Presidency not on his own but as second rider on the presidential horse, where is he: If he is a first-rate man, his nerve and confidence will be shaken, his talents wasted and soured, even as his publicity urges him on toward the ultimate office for which, the longer he serves in the second place, the less ready he may be. If he is not a first-rate man he should not be in a position to inherit or claim the presidency. Why not therefore abolish this mischievous office and work out a more sensible mode of succession?[58]

The Throttlebottom complex would be solved. We would stop elevating to the presidency individuals who have not really won election. Presidents could appoint one or more vice presidents. "If the principle be accepted—the principle that if a president vanishes, it is better for the people to elect a new president than endure a vice president who was never voted for that office, who became vice president for reasons other than his presidential qualifications and who may very well have been badly damaged by his vice presidential experience—the problem is one of working out the mechanics of the intermediate election. This is not easy but far from impossible."[59]

At the heart of the matter is whether we want to be a government of and by the people. It is widely acknowledged that we sometimes choose vice presidents for the wrong reasons. It is equally clear, however, that today the vice presidency has become a stepping stone either to the presidency directly or at least to presidential nomination. Nearly 25 percent of our presidents did not serve out the terms they were serving. And nearly a third of the presidents were vice presidents first (Table 10.2).

The issues at stake, especially in this proposal, are issues of democratic procedure and political legitimacy. Many say that the nation could not afford a special election after a presidential death, as in 1945 or in

TABLE 10.2
Vice Presidents Who Became President

Year	Vice President	President served	Reason for Obtaining office
1797	John Adams	George Washington	Elected to office
1801	Thomas Jefferson	John Adams	Elected to office
1837	Martin Van Buren	Andrew Jackson	Elected to office
1841	John Tyler	William Henry Harrison	Harrison died
1850	Millard Fillmore	Zachary Taylor	Taylor died
1865	Andrew Jackson	Abraham Lincoln	Lincoln killed
1881	Chester A. Arthur	James A. Garfield	Garfield killed
1901	Theodore Roosevelt	William McKinley	McKinley killed
1923	Calvin Coolidge	William G. Harding	Harding died
1945	Harry S Truman	Franklin D. Roosevelt	Roosevelt died
1963	Lyndon B. Johnson	John F. Kennedy	Kennedy killed
1968	Richard M. Nixon	Dwight D. Eisenhower	Elected to office
1974	Gerald R. Ford	Richard M. Nixon	Nixon resigned
1989	George Bush	Ronald Reagan	Elected to office

1963, but many other Western nations do so regularly. Others say intermediate elections violate the tradition of quadrennial elections. But what is so compelling or virtuous about waiting until that fourth year if what is at question is the quality and character of the nation's leadership? Moreover, the special election would merely be to fill out the remainder of the departed president's term. Would this new arrangement be a departure from the intent and spirit of the founding fathers? Although strictly speaking they opposed direct popular election, it would reaffirm their general view. They opposed selection of the president by the Congress. As much as possible they wanted the people to have the main say. What could be more sensible for a self-governing democracy than that the president must, except for the briefest periods, be a person elected to that office by the people?

There are, however, several disadvantages to the idea of abolishing the vice presidency and conducting a special election to fill a presidential vacancy. First, it would invite instability and turmoil during the succession period. Second, it is doubtful that a sixty- or ninety-day campaign in a nation this large would allow for the clarification of competing issues. It is one thing for Canada or France to conduct a short campaign, but the United States is substantially larger and has weaker political parties, and it usually takes longer here for issues to get sharpened. Finally, most people, rightly or wrongly, are not displeased by the existing system. Neither are they displeased by those who in the twentieth century have become vice presidents through the current arrangements.

In sum, the vice presidential nomination process has had a mixed history. We now have had nine examples of what happens when a vice president is forced by necessity to become president. Several became respected presidents. Yet at least half were major disappointments. More disturbing, however, is that an assessment of those who have mercifully not become president, from Aaron Burr to Spiro Agnew to Dan Quayle, is not in the least reassuring.

None of the reform proposals suggested in the past two decades is without liabilities. Even the most casual of political science students can discern most of the trade-offs and the deficiencies. Nor has any of them captured a following. Most people are resigned to living with the existing system.

However, even if we stick with the present system for selecting vice presidential nominees, several sensible modest improvements can be made. None of these requires a constitutional amendment and most are merely the product of learning from past mistakes. In a condensed form, these are the recommendations most frequently made:

- Presidential candidates should have their staffs recommend and consider several vice presidential candidates, well in advance of the convention. Whenever possible the presidential nominee should interview these candidates, also well in advance.

- Presidential candidates should give highest priority to the need for ideological compatibility, treating "electoral balancing" as a secondary factor. Note, however, that the "two are neither mutually exclusive nor naturally contradictory."[60]

- National parties should provide for a flexible convention schedule, permitting an extra day if necessary for presidential nominees to deliberate over the vice presidential nominee choices. Provisions should be made, in special cases, for a delay of a few weeks and the approval of the running mate at a miniconvention. Such special cases might be on such occasions as when a classic "dark horse" or last-minute compromise candidate wins the presidential nomination and he or she had not expected to win and hence was unprepared for the vice presidential selection responsibility.

- The candidates should make a list of serious preferences for the vice presidency before the convention, to facilitate media and public examination; and they are encouraged to initiate direct contact and staff liaison with potential running mates.[61]

Common-sense suggestions such as these will not guarantee an error-proof selection process, but they should reduce the risks.

Walter Mondale outlined a helpful list of job qualities he thought were essential to keep in mind as presidential candidates select a vice presidential running mate. He provided this at the time he was engaged in selecting his own running mate in June of 1984. "My choice" he said "will be guided by the need to select someone totally qualified to assume the office of the President should that be necessary." Then he went on to say there were many ways a vice president can help a president govern. For example:

1. The vice president, the only other official elected nationwide, can advise a president from a national perspective.

2. Without specific ties or allegiances, the vice president can help break through the bureaucracy and solve complex problems.

3. A vice president can speak for the president on Capitol Hill and help advance the President's legislative program.

4. The vice president can be the president's most effective spokesperson with the public.

5. The vice president can extend the president's reach in foreign policy by conducting negotiations at the highest levels and serving as the president's eyes and ears abroad.[62]

This list will vary from presidential candidate to presidential candidate but is a good beginning, and journalists and party leaders should press candidates to acknowledge these qualities and possible roles whenever the selection process is taking place. In the last analysis, however, presidential nominees are still likely to choose their running mates with their probable impact on the electoral outcomes in mind. Nor is this concern for political considerations necessarily unreasonable. "A party facing an election of such great consequence as the race for the White House would be irrational not to weigh heavily the probable political impact of prospective running mates," wrote Joel Goldstein. He added, "A two-party system provides many advantages to a democracy. It simplifies voter choice, promotes consensus, adds stability to a society, and allows majorities to form. But every virtue has its costs. Compromise is part of the price. The use of the vice presidential nomination to placate disgruntled factions is part of the mortar that holds the political system together."[63]

SUCCESSION

Despite all of the other problems associated with the vice presidency, the office has done one thing reasonably well—solving our succession problem. Many nations have faltered or even been torn apart by civil upheavals because they failed to provide for a prompt, orderly means of transition. Our citizens, like most people, long for system stability and for a heightened sense of continuity with an understood heritage.

Americans used to joke about vice presidents who because of some emergency became "His Accidency" or president-by-chance. However, most Americans are grateful when the leadership vacuum is replaced and the new leader, endowed with legitimacy, acts with confidence and dispatch. Our vice presidential succession process may be imperfect, but one of its virtues is the stability it brings to our governance arrangements. Trauma, chaos, and instability are avoided, as people know well in advance that there is a plan, a legitimate plan that will be put into effect. Whatever one may think of the succession plan, the reassurance rendered is somewhat remarkable.

The framers provided that the vice president should be first in the order of succession, but they left it up to Congress to establish the rest of the order of succession. Three statutes and one constitutional amendment speak to this matter. In the first presidential succession act, passed in 1792, the president pro tempore of the U.S. Senate (by custom the senior member of the majority party in the Senate) was named first in line behind the vice president, followed by the speaker of the House of Representatives. If a double vacancy occurred (that is, if something happened to both the president and vice president), this same 1792 act provided that the president pro tempore of the Senate would become acting president "until a president be elected." Hence an immediate special election of some kind would have been called to elect a new president, unless the vacancy occurred in the last months of the presidential term.

This first statute was superseded in 1886 when the heads of the cabinet departments, beginning with the secretary of state, were named as the line of succession. This may have been done to avoid violating the principle of separation of powers. Many people, however, believe an elected official should stand first in line to fill the presidency. Harry Truman, while serving out FDR's fourth term, urged a return to succession by elected congressional officials. In 1947, Congress approved this change, placing the speaker of the House first behind the vice president, followed by the Senate's president pro tempore, followed by the cabinet officers in terms of the departmental seniority—secretary of state and so on.

Questions persisted, however, regarding cases of a vacancy in the vice presidency. Before the Twenty-fifth Amendment to the Constitution, the vice presidency had been vacant eighteen times for a total of about forty years. Fortunately, during these periods there had been no need to go down the line of presidential succession. Section 2 of the Twenty-fifth Amendment, ratified in 1967, now provides that "whenever there is a vacancy in the office of the vice president, the president shall nominate a vice president who shall take office upon confirmation by a majority vote of both Houses of Congress."

Some critics of this provision say it gives too much power to the president. Defenders of the provision counter the criticisms as follows:

> In giving the President a dominant role in filling a vacancy in the vice presidency, the proposed amendment is consistent with present practice whereby the presidential candidate selects his own running mate who must be approved by the people through their representatives. It is practical because it recognizes the fact that a vice president's effectiveness in our government depends on his rapport with the president. If he is of the same party and of compatible temperament and views, all of which would be likely under the proposed amendment, his chances of becoming fully informed and adequately prepared to assume presidential power, if called upon, are excellent.[64]

Some analysts say this provision is overly vague on certain points. Thus some would amend it to add deadlines against stalling by either the White House or Congress. Others wondered what would happen if a president resigned or died while his nomination of a vice president was still pending. Would the nominee still stand? Or would the speaker of the House, who had just become president, have the right to nominate his choice? "Would Congress have the right to choose between the two nominees? The fact that there are any open questions about presidential succession is, of course, dangerous, for there is nothing more threatening to a constitutional democracy than doubts about who has the legitimate right to govern."[65]

When Nixon was Eisenhower's vice president they had come to an informal agreement about when the vice president would take over in cases of emergencies; obviously it would sometimes be up to the determination of the president, or Nixon along with the president's doctors. By the same token, Eisenhower said, "I will be the one to determine if and when it is proper for me to resume presidential responsibilities."[66]

Section 3 of the Twenty-fifth Amendment provides for presidents to delegate presidential powers and duties to the vice president for a specified period. Debates had long raged over questions of presidential inability or disability. Many presidents have been temporarily unable to fulfill their responsibilities, and two presidents had serious disabilities—Garfield for two and a half months before he died, and Wilson for about a year. In each case their vice presidents did not assume presidential duties, in part because they wished to avoid usurping the powers of that office. Now a president can temporarily step aside. Section 3 reads:

> Whenever the President transmits to the President pro tempore of the Senate and the Speaker of the House of Representatives his written declaration that he is unable to discharge the powers and duties of his office, and until he transmits to them a written declaration to the contrary, such powers and duties shall be discharged by the Vice President as Acting President.

But will a president invoke this section, save in extraordinary occasions? Reagan's aides chose not to do so during his operations for bullet wounds after the March 1981 attempt on his life. Reagan's aides did not feel the president would be totally incapacitated for more than a few hours—which was the case, although the first couple of days were rough. But there are political implications as well. Invoking this section of the Twenty-fifth Amendment might alarm the public, might dramatize the president's vulnerability, might weaken the image of strength of the incumbent.

In July of 1985, President Reagan signed a letter which read in part: "I have determined and it is my intention and direction that Vice President George Bush shall discharge [presidential] powers and duties in my stead, commencing with the administration of anesthesia to me." With these words, Reagan—without formally invoking Section 3 of the Twenty-fifth Amendment—made Vice President George Bush acting president. Reagan would be operated on for colon cancer, and from the time he was anesthetized until seven hours and fifty-four minutes later, power was transferred to his vice president.

While this was a controversial step, the president wanted to avoid repetition of the chaos that resulted after he was shot in 1981, when Secretary of State Alexander Haig marched before television cameras to announce "I am in control here." During his hours as acting president, George Bush wisely played a low-key role, and that evening when President Reagan signed another letter stating that he was prepared to once again discharge his duties as president, all power was returned to the president. This is the only time Section 3 of the Twenty-fifth Amendment has been invoked (however informally), and it ended up raising more questions than it answered.

President Reagan, wisely if reluctantly, followed the sensible procedures provided in the Twenty-fifth Amendment to permit the vice president to serve as acting president when he underwent surgery. Yet Reagan acted in a way that raises questions about Section 3. Reagan's letter to the president pro tempore of the Senate and the speaker of the House of Representatives included the following curious sentence: "I do not believe that the drafters of this amendment intended its application to situations such as the instant one." Reagan was probably wrong. No doubt he and his aides feared that Reagan might have further illnesses and operations and that frequent invoking of this section might diminish Reagan's leadership image. Anyone who has ever worked in the White House appreciates how much a president's influence rests on how the Congress and the public perceive the president's strength, energy, and ability to persuade.

Bush, by the way, exercised no presidential responsibilities during the nearly eight hours of that historic July thirteenth in which he was, in effect, president. Curiously, Reagan shied away from referring to Bush as the "acting president," yet that is the term specified in the amendment. Bush apparently was not informed of the exact time he was assuming (or was supposed to assume) his responsibilities. For practical purposes, the then White House chief-of-staff, Donald Reagan, was the man in charge during this interval, and, perhaps, for most of the time Reagan was undergoing surgery. Bush remained at the vice presidential residence ready to act if needed, yet dependent on the chief-of-staff who, along with Reagan's wife, remained at the president's side.

There was confusion too about when the vice president was formally put in the position of discharging the powers and duties of the presidency. Was it when the White House staff telephoned him? Was it when the President signed the required letters to the leaders of Congress? Was it when those leaders received the letters?

Less than eight hours after his surgery, Reagan signed a second letter to the congressional leaders reclaiming his responsibilities. Bush was presumably also notified at the same time. Should these notifications be by letter, telephone, fax, or e-mail? White House aides inevitably will play the intermediary role in most of these circumstances, but this first test suggests there are additional questions and procedures to be ironed out.

Confusion only increased when in May of 1991, President Bush announced that, in the event his irregular heartbeat required electric shock therapy, power would be turned over to Vice President Quayle. While such a transfer of power was never required, the process again raised troubling questions.

What if there is contention about whether the president is disabled or fit? Section 4 of the Twenty-fifth Amendment speaks to this, though here again, the provisions invite ambiguous interpretation. The beginning of this section reads:

> Whenever the Vice President and a majority of either the principal officers of the executive departments or of such other body as Congress may by law provide, transmit to the President pro tempore of the Senate and the Speaker of the House of Representatives their written declaration that the President is unable to discharge the powers and duties of his office, the Vice President shall immediately assume the powers and duties of the office as Acting President.

The "principal officers of the executive departments" refers implicitly, if not explicitly, to the members of the cabinet. However, nowhere in the Constitution or anywhere else is there a formal definition of who is in the cabinet. It varies not only from president to president but from season to season, depending on whom the president wishes to include.

Another problem in this section that may raise uncertainty and even confusion in the future involves the provision of how a president may regain his powers. A president, so the amendment provides, can inform the leaders of Congress that no inability exists and ask to resume the powers, but the vice president, together with a majority of the cabinet, can contest the president's assertion. What then? Well, says the provision, Congress has to decide very quickly whether the president is able or unable. Complications can plainly arise in the vague wording of this section. A president could conceivably keep stating he was able to continue. Meanwhile, while Congress would be debating the issue, two persons could both be attempting to exercise the powers and duties of the presidency.

These problems can be cleared up without resorting to any changes in the Twenty-fifth Amendment. Informal agreements and unambiguous precedents will help diminish the confusion. The amendment is plainly a compromise of many different proposals, none of which is entirely satisfactory. Constitution writing is always hard, and to try to accommodate every possibility is impracticable. Not every contingency can be controlled by law or constitutionally specified protocol. The amendment is an improvement, although not definitive for every situation. Judgment on its overall effectiveness must await the inevitable future tests of its various provisions.[67]

CONCLUSION

What of the nature of the vice presidency? Ambiguous and fragile though it may be, this office is here to stay, paradoxes, uncertainties, tensions, and all. The office solves our nation's succession problems and can be made into a useful learning and advisory position. The job will remain attractive to aspiring politicians precisely because it is one of the major paths to the presidency. The Mondale, Bush, and Gore experiences, on balance, have added to its prestige.

Vice presidents will continue to have, and perhaps should have, a more or less undefined set of troubleshooting and advisory functions. Different incumbents will bring different skills and strengths to the office. Some will be better at diplomatic functions. Some will be gifted political negotiators and coalition builders. Some will be excellent liaisons with mayors and governors. A certain flexibility is needed.

Given all its problems and all the ambiguities of the position, a case can be made to abolish it. However, this is not going to be done. Our tradition, Constitution, and political tendencies all point toward retaining the office. Moreover, amending the Constitution to make more explicit that vice presidents should now be executive and not legislative branch officials would not be worth the effort. However, if there is ever a constitutional convention or a comprehensive constitutional amendment concerning the presidency it would be prudent to relieve vice presidents of their senate responsibilities—including the presiding *and* the casting of tie-breaking votes. Those responsibilities, as noted earlier, no longer seem relevant for vice presidents.

Efforts in the past several presidencies to make the office function more effectively should continue in the twenty-first century. An obvious way of upgrading the vice presidency is to enhance the quality of the vice presidential selection process. Although no job can fully prepare an individual for the presidency, recent experience suggests the once ridiculed office of the vice presidency can often be refashioned, at least partially, to accomplish this task.

CHAPTER 11

"If Men Were Angels . . .": Presidential Leadership and Accountability

If men were angels, no government would be necessary. If angels were to govern man, neither external nor internal controls in government would be necessary. In framing a government which is to be administered by men over men the great difficulty lies in this, you must first enable the government to control the government and in the next place oblige it to control itself. A dependence on the people is no doubt the primary control on the government. But experience has taught mankind the necessity for auxiliary precautions.

James Madison, *Federalist Papers* No. 51 (1788)

A strong president is a bad President, a curse upon the land, unless his means are constitutional and his ends democratic, unless he acts in ways that are fair, dignified, and familiar, and pursues policies to which a 'persistent and undoubted' majority of the people has given support. We honor the great Presidents of the past, not for their strength, but for the fact that they used it wisely to build a better America.

Clinton Rossiter, *The American Presidency* (Harcourt, Brace and World, 1956), p. 257

People with vast power at their disposal get cut off from reality, and their power is inevitably misused. One Administration will have its Watergate, another its Vietnam. Clearly, there is a need for Congress, the courts, the media and the general public, each in its own way to work to lessen both the power and the aura of divine right that now surround our President.

Jeb S. Magruder, convicted Nixon White House aide, *Los Angeles Times* May 22, 1974, p. II-7

Novelist W. Somerset Maugham once said "There are three rules for writing a novel. Unfortunately, no one knows what they are." In the same vein, we are tempted to conclude that there are three rules to being an effective president, yet no one knows exactly what they are.

As we have discussed, the presidency changes from season to season, occupant to occupant, issue to issue. We may never unravel most of the paradoxes of the American presidency. Yet there are things the American people and presidents can do to encourage realistic and effective presidential performance.

As noted at the beginning of this book, we want to be led, yet we cherish our independence and freedom. We want a "take charge" leader in the White House, yet we demand accountable and responsive leadership. Many Americans are now less content to hold presidents to account only every four years when they go to the polls, they insist on daily accountability. Our system is built on distrust of powerful leaders and the need for their accountability.

Compared with the heady days of FDR, and later the national security state of the Cold War era, the presidency of today is a more constrained office. If FDR invented the modern presidency during the depression and World War II, the demands of the Cold War further enlarged and empowered the office. But in this post–Cold War end of the century age, the demand for a strong, centralized presidency seems less pressing.

While we may on occasion need heroic leadership, we are less in need of presidential dominance than in the past sixty-five years. The end of the Cold War has liberated us from the need for deference to the powerful presidency model, which proved effective on occasion yet dangerous at other times. However, as the United States enters the twenty-first century, we are rightly still concerned with how best to keep presidents effective and honest.

This chapter will examine the need for holding presidents accountable, and will survey some of the most debated existing and proposed checks on the president.

HOLDING PRESIDENTS TO ACCOUNT

As our theme in these chapters suggests, any discussion of presidential leadership and accountability must take into account the ever-present paradoxes of the presidency. Some part of us wants a larger-than-life, two-gun, charismatic Mount Rushmore leader. Harrison Ford in the film *Air Force One* (1997) vivified this yearning. Still, there is the remarkably enduring antigovernment, antileadership, chronic-complainer syndrome. We want strong, gutsy leadership to operate on alternate days with a "national city manager." We want presidents to have a wealth of power to solve our problems, yet not so much they can do lasting damage.

Accountability implies not only responsiveness to majority desires and answerability for actions but also taking the people and their views into account. It also implies a performance guided by integrity and character. Accountability implies as well that important decisions could be explained to the people to allow them to appraise how well a president is handling the responsibilities of the office.

To whom is accountability owed? No president, it would seem, can be more than partially accountable to the people, for each president will listen to some people and some points of view more than to others. If we have learned anything in recent years, however, it is that the doctrine of presidential infallibility has been rejected. Arbitrary rule by powerful executives has always been rejected here. But what should be done when there are sharp differences between experts or when expert opinion differs sharply from the preponderance of public opinion? How much accountability, and what kind, is desirable? Is it not possible that the quest for ultimate accountability will result in a presidency without the prerogatives and independent discretion necessary for creative leadership?

The modern presidency, in fact, may be unaccountable because it is too strong and independent in certain areas and too weak and dependent in others. One of the perplexing circumstances characterizing the modern presidency is that considerable restraints sometimes exist where restraints are least desirable and inadequate restraints are available where they are needed. Also, presidential strength is no guarantee that a president will be responsive or answerable. Indeed, significant independent strength may encourage low answerability when it suits a president's short-term personal power goals.

THE PRESIDENCY AND DEMOCRATIC THEORY

How do you grant yet control power? Can leaders be empowered yet also democratized?

These are classic questions our framers faced and these have been central to debates in democratic political theory. Leadership implies power, accountability implies limits. Contradictions aside, accountability is a fundamental piece of the democratic puzzle. In essence it denotes that public officials are answerable for their actions. But to whom? Within what limits? Through what means?

There are essentially three types of accountability: *ultimate accountability* (which the United States has via the impeachment process), *periodic accountability* (provided for by general elections and occasional landmark Supreme Court decisions), and *daily accountability* (somewhat contained in the separation of power).[1] James Madison believed elections provided the "primary check on government" and that the separation of powers ("ambition will be made to counteract ambition") plus "auxiliary precautions" should take care of the rest.[2]

There *are* times when presidents abuse power or behave corruptly.

But even in the two notable bouts with presidential abuses, Watergate and the Iran-Contra scandal, the president was stopped by the countervailing forces of a free press, an independent Congress, an independent judiciary, and a (late to be sure) aroused public.

We may hold presidents accountable, but can they be held responsible? That is, can they muster enough power to govern? One means to improve accountability and also empower leadership is to strengthen the party system in America. Our parties are, at least by European standards, relatively weak, undisciplined, and nonideological. A stronger party system could organize and mobilize citizens and government, diminish the fragmentation of the separation of powers, and mitigate against the atomization of our citizenry. If the parties were more disciplined and programmatic, the government's ability to push through its programs would doubtless be enhanced.

A more responsible party system would also ground presidents in a more consensus-oriented style of leadership, and thereby diminish the independent, unconnected brand of leadership so often attempted by recent presidents. A more robust party system can help join the president and Congress together in a more cooperative relationship.[3]

All presidents work to be successful, but what does it mean to be a success? Great popularity? A good historical reputation? Achieving your policy goals? A high congressional box score? Getting your way?

If success is measured merely by getting one's way, then many bullies would be judged successful. But success means more than getting what one wants. In determining success, we must always ask "power for what *ends?*," because power divorced from purpose is potentially dangerous and democratically undesirable.

Presidential politics should always be concerned with central issues and values. Candidates who run for the White House and thereby seek the power to influence the lives of millions of Americans ought to do so because they have a vision of building a better and more just America. If this is not the case, those candidates who are merely seeking power for its own sake should be smoked out in the election process.

If we look on government as the enemy and politics as a dirty word, our anger turns to apathy, and power (but not responsibility) slips through our hands. We often look at politics not as a means to achieve public good, but as an evil; we see elections as the choice between the lesser of two evils; we presume that our democratic responsibilities are satisfied merely by the act of voting every so often, or we drop out of politics. Consequently many people abandon politics altogether.

This is why politics and elections matter so much. People who give up on politics in effect abdicate the possibility of implementing their most cherished policy ideas. People who give up on politics and parties are essentially giving others power over their lives.

In a democracy, a successful president pursues and uses power, not for selfish ends, not to aggrandize his or her own status, but to help solve problems and help citizens enjoy the blessings of freedom and opportunity.

The best of democratic leaders are teachers who both understand and educate all of us about the promise and mission of America. They move the government in pursuit of the consensus generated from the values of the nation. They appeal to the best in citizens and attempt to lead the nation towards its better self.[4]

Franklin Roosevelt suggested that the presidency "is preeminently a place of moral leadership."[5] Thus, presidents may use their office as a "bully pulpit" to, when at their best, appeal to our better instincts, and lead democratically. It was through politics and government that the progressive social movements of this century helped move us toward greater racial and gender equality, devised policies to expand education and opportunities to a wider segment of the population, attempted to protect and expand the rights of citizens. These battles are far from over. As a nation we have a long way to go before we can truly grant the blessings of liberty and prosperity to all citizens, yet it is through politics—and only through politics—that we can hope to achieve these goals. And if we want our politics and our government to succeed, we must find ways for citizens to guide and encourage responsible presidential leadership.

Presidents may and have also used the powers of the presidency to promote economic stability and economic growth in America. The Roosevelts, Wilson, Kennedy, Reagan, and Clinton all strived to stimulate the economy and promote favorable trade programs that in turn created jobs and economic security. Presidents generally know what is expected of them as promoters of economic development, yet, here again, they are likely to respond to the yearnings and lobbying of those who become actively engaged in the political process and party politics.

The ends power serves are important, but in presidential terms, virtue is not enough. A successful president must have *character* and *competence*. Character without competence (resources, skill, power) gives us noble but ineffective leaders; competence without character may lead to government by demagogues.

Presidents who lead in the democratic spirit can encourage leaders, foster citizen responsibility, and inspire others to assume leadership responsibilities in their communities. Democratic leaders establish a purposeful vision, pursue progressive goals, and question, challenge, engage, and educate citizens.

The United States needs a strong presidency and a democratically controlled presidency, and this in turn necessitates a strong civic culture.[6] Political theorist Benjamin R. Barber notes the challenge inherent in such a quest:

> At the heart of democratic theory lies a profound dilemma that has afflicted democratic practice at least since the eighteenth century. Democ-

racy requires both effective leadership and vigorous citizenship: yet the conditions and consequences of leadership often seem to undermine civic vigor. Although it cries for both, democracy must customarily make do either with strong leadership or with strong citizens. For the most part, depending on devices of representation in large-scale societies, democracy in the West has settled for strong leaders and correspondingly weak citizens.[7]

The American presidency operates within a system of shared power, one in which the claims of many groups constantly compete. Presidential struggles with other governmental and extragovernmental centers of power stem from the larger societal conflicts over values and the allocation of wealth and opportunity. As a result, the presidency becomes a place in which few radical decisions are made; most of its domestic policies are exploratory, remedial, or experimental modifications of past practices.

Limitations on a president's freedom of action are, to be sure, often desirable. Many of the checks and balances that are still at work today were deliberately designed by the framers of the Constitution. In some measure, presidents should be the agents of their campaign commitments, their parties, and their announced programs. They should be responsive most of the time to the views of the majority of the American people. Presidential behavior should be informed by the Constitution, existing laws, and the generally understood, albeit hazier, values that define democratic procedure. The notion that party programs, spelled out in campaigns, allow the public some control over policy through the election process is a valuable brake, one that needs, if anything, to be revitalized. Other brakes that limit presidential discretion may be viewed as positive or negative, depending on an individual's political and economic views. The constraining of a president by the bureaucracy and by special interests is implicitly, if not explicitly, a kind of accountability, even if it is not exactly the kind we want as our prime constitutional safeguard against the abuse of power.

The American political system is deliberately designed to enhance the chances of special interests to veto policies that affect them. Although the various economic and professional elites may not be as cohesive and omnipotent as the power-elite school suggests, the wealthier interest groups have perpetuated decidedly favorable governmental privileges to advance their business and professional goals. Although at times of crisis there are substantial incentives for subordinating special claims to the nation's well-being, such times are a presidential luxury. Under normal conditions, an elaborate network of influences and obligations may frustrate presidential objectives, especially in the area of domestic policy.

A president's leeway for achievement can be determined by the degree to which consensus or conflict exists among elite interest groups within a particular arena of public policy. If the policy elite of a given profession or industry share wide agreement on a particular issue, it is very difficult for a president to effect an opposing point of view. Occasional exceptions such as Medicare, automobile safety devices, and antipollution legislation are not persuasive, because the profession or industry in question seldom lost much and the costs for such programs were in most cases passed on in some way to the consumer or taxpayer. If, however, cleavage or confusion occurs over substantive or procedural matters, a president has some independent influence; although even then, the scope and type of his influence will be shaped by the character of the conflict among these elite. Thus, Johnson's efforts to create model cities as demonstrations of how social and physical planning could produce decent and livable cities soon was heavily influenced by pressures from home builders, developers, real-estate associations, big-city mayors, and other strategically positioned interests. Likewise, despite widespread public support for rapid progress on the environmental front, Carter's environmental protection recommendations soon became influenced by the views of the automobile manufacturers as well as by the unions potentially affected by stringent standards and too rapid implementation. More recently, President Clinton's efforts at health care reform met with fierce opposition from insurance companies and medical care providers who felt threatened by the proposed reforms. They were able to mobilize the public and Congress and prevented Clinton's proposals from being enacted into law. Sometimes a consensus among policy elites may be the product of presidential commitment, but the reverse is more likely to be the case.

Prior commitments to special interests inhibit planning, brake a president's capacity to focus on new problems, and help to exhaust his political credit. Despite high expectations, presidents may find themselves merely a strategically situated broker for their own party, able only in a limited way to affect existing patterns of grants or subsidies.

Every grant program generates concrete benefits to a particular group, and possessiveness characterizes nearly every group that has participated in the growth of federal aid programs since the New Deal. According to the doctrine of interest groups, the unorganized are left out of most policymaking equations. In fact, seldom does an interest group emerge that has as its aim the promotion of the public benefit, a program that would benefit everybody. At the same time, the standards of justice and respect for law deteriorate amid informal, frankly feudal negotiations among those stronger interests who can adjust the laws to their own advantage and profit.

In the end, all three branches of government and the bureaucracy listen more attentively and usually yield to the ideas from those segments of society able to represent themselves, able to shape the character of those branches, and able to supply precisely that information and argumentation needed to make the system move. So it is that the many well-heeled interests continue to enjoy a special advantage in any contest with a president who is a genuine progressive.

The founders would most certainly have been pleased to see how the system of checks and balances has thwarted executive tyranny. But they would have perhaps been less pleased with the gridlock that so often characterizes relations between the president and Congress.

"How," political scientist Bert Rockman asked in *The Leadership Question,* "can leadership be exerted yet restrained?"[8] It is a question that confounded the founders and continues to trouble us today. Political gridlock is often the reality. Is the presidency broken? Does it need to be fixed? Would-be reformers must be cautious to ensure that reform would not deform.

Is the separation-of-powers model *the* problem? Does it create deadlock and paralysis? If you are the president, there must be times when it seems so. Woodrow Wilson, writing in 1884 long before occupying the White House, saw the separation as creating a massive political escape clause for blame and responsibility. Wrote Wilson:

> Power and strict accountability for its use are the essential constituents of good government. . . . It is, therefore, manifestly a radical defect in our federal system that it parcels out power and confuses responsibility as it does. The main purpose of the Convention of 1787 seems to have been to accomplish this grievous mistake. . . . Were it possible to call together again the members of that wonderful Convention . . . they would be the first to admit that the only fruit of dividing power had been to make it irresponsible.[9]

Upon reflection we are reminded of the positive benefit of separating, sharing, and overlapping power. If one values, as we do, deliberation, discussion, and debate; if we accept a model of democratic governing based on consensus and cooperation, then the reform agenda will be short. But some see the separation as the likely suspect in the crime of stalemate and gridlock.

Americans are great fixers. We have proposed hundreds of amendments to our Constitution in the hopes of improving our republic. The presidency is the subject for a large number of these proposed reforms. It would require another book to discuss and evaluate even the most plausible proposed checks and balances on the presidency. But it is important to discuss at least a few proposed reforms *that we do not believe would be helpful.* We will follow this with a discussion of some existing checks that absolutely need to exist and in same cases need to be strengthened.

THE SIX-YEAR NONRENEWABLE TERM NONSENSE

One of the more curious remedies persistently suggested in discussions of reforming the presidency is the idea of a single six-year presidential term. This is certainly not a new idea, having been originally proposed in Congress as early as 1826. It has been reintroduced more than 150 times since then and has won backing from at least ten presidents, including

Johnson, Nixon, and Carter. This reform could only be achieved by amending our Constitution.

In April 1979 President Carter allowed that the press would ascribe purer motives to his policies if he did not have the option of seeking re-election. And so he endorsed the six-year term as a sensible reform. Carter's attorney general, Griffin B. Bell, called the six-year term an idea whose time had come:

> This change will enable a President to devote 100 percent of his or her attention to the office. No time would be spent in seeking reelection. Under the present system, the President serves three years and then must spend a substantial part of the fourth year in running for reelection. . . . Moreover, the current four-year term is actually too short to achieve any of the major changes and improvements that a President should accomplish. The funding cycles are so long that it is well into a President's third year before his own program changes take effect. This leaves the bureaucracy in control.
>
> A single six-year term would permit the long-term study, planning, and implementation that our government needs, plus saving that fourth year now lost to campaigning.[10]

Proponents contend that the single six-year term would remove a president from the negative kind of partisan politics. The assumption is that once elected to such a term, with no possibility of reelection, presidents would cast aside partisan calculations and provide leadership for all of the people. Presidents would do what is right even if this meant that their party would lose votes, their friends would suffer financial losses, or their own political future would be damaged. Former Johnson aide Jack Valenti wrote that "if the Watergate mess tells us anything it is that the reelection of a President is the most nagging concern in the White House. . . ." Further, he asserted that "Watergate would never have occurred if Presidential aides were not obsessed with reelection. If they had been comfortable in tenure, knowing that in six years they would lose their lease—and in that short time they must write their record as bravely and wisely as possible—is it not possible that their arrogance might have softened and their reach for power might have shortened?"[11]

Advocates of the single six-year term see it as a means of making the presidency more objective, "neutral," and "reasonable." They want to deemphasize the divisive aspects of electoral and partisan politics, to elevate the presidency above selfish or factional ambitions. Some of those who favor this idea see it also as a means of making sure that no president succeeds himself (or herself). They prize the concept of citizen-politicians assuming the office of president for a fixed term and then retiring. Some say, too, that a term of six years would strengthen a president's hand in recruiting top managers to the executive departments. Implicit in all these arguments is the hope that the dignity of the office can be enhanced by encouraging presidents to act so as to never favor one party over another, one region over another, or one class over another. Also implicit is the verdict that the roles of politician and statesman are incompatible. Presidents look unstatesmanlike to some people when they appear at party fund-raising dinners or intervene in state and congressional elections.

Arguing that we must liberate the presidency from "unnecessary political burdens," Senator Mike Mansfield said in 1971 that it is intolerable that a president "is compelled to devote his time, energy and talents to what can be termed only as purely political tasks. . . . A president facing reelection faces . . . a host of demands that range from attending the needs of political office holders, office seekers, financial backers and all the rest to riding herd on the day-to-day developments within the pedestrian partisan arena." [12] Others also feel that the country's chief executives should be more businesslike and that reducing their reelection activities would ensure more time and energy for substantive planning and systematic program implementation. Some hope, moreover, that the six-year term would enable a president to overcome both a deference to special interests and the timidity that results from having to keep an eye on the forthcoming election.

In short, the case in favor of the single six-year term is based on several expectations, namely, that it would:

- reduce the role of politics in the White House
- liberate a president from the worries and indignities of a reelection effort
- allow more time to concentrate on policy planning and program implementation
- liberate a president from the pressures of special-interest groups and party-line politics, allowing him to exercise greater independence of judgment and nonpartisan leadership
- eliminate the advantages of incumbency from presidential elections
- allow a president to make decisions free from the temptation of political expediency
- enforce the common-sense idea that a period of six years is enough even for the most robust individual

Despite a few attractive features, the six-year term would cause more problems than it would solve. The required reelection after four years is one of the most democratic aspects of the presidency. It affords an opportunity for assessment. It enhances the likelihood that a president will carefully weigh the effects of whatever he or she does on his reelection chances. At the core of our system is the belief that our president should have to worry about reelection and be subject to all the same vicissitudes of politics as other elected officials. Moreover, a political party should retain the threat of dumping a president as a check on the incumbent and the office, especially on a president who refuses to honor his or her party's pledges.

When the U.S. Senate in 1913 passed a resolution in favor of the single six-year term, Woodrow Wilson argued against it and his reasoning still seems valid: "The argument is not that it is clearly known now just how long each President should remain in office. Four years is too long a term for a President who is not the true spokesman of the people, who is imposed upon and does not lead. It is too short a term for a President who is doing, or attempting, a great work of reform, and who has not had time to finish it." Wilson also contended that "to change the term to six years would be to increase the likelihood of its being too long without any assurance that it would, in happy cases, be long enough. A fixed constitutional limitation to a single term of office is highly arbitrary and unsatisfactory from any point of view."[13]

The proposed divorce between the presidency and politics presupposes a significantly different kind of political system from that of the United States, which is glued together largely by ambiguity, compromise, and the extensive sharing of powers. In light of the requisites of democracy, the presidency must be a highly political office and the president an expert practitioner of the art of politics. Quite simply, there is no other way for presidents to negotiate favorable coalitions within the country, Congress, and the executive branch and to gather the authority needed to translate ideas into accomplishments. A president who remains aloof from politics, campaigns, and partisan alliances does so at the risk of becoming the prisoner of events, special interests, or his own whims.

The very means for bringing a president in touch with reality is the process of political debate and political bargaining, with all of the necessary uncertainty, changes of course, arguments, and listening to other points of view. What makes domestic politics so distasteful to presidents, that it is full of groups to persuade and committees to inform, is precisely its virtue; indeed, it is the major hope for maintaining an open presidency, one neither bound by its own sources of information nor aloof to the point that it will no longer listen.

By calling the president "more presidential" whenever a president ignores partisan politics, citizens encourage that president to even greater isolation. By turning up their noses at politics in the White House and urging the president to get on with the real business of guiding the nation, they also help to establish the two important conditions for secrecy and duplicity, with which the nation has become so familiar. First, with all the apparatus and technology for secret statesmanship at hand, presidents can more easily call on aides when something needs fixing than persuade the public or Congress to their point of view. Second, because presidents will look unpresidential if they participate in normal party politics, their aides must go through grotesque contortions to prove that their boss has never thought about anything except being president of all the people. The tactic of secrecy, so tempting to those who have it within their grasp, amounts to insulating the president from the normal checks and balances of the political system. New bait will be needed to lure presidents out of this comfortable sanctuary and into the morass of open politics, for the present enticements are small.

One way to prevent future abuses of presidential power, as others have noted, is to make the White House more open; and one way to do that, as has not been suggested so often, is to begin regarding a president as a politician once again. *Politics,* in the best sense of that term, is the art of making decisions in the context of debate, dialogue, and open two-way conversations, the art of making the difficult and desirable possible. This kind of politics at the White House should not be diminished. Indeed, as pointed out above, it is highly desirable that presidents be great practitioners of the craft of politics. They, as well as Congress and our parties, would profit from more politics, not less.

Most of the effective presidents have been highly political. They knew how to stretch the limited resources of the office, and they loved politics and enjoyed the responsibilities of party leadership. The nation has been well served by sensitive politicians disciplined by the general thrust of partisan and public thinking. Many of the least political presidents were the least successful and seemingly the least suited temperamentally to the rigors of the office. The best have been those who listened to people, who responded to majority as well as to intense minority sentiment, who saw that political parties are often the most important vehicle for communicating voter preferences to those in public office, and who were attentive to the diversity and intensity of public attitudes even as they attempted to educate and to influence the direction of opinion.

Everything a president does has political consequences, and every political act by a president has implications for the state of the presidency. The nation must fully recognize that presidents will and must be political, that they ought to be vigorous partisan leaders. Bipartisanship has often been overrated. James MacGregor Burns aptly noted that "almost as many crimes have been committed in the name of mindless bipartisanship as in the name of mindless patriotism."[14] If patriotism in an autocratic system implies blind loyalty to the regime, then patriotism in a democracy must include a responsibility and even obligation to speak out as a citizen whenever one believes that the government is following an unjust or misguided course of action.

If national leaders do become isolated or insulated from the mood of the public, then electing presidents for longer terms would only encourage this tendency.[15] Frequent elections necessarily remain a major means of motivating responsive and responsible behavior. An apolitical president, disinterested in reelection, motivated by personal principle or moralistic abstractions, and aloof from the concerns of our political parties, could become a highly irresponsible president. Elections customarily force an assessment of presidential performance. They are welcomed when promises have been kept and feared when performance has been unsatisfactory. Was it, for example, a mere coincidence, or was President Nixon's Vietnam troop-withdrawal rate calculated with the election of 1972 in mind? Was the Johnson-Humphrey bombing halt of 1968 aimed toward that year's election? Nixon's economic game-plan reversal in 1971, Johnson's vain efforts at peace negotiations in 1967 and 1968, and Bill Clinton's 1996 embrace of several of the House Republicans' key issues were unmistakenly related to the positive, constructive, and dynamic character of American elections.

Although change in important national policy is a slow process, a six-year term is not necessarily an appropriate remedy for this. Frequently, policy changes whose pace has frustrated the White House have come slowly because they have been highly controversial and adequate support had not yet been assembled. Mobilizing support is just as much a presidential responsibility as proclaiming the need, and support would be no less crucial with a seven-year or a seventeen-year term. Only a shrewdly political president who is also his party's leader, who is sensitive to political moods, and who is allied with dozens of the political party elite can build those coalitions able to bridge the separation of powers in Washington and to offset the strong forces bent on thwarting progress.

A president who cannot be reelected after four years is unlikely to accomplish anything of value if given a free ride for another two. What was true in the past remains true today: effective national leadership requires what the Constitution actually tried to discourage, that a party or faction disperse its members or its influence across the branches of government. Under normal circumstances, a president who ignores this maxim or retreats from these partisan and political responsibilities is unlikely to achieve much in the way of substantive policy innovation. A president elected to a single six-year term would be a president inescapably confronted with a bureaucracy as well as senior political appointees even less responsive than now. Even when presidents are both popular and eligible for reelection, they depend on senior and mid-career civil servants, a situation summed up in the wry Washington saying that "the bureaucracy eats presidents for lunch." When it is known that a chief executive is to leave by a certain date, bureaucratic entrepreneurs suddenly enjoy wider degrees of discretion and independence. Re-eligibility, used or not, is a potentially significant political resource in the hands of a president; denying that resource, even in the more limited way that the Twenty-second Amendment has done, will diminish the leadership discretion of future presidents who desire to be activist initiators of policy. President Truman spoke to this point: "You do not have to be very smart to know that an officeholder who is not eligible for reelection loses a lot of influence. . . . It makes no sense to treat a President this way—no matter who he is—Republican or Democrat. He is still President of the whole country and all of us are dependent on him; and we ought to give him the tools to do his job."[16]

There is yet another practical consideration about the six-year term: If we elect a truly outstanding president under this proposed reform we have him or her in the White House for two years less than under the present practice that would have provided eight years. If we elect a really bad president we are stuck for two years more than under the present system that provides for getting rid of such types at the end of four years.

The single-term proposal has a comforting ring of good, old-time government and nonpartisanship to it. Yet it represents the last gasp of those who cling to the hope that we can separate national leadership from the crucible of politics and of those who contend our presidency is too beholden to the workings of a patronage or spoils system. Neither is the case: the former remains an impossibility—it is impossible to take the politics out of public leadership in a democracy—whereas the latter is a problem whose time largely has passed. Equally undesirable is the notion that intense conflict over policy choices, that is, intense political activity, somehow can be removed from the presidency. The conflicts that surround the presidency and require a president to act as a public mediator, mirror those existing and potential conflicts over values that exist within the American society at large. If presidents were not required to resolve political conflicts by making political choices, they would not be fulfilling those responsibilities we rightly associate with democratic leadership.

A PLURAL EXECUTIVE IS ALSO NOT NEEDED

The job of the president, some think, has now become too complex and its reach too extended to be entrusted to one individual's fallible judgment. What can be done to lessen our excessive dependence on the accident of presidential personalities? One solution is to have several presidents.[17] The idea of a plural executive received consideration and some support at the 1787 Constitutional Convention.

The vast responsibilities of an executive who is also chief-of-state and top party leader are expected nowhere else. Political scientist Herman Finer was sufficiently alarmed by the growth of presidential power that he came out strongly in favor of introducing some kind of plural or collegial element to our national leadership: "The burdens . . . are necessarily so multifarious that to avoid a fatal collapse of efficiency and responsibility the President would have to be a titan and a genius. A collective Presidency might have these qualities, but not a solitary man. A solitary President is a gamble this nation cannot afford."[18] Hence, Finer suggested having one president and eleven vice presidents.

There are numerous variations of the plural executive concept. Some persons have advocated the election of two or three presidents, separating domestic and foreign responsibilities, or separating ceremonial from policy duties, or dividing up the policy formulation and implementation tasks of the presidency. Former Republican Senator Mark Hatfield suggested that we have an elected vice president, an elected attorney general, and elected heads of the major domestic departments. The Hatfield proposal suggests a plural executive similar to that found in most of the states.[19]

A more modest suggestion, made by former FDR adviser Benjamin Cohen, calls for the creation of a small executive council of not less than five or more than eight distinguished citizens who would be consulted by a president prior to crucial policy decisions. The Cohen proposal seeks to achieve some of the aims of a plural presidency while leaving our existing singular presidency almost intact. His executive council members would be nominated by a president but subject to Senate confirmation. The council members would be persons of independent political position, widely respected in and out of Congress.

Membership in this executive council would be a full-time job, according to Cohen. The idea is to oblige a president to consult this group of eminent persons before the president acts on critical national security matters, although the president alone would have the ultimate power of decision and the last word.

> The Executive Council should constitute a small super-cabinet with authority to participate in the decision-making process before important or potentially important Presidential plans, programs, and policies are finalized. . . .
>
> The members of the Executive Council, individually or collectively, should also have adequate authority within the limits prescribed and guidelines set by the President to act for him in monitoring, approving, and in coordinating the policies and programs of various departments and agencies in order to keep them within and abreast of the Presidential guidelines. But the members of the Executive Council should scrupulously avoid involvement in the minutia of departmental or agency operations. They should operate with very limited staffs of their own and avoid duplicating the staffs of the departments and agencies, although they should have authority to request permission of the heads of departments and agencies to borrow qualified persons for work on particular assignments. . . .[20]

Cohen's proposal emphasized that one of the critical dangers to presidential leadership comes from isolation from peers. Presidents are human beings subject to human frailties. They have their off days, their blind spots, their periods of emotional anxiety, their occasional needs to display a macho aggressiveness.[21] "Quiet consultation by our Presidents before they make their momentous decisions with a small Council of wise and respected persons may protect our Presidents, our nation and our world from much of the hazards of fateful decisions which ultimately must be made by one man."[22]

The Cohen proposal is one of the few plural executive ideas that might be carried out through legislation rather than through constitutional amendment. It might yield some of the benefits of the plural presidency without inviting most of the liabilities. Although it was one of the few post-Vietnam, post-Watergate reform suggestions to warrant careful consideration, it received little attention and no support. Americans clearly retain the idea that a single executive has to be in charge. The Cohen proposal would if enacted, further confuse the role of the cabinet and the vice president. Inevitably, too, many in Congress would contend that such an executive council would be a further aggrandizement by the executive of legislative and policymaking powers that rightly belong to Congress—a contention that has some merit.

The defects of most plural executive schemes are fairly self-evident: competition and conflict within the executive, a further swelling of the size of the presidency, and a diffusion and confusion of responsibilities. In times of crisis the members of a truly collective presidency would be expected to show unanimity. But what if they did not come to an agreement? Would the public have confidence in a major policy arrived at by a three or two to one decision? Surely the bureaucracy might heed such decisions less assiduously. Might not a collectivized presidency court paralysis or indecision, or both, in a nuclear attack or in an international monetary crisis, when swiftness and decisiveness would be most needed?

Plainly, a plural executive would have more difficulty in establishing priorities for the nation. For better or worse, the nation and especially Congress often look to the White House for its public policy agenda. The president must possess the ability to integrate, synthesize, and especially assess the relative merits of one policy with respect to others. How well can this be performed by a collective presidency, where each executive is in charge of just a part or even a fraction of the whole? Our existing system of checks and balances and limited government already seems, according to many people, designed more for paralysis than leadership. The plural executive would just accentuate this characteristic.

Perhaps the most frequent complaint about the plural executive idea is that it would create great difficulties in assigning accountability. Alexander Hamilton argued that the restraints of public opinion on the presidency would lose their efficacy if there were several executives rather than one. Whom should the people blame when things went wrong? Which one or set of executives should be impeached? Far better that there should be a single person for the people to oversee. Hamilton's 1788 brief in behalf of a single executive remains compelling:

> . . . the plurality of the Executive tends to deprive the people of the two greatest securities they can have for the faithful exercise of any delegated power, first, the restraints of public opinion, which lose their efficacy as well on account of the division of the censure attendant on bad measures among a number, as on account of the uncertainty on whom it ought to fall; and, second, the opportunity of discovering with facility and clearness the misconduct of the persons they trust, in order either to their removal from office, or to their actual punishment in cases which admit of it.[23]

Under the existing system, presidents already have substantial leeway in creating subordinate positions for sharing their burden. If they so desire, they can appoint a chief-of-staff or create a de facto assistant president to assume major responsibilities. Many presidents have done precisely this. From a president's point of view, one of the advantages of the current system is that all such assistants or deputy presidents are on a temporary assignment and serve at the pleasure of the president. They are expendable—they can be shifted to some independent agency or even used as a scapegoat when things go wrong. Under the existing system, a president need not be saddled with permanent deputies with whom he or she may have major disagreements.

In general, suggestions for institutionalizing more help for presidents by fixing into law or grafting onto the Constitution any plan for a plural executive should be rejected. The presidency, as opposed to the president, is already a collective entity, and individual presidents usually have ample discretion to organize their own office as they please.

NO TO "VOTES OF NO CONFIDENCE"

One of the boldest and perhaps most dramatic responses to Watergate and the abuse of power by Richard Nixon was a proposed constitutional amendment providing for a vote by Congress of no confidence in a president. The effect would be similar to the recall, now provided for in about sixteen states. Introduced in Congress by Representative Henry Reuss on August 15, 1974, H. J. Resolution 1111 gained little attention and little support.[24] But it merits some attention in our discussion of presidential accountability, if only because it poses a rather strong alternative to current practices. Indeed, the vote of confidence, if enacted here, would have far-reaching consequences, so much so that it might substantially transform the system of government in the United States as we now know it.

How would the no-confidence amendment work? A three-fifths vote of the members of each House present and voting would be necessary. Such a resolution would take priority over any other pending issue before Congress. If adopted, Congress would fix a date, between 90 and 110 days, for a special election for the president and vice president as well as for members of Congress. If it occurs near the regular congressional election date, that date would be used. Note that the incumbent president is eligible to stand for reelection even though he was the target of the no-confidence vote.

A reason put forward in defense of the no-confidence proposal is that the presidency in modern times has grown too powerful, especially in crisis contexts and in foreign affairs. Presidential power, it is argued, has risen above the level where the system of checks and balances can be effective in countering presidential actions. More bluntly, however, advocates of the no-confidence or similar national recall proposals believe that the four-year fixed term is a liability if and when we have incompetent presidents who lose the confidence of the nation. Incompetence, they point out, is not an impeachable offense. A "no-confidence" vote might also be a way to reconstitute a hopelessly deadlocked or failed government.

A goal of the proposed vote of confidence is to make future presidents more accountable to Congress, as well as more accountable to the American people. A president, it is assumed, would realize that he or she is accountable for his actions, proposed programs, negotiations, policies, and decisions and would have to face up to criticisms by Congress. The commission of high crimes and misdemeanors would no longer be needed to justify a president's dismissal. Maintaining the confidence of Congress and the general public would be an ongoing necessity. Matters such as Vietnam policy, for example, would have to be discussed in greater detail with congressional leaders to ensure that a president had the support of Congress. The vote of no confidence would be a means of retaliation against a president who too often worked behind the scenes or otherwise manipulated the spirit of checks and balances. Advocates reason that major decisions would have to be made by consultation, instead of by one person. To some extent, then, this constitutional amendment would introduce a certain amount of plural or shared decision making in our national government. The fundamental dangers of decisions by a Lone Ranger president would be thereby reduced.

A primary advantage of the Reuss amendment, of course, was that it did not require proof of illegal conduct. With the impeachment process, high crimes and misdemeanors must be proved. But as James Sundquist wrote, "In today's meanings of those words, a president who has simply lost his capacity to lead and govern because of bungling, betrayal by ill-chosen subordinates, or any of the other weaknesses that can lead to misuse of presidential power, cannot for that reason be relieved of power."[25] The Reuss proposal, however, could be used as a clear statement of dissatisfaction with a president's leadership abilities or his policies or lack thereof.

In a nutshell, the chief arguments in favor of the vote of no confidence are these:

- Impeachment is an inefficient check. At best it protects against gross criminal violations of the public trust but not against presidential incompetence.

- The president would be more disposed to working with Congress, explaining his policies and educating Congress and the general public about his plans and conduct in office.

- This proposal would force more presidential consultation with Congress and the leaders of the major parties and should lessen the secrecy surrounding presidential policymaking.

- The proposal does not take power away from the president; it only makes him more responsible for how he uses his power. It might prevent a president from trying to be "above politics" and from isolating himself from criticism.

• It would allow Congress to act in the face of negligence, gross incompetence, and disastrous policies, not just explicitly criminal
  abuses.

Critics viewed the Reuss amendment as an example of "good intentions, bad policy." They pointed out that Congress has plenty of resources
with which to check a president if only they would use them. Further,
even though presidents have become paramount in the conduct of foreign
affairs, Congress has numerous means at its disposal to oversee this exercise of power.

In certain situations it would seem that the vote of no confidence
would give Congress the power to frustrate continually a president with
whom it disagreed. The alternatives are a government of continuous presidential elections and overall paralysis or a government in which president
and Congress are so close as to defeat the basic concept of the separation
of powers so fundamental to our system. Perhaps this latter condition of
power centralized in an overly united president and Congress would, under certain circumstances, be more dangerous to the basic freedom and
spirit of the United States than any single "imperial" president.

There is also the argument that a vote of no confidence could actually be
used to strengthen the hand of an already strong leader, much as Hitler and
de Gaulle used plebiscites to weaken their opposition. If timed properly, a
president who wins a vote of no confidence with a strong show of votes can
use it as a false rally of support. One imagines, for example, that Lyndon
Johnson could have won a vote of no confidence on his Vietnam policy in
1966. Or Richard Nixon might have won a vote of no confidence during the
early stages, say the spring or summer of 1973, of his Watergate crisis. In
both cases, these "victories" might have strengthened the hand of those men
in defiance of a more responsive, accountable posture.

A vote of no confidence arrangement might lead presidents to avoid
making significant changes in policy that would antagonize Congress. Innovative leadership would be thwarted, as presidents might gear most of
their actions to public opinion polls or to the wishes of the majority at the
expense of minority rights. The proposal could lead a president to concentrate on short-term or immediately popular initiatives to "create" favorable
public approval at the expense of long-term planning.

In short, the cure would doubtless be worse than. the occasional ail-
ment.[26] The vote of no confidence procedure would not necessarily improve the quality of presidential leadership, nor would it enhance accountability in any significant way. Presidents do not intentionally make poor
decisions. What could result might be far worse than the rare arrogant
president we have had to endure. This measure might make presidents
too dependent on Congress, or conform too closely with popular opinion.
This measure might give us an endless line of unsuccessful short-term
presidents and as a result a paralyzed nation. We risk the unwise weakening of the presidency by such an amendment. Abuses of presidential

power do need to be curbed, but this proposal is not the way to do it. We need better and wiser leadership, not a weakened presidency.

THE ILLUSION OF THE THIRD PARTY PANACEA

One of the more appealing (on the surface) political reforms is the idea that perhaps a major third party could, if it were able to organize and win a respectable following, produce better presidential candidates and straighten out the existing two parties.

Congressman John Anderson preached this reform in 1980 when he challenged Reagan and Carter in the presidential general election. Businessman Ross Perot, who won nearly 19 percent of the popular vote in 1992 and 8 percent in 1996, tried in vain to build a third party he called the Reform Party. In both Anderson's and Perot's cases, they faulted the two major parties for causing the huge national debt and for failing to control the sprawling out-of-control entitlement programs. Both Anderson and Perot deserve credit for raising important issues and for forcing the Democrats and Republicans to explain their policies. Third parties in America are at their best when they call attention to controversial issues and criticize the establishment parties for neglecting to do what they have promised to do.

It is true that our seemingly entrenched two major parties have made it extremely difficult for an existing or future third party to become a major party. The last time this happened was in the late 1850s! And, of course, no third party has won the presidency since 1860. Third parties have never shaped national policy from inside the government. And, although their points of view often get absorbed into one of the major parties, third parties and their candidates have rarely had much influence on national policy.

Your authors are not opposed to third parties or the efforts to create a viable future third party. But the political realities are that third parties do not have bright prospects in the next several years, and past experience is rather clear in demonstrating that serious third parties are readily put out of business when they do not have compelling policy ideas. Simply put, one of the major two parties will steal, modify, and adapt such ideas into their own platforms and promises. That's what Nixon did with some of George Wallace's ideas in 1968 and 1972. That's what Democrats and Republicans are trying to do now with some of Ross Perot's better ideas.[27]

Doubtless our two-party system needs to be challenged and shaken up. There is little doubt too that a serious multiparty system in America would stir more people to participate in politics and elections, and this would be a healthy development (for the most part). Yet the two-party system has also served this country well throughout the twentieth century. It needs challenge and revitalization, but this can be done as well— and perhaps better—from within the parties as outside of them.

Strengthening, reforming, and improving the two-party system offers more hope and is more realistic than the search for the magical third party. It is highly unlikely that a third party will provide the answers to most of the pressing economic, social, and international problems facing the United States as we begin the twenty-first century.

In contrast to the six-year term idea, the idea of a plural presidency, notions of votes of no confidence, and the illusions of the third party panacea, there are existing checks that work reasonably well. These need better appreciation. There are also additional ways to make our system of governance and the presidency act in a more accountable and responsible manner.

THE ULTIMATE CHECK: IMPEACHMENT AND REMOVAL

Impeachment is obviously one of the most potent checks against the abuse of executive power, yet over the nation's history it has been the least used check. For practical purposes, it is a political action, phrased in legal terminology, against an official of the federal government. The Constitution deals with the subject of impeachment and conviction in six places, but the scope of the power is outlined in Article II, Section 4.

> The President, Vice President and all civil officers of the United States, shall be removed from Office on Impeachment for, and Conviction of, Treason, Bribery, or other high Crimes and Misdemeanors.

In the impeachment process, the House of Representatives acts as the prosecutor and the Senate serves as judge and jury. Any member of the House may initiate impeachment proceedings by introducing a resolution to that effect in the House. The House Judiciary Committee may conduct hearings and investigations. The committee then decides either in favor of or against an impeachment verdict and sends its conclusions on to the full house. A 50-percent vote in the House is needed to impeach. Select members of the House, if an impeachment is enacted, would then try the case before the U.S. Senate. In the Senate a two-thirds vote of those members present is needed for conviction and removal.

Only thirteen national officials have been impeached by the House since 1789. Of these, eleven were tried in the Senate. Four were convicted, six were acquitted, and one resigned before the Senate took action. In the two remaining cases, the charges were dismissed after the person had been forced to resign national office. Nine of these cases involved federal judges, one involved a senator, one a secretary of war, and one a president, Andrew Johnson, who was overwhelmingly impeached by the House in February 1868 but missed conviction in the Senate by one vote (35–19) in May of that same year.

The impeachment and removal of a president has been a much misunderstood and an obviously cumbersome means of accountability. Its use is fraught with emotion and hazardous side effects, and it necessarily remains a device to be used only as a last resort.

One of President Nixon's involuntary contributions to our understanding of presidential politics is that he provided the occasion for clarifying the character and usage of the impeachment and removal power, for the most significant controversial constitutional question about impeachment had been what were the appropriate grounds for this action. *Treason* and *bribery* are clear legal terms and cause no problem. It was the phrase *other high Crimes and Misdemeanors* that raised so many hard-to-answer questions. As one group of lawyers put it:

> Is that phrase limited to acts which would be indictable as criminal offenses, or was it intended to reach abuses of office or breaches of trust not constituting criminal acts? If impeachment and removal may properly rest on activities which do not constitute crimes, are there any limits in principle on the type of conduct which can be the basis for impeachment and removal, or should the exercise of these powers be governed solely by the free play of our political system?[28]

Some analysts, especially lawyers defending the potentially impeached, have argued that a person can be impeached and removed from office only on criminal charges or on offenses that would be indictable in the criminal court. The weight of most recent scholarship, however, Nixon's defense to the contrary notwithstanding, supports the construction of "high crimes and misdemeanors" as not limited to offenses under ordinary criminal law.[29]

One possible view is that an impeachable offense is whatever a majority of the House say it is. The extreme opposite view, or the Nixon defense view, holds that impeachment should be voted only on proof of serious, indictable crimes. In the celebrated Richard Nixon impeachment proceedings, the House Judiciary Committee adopted a middle stance, one that is likely to have a controlling, if not legal, influence on future impeachment efforts. The Judiciary Committee in 1974 held that violation of a criminal statute is not a prerequisite for impeachment as long as the offense is a serious one. Committee members were well aware that an impeachable offense should not be taken to mean anything around which political expediency might organize a majority in the House and two-thirds in the Senate. In effect, the majority of the House Judiciary Committee believed that a gross breach of trust or serious abuse of power was necessary before passage of an impeachment resolution.

One of the lawyers for the House Judiciary Committee, in particularly probing analysis of the impeachment power, concluded that there are four general ways a president can violate his or her constitutional duty to see that the laws are faithfully executed—all of which could justify impeachment investigations: (1) If the president attempts, either directly or indirectly, to induce a subordinate to engage in illegal activity; (2) if the president believes that some form of misconduct on the part of a subordinate within the executive branch has occurred and the president takes no action to correct it or even to respond to it properly; (3) if the president asks an official or assistant to take ethically questionable steps and does not supervise his or her activity; and (4) if a president does not oversee adequately the actions of his subordinates. It is suggested in this latter case that a president is accountable for any misconduct even if it takes place without his immediate knowledge.[30]

President Nixon resigned before the full House impeached him. The House Judiciary Committee, however, had voted to approve impeachment on three counts. It was the view of most observers that Nixon would have been impeached, then convicted and removed from office by the Senate shortly thereafter. He chose to resign, in part, many feel, because he would have lost the right to his pension and other post-presidency fringe benefits had he been convicted in the U.S. Senate. No doubt he decided to resign as well because he had lost the backing of the country and the public support needed to function even at a minimal level in that office.

Many observers note that impeachment is a time-consuming and highly traumatic instrument to deploy. True. But there is no reason to suppose in most instances that impeachment would be any more traumatic than having a person such as Richard Nixon continue in office for another two years when he has long since lost the ability to govern effectively. The impeachment and removal powers may well be an elaborate and difficult-to-use means to hold presidents to account, but as the Andrew Johnson and Richard Nixon cases attest, each in its own way, it can be used. More important, its availability is an ever-constant reminder to presidents that their power must be exercised in an accountable manner.

THE NEED FOR OCCASIONAL INDEPENDENT COUNSELS

Only a handful of the more than one hundred Watergate-related reform bills ever made it through Congress. One of the more controversial ones that did was a provision that would establish, under certain circumstances, a temporary special prosecutor to investigate and prosecute executive branch conflict-of-interest allegations. Later legislators would rightly substitute the name *independent counsels* for the more biased term *special prosecutors.*

Watergate demonstrated that the executive branch could not investigate itself lest there be the appearance—if not the reality—of a conflict of interest. After considerable debate Congress passed the Special Prosecutor Act in 1978. This law allowed the attorney general to appoint a special investigator under certain circumstances.

Under this legislation it is up to the attorney general to determine whether a case of conflict of interest is valid enough to warrant further investigation and prosecution. This would apply in the cases of about one hundred top executive office and Justice Department posts and top officers of a presidential election campaign. The statute mandates the appointment by a designated panel of three federal judges of an independent counsel whenever charges of a violation of federal criminal law arise in connection with these top officials.

Did the Congress overreact in creating the special prosecutor law? No. Presidents, like all executives, set the moral tone for their administrations, give clues as to acceptable and unacceptable behavior, establish norms and limits. Presidents demonstrate by words and deeds the kind of behavior that will be tolerated and the standards applicable to the entire administration. Independent counsels, while perhaps overused in recent years, do serve a useful purpose when employed in appropriate circumstances.

If politics is, or appears to be, scandal ridden; if self-interest and not pursuit of the public interest appears to be the chief motivating force in politics, then it should not surprise us that the general public sees "politics" as a dirty word. But this may undermine democracy and the respect for the rule of law, as well as drive people out of the political process. Thus, it is vital to the health of constitutional democracy that corruption be exposed and the wrongdoers punished.

In the history of the presidency, the administrations considered most corrupt are the Grant, Harding, and Nixon presidencies.[31] The Reagan and Clinton administrations, sadly, have also had major ethical and criminal problems. Ulysses S. Grant, inexperienced in the world of politics, was taken advantage of by avaricious associates. Likewise, Warren G. Harding's lax management style allowed greedy underlings to take advantage of him.

The Nixon administration was markedly different from the Grant and Harding presidencies in that here the president was an active "co-conspirator" in the crimes of Watergate.[32] And during the Reagan years, several types of corruption were exposed, some of which were linked directly to the president. With both what was referred to as the "sleaze factor," in which over two hundred White House officials were accused of criminal behavior[33] (among them Richard Allen, the president's first national security adviser; Labor Secretary Raymond Donovan, who was indicted but not convicted; Rita Lavalle of the EPA; Attorney General Ed Meese; aide Michael Deaver, and White House political director Lyn Nofzinger) and the Iran-Contra scandal, the Reagan administration faced an unprecedented number of accusations of wrongdoing.

Charges of wrongdoing have hounded President Clinton. In January of 1992, then candidate Clinton faced accusations of sexual misconduct, and shortly after he was elected, charges of shady financial dealings surfaced in the Whitewater affair. When aide Vince Foster committed suicide, charges that the Clintons had a hand in the death were leveled. And charges of fund-raising improprieties stemming from Clinton's 1996 reelection bid hounded him in his second term.

Several independent counsels were appointed to investigate Clinton and his associates. Kenneth Starr, appointed to investigate Whitewater-related accusations, sparked controversy for his public association with anti-Clinton organizations, and for the lengthy and expensive investigation against the president. Clinton's prepresidential as well as presidential behavior certainly provided a lot for investigators.[34]

There were seventeen independent counsel investigations between 1978 and 1998. The process has not worked perfectly, and indeed for a two-year period from 1992 to 1994 the law was allowed to lapse.

Critics have argued that some independent counsels have politicized their investigations and that in some cases, such as the Iran-Contra affair and in the more recent Whitewater and campaign financing investigations, the investigations have gone on for too long.

Some critics think too that the independent counsels are too independent. But this independence is necessary if the process is to have integrity. Legal scholar Peter Shane offers this useful perspective:

> The relative independence of the independent counsel is secured not only by his or her judicial appointment but also by statutory protection against discharge. Independent counsels ordinarily have job tenure until they determine that all matters within their jurisdiction are adequately resolved. They may be removed only by personal action of the Attorney General, and then only for "good cause, physical disability, mental incompetency, or . . . other condition that substantially impairs" the independent counsel's job performance. As a consequence, the Attorney General—and, thus, the President—is without power to threaten an independent counsel with removal because of disagreements about the conduct of an investigation or prosecution. Decisions regarding the sufficiency of evidence for an indictment, the appropriateness of alternative investigative techniques, witness selection and so on, are entirely within the independent counsel's discretion.[35]

What is now called the *Independent Counsel Act* will always be controversial. Republicans complained bitterly when Iran-Contra and other Reagan-era scandals were investigated. Democrats complained almost as loudly as investigations of the Clinton administration dragged on. But the independent counsel process is essential to keeping presidents and other cabinet-level executive officials accountable.

We agree with former Watergate special prosecutor Archibald Cox that "In the rare national crises that arise when there is serious evidence of criminal misconduct by a president or other high executive official, only an independent counsel, chosen by a judicial panel, can provide the best assurance of thorough and impartial investigation followed by fair-minded prosecution or public dismissal of the charges."[36]

Cox believes the potential for abuse of this process can be reduced by a number of changes. He suggests the independent counsel provisions should be limited just to the president and cabinet-level officials. It should also, he believes, be limited to crimes committed in national office. "The benefits of allowing investigation of earlier wrongdoing are outweighed," Cox writes, "by the risk of encouraging opposing politicians to dig for stories of past misdeeds."[37] The process should probably also have a time limit of up to a year. Most important, Cox adds, "independent counsels must see their function not as pursuit of a target to be wounded or destroyed, but as an impartial inquiry with as much concern for public exoneration of the innocent as for indictment of the guilty."[38]

The independent counsel law was challenged in *Morrison v. Olson,*[39] but in 1988 the Supreme Court upheld its constitutionality. With all its potential faults and possible abuses, the independent counsel process needs to be kept.

THE NEED FOR GREATER CAMPAIGN FINANCE INTEGRITY

As stated in Chapter 2, running for president is a very costly enterprise. Candidates must raise vast sums of money, and the sheer magnitude of this fund raising poses serious questions for democratic accountability and integrity. The dependence of candidates on the largess of big donors raises the appearance, if not the reality, of corruption.

In the 1960s, presidential campaigns were funded by less public means than today. Joseph Kennedy, father of John Kennedy, once scolded his son for spending too much on the 1960 race, complaining that he was interested in financing a victory, not a landslide. During the 1960s the costs of presidential campaigning began to skyrocket. The main cause of this was television. The costs of television caused candidates to spend more and more time raising money to pay for costly TV advertising.

The scandals of the 1972 Nixon reelection campaign caused a backlash, prompting the Congress to attempt to reform the method of financing presidential elections. A series of laws and amendments to the laws sought to limit the impact of money on the selection process, but the laws had loopholes that were soon exploited by various campaigns. One of the biggest loopholes was the "soft money" contribution, where money was given to the political party and not directly to the candidate, thus allowing candidates to bypass the legal limits imposed on their campaigns.[40]

In 1992, Democrats spent $85 million in "hard" money to the Republicans' $164 million. In soft money the Republicans outspent the Democrats $46 million to $31 million. By 1996, costs soared. Republicans spent $278 million in hard money to the Democrats' $146 million. In soft money they spent $121 million to the Democrats' $106 million.[41]

President Clinton suffered a number of embarrassments when disclosures of his 1996 fund-raising efforts revealed questionable judgment relating to money taken from a number of foreign sources and reports that for $25,000 a donor could attend an "event" at the vice presidential mansion, for $50,000 could meet with the president at the White House, and for

$100,000 or more, could dine with the president at the Hay Adams Hotel or be an overnight guest at the White House.[42]

Clearly the reform efforts of the 1970s had not worked as intended. Is there a solution to this continuing problem? Some call for tightening the loopholes in current law; others suggest stricter limits on donations; and others want stiffer penalties for violations of current law. Still others suggest that television stations be required to give free prime-time spots to candidates in hopes that this will diminish the need for such large sums of money.[43]

As the 1970s campaign finance reform efforts indicate, creative and determined candidates are usually able to find a loophole to frustrate reformist intent. Some combination of major limits on soft money contributions and free television time for major candidates is now needed to guarantee the integrity of our political process. It is probably not wise to rely on legal prohibitions alone to control the campaign finance mess. Here, as in many cases, the most hopeful corrective is an aroused, educated, and demanding public. Yes, full disclosure of sources of funding and spending; and yes, closing loopholes offers some help. But in the end, it must be the voters who demand better of their candidates. Laws alone will not do the job; laws backed by an informed public may.

THE NECESSITY FOR POLITICS AND HEALTHY POLITICAL PARTIES

A basic question throughout this book is not whether government by the people is possible or even desirable in the modern world but rather how the political system and the relationship between leadership and citizens can be transformed so they will approach more closely the ideals of democracy. This argument explicitly rejects the view that things must remain as they are because that is the way underlying forces make them. These pages have often emphasized the need for a strong yet also a lean and accountable presidency, a presidency that could achieve peace, economic prosperity, and the reforms and innovative changes that would broaden the economic and political opportunities of everyone.

The simplistic notion that returning to the drawing boards and coming up with a new charter or a new constitution will provide the needed solutions is rejected. A new constitutional convention is not needed. Solving major policy problems and keeping presidents honest and responsible are more likely to be accomplished by *political* than any additional *constitutional* means. No single institutional innovation we have ever heard of could guarantee a commitment to truth, compassion, and justice. Formal constitutional provisions to guard against presidential isolation, such as the institutionalization of government-sponsored votes of confidence or a lengthened presidential term, are not sensible ways to increase accountability.

That the American political system and leaders are asked to undertake much of what the rest of society refuses to do is a continual problem in this nation. The promise of the presidency symbolizes the hopes of the people, and certainly there is nothing wrong in calling on our president to summon up exalted national commitments. But the attempt to reconstitute any single institution in a large, complex society may be rather futile if the fundamental purpose of that institution is to represent and respond to the dominant values of the society. There is little doubt that this society's values are rooted in a strong faith in political and social gradualism, in a deep fear of revolutionary change, and in a steadfast devotion to most of what constitutes the existing order.

Political controls, however, do need to be sharpened and strengthened to ensure a continual public and congressional scrutiny of presidential activity. Openness and candor often have been lacking. Presidents and their aides sometimes supply disappointingly little information to the press, to Congress, or to the public on matters of executive agreements, vetoes, executive orders, complex arms sales, and how they raise campaign contributions. In the seemingly endless attempt to accentuate the positive, White House spin controllers too often have distorted news and thereby aggravated difficulties in credibility by claiming too much credit for fortuitous events or for policy initiatives that may or may not achieve sustained or desirable ends and by projecting the appearance of boldness, usually at the expense of candid discussions of the complexity of problems, the modesty of proposed solutions, and the realities of who must pay and how much.

A free society must mean a society based explicitly on free competition, most particularly competition in ideas and opinions, and by frank discussions of alternative national purposes and goals. Elected leaders and a vigorous press must ceaselessly attack ignorance, apathy, and mindless nationalism—the classic enemies of democracy. The citizen must resist sentimental and rhetorical patriotism that espouses everything as a matter for top priority but in practice eschews the tough political programs that must be begun and implemented. Needed is a far more thoughtful way of looking at the presidency, leadership, and citizen responsibilities.

Also, Congress, the press, and the public must use all existing political controls as a means to inspire as well as to check presidents. Citizens must insist that presidents lend their voice and energies to those who are not represented by well-heeled lobbyists. Strengthening the have-not sectors of society and giving a fair hearing to minorities will always remain major presidential responsibilities and an essential part of the legitimacy of the modern presidency.

We may want to change the presidency, yet *we* must change also. We must make the presidency work, but we must also strengthen Congress, modify public expectations and demands, and strengthen the party system. The interconnectedness of the American system—not the reform of one branch, but the reform of the American governmental system—

should be the goal. And while ours is primarily a presidentially driven system, it is also a system driven by cooperation and engagement between the separate institutions that share power.[44]

A precondition of making the separation of powers work better is the need to develop a *consensus* around particular governing ideas. This does not mean that all Americans must march in lockstep behind the will of the majority, nor does it mean that we must, or even can, come together on such issues as abortion, welfare, or prayer in school, but it does point to the need for Americans to feel and act like a nation; a people joined together in common pursuit of the common good; a nation, not merely an aggregation of individuals. The routine division between Congress and the president reflects not merely the institutional battles between two branches of government but also reflects the deeper conflicts and divisions that divide the nation and its people.

The ideal conditions for presidential accountability are difficult to spell out, but the public should know what a president's priorities are, how they will be financed, who will gain and who will lose, and what the alternatives are. The public and its representatives in normal circumstances should be given a chance to evaluate presidential priorities and give their views. Where strong majorities exist, Congress should be able to compete with a president in shaping the nation's policy agenda. Substantial controversy exists, however, over the extent to which public opinion should shape or dictate presidential choices. Government by public opinion, however it is devised, can never guarantee justice or wisdom.

We can complain about the paucity of formal means and the decline in the effectiveness of informal checks for making a president accountable. But we should also appreciate that presidents need flexibility, perhaps today more than ever. No one seriously proposes that a president's decisions should merely reflect majority opinion. The structure of the office in part reflects the desire of its designers to prevent presidents from being threatened or rushed into action by the shifting gusts of public passion. In practice, the definition of acceptable limits for presidential accountability will vary over time. If the standards for presidential accountability tilt too far in the direction either of public opinion or of independence and isolation, a president is less able to provide those subtle accommodating and mediating elements of leadership that are essential for effective democratic government. No task defines the essence of presidential leadership in a pluralistic society better than that of devising a workable and purposeful adjustment of the conflicting views of experts, elite groups, and the people as a whole.

With the proliferation of public opinion polls, presidents are barraged with nearly minute-by-minute CNN-style reports on everything from "How is the president doing?" to the most trivial of minutiae. The time frame for presidential decision making and problem solving has been dramatically shortened. Issues cannot percolate up or ripen; problems cannot be debated, discussed, analyzed. Our instant-gratification culture wants answers *now*. Thus presidents cannot allow solutions to emerge; they must have answers, not questions.

Although certain presidents have tried on occasion to govern without the benefit of public and partisan support, they seldom have succeeded. Presidents are in fact usually heavily influenced by their anticipation not only of the next election but also of tomorrow's headlines and editorials, next week's Gallup poll, next month's congressional hearings, and possible reprisals against their programs by Congress, the Supreme Court, the opposition party, and other institutions. In what other nation can a chief executive be overruled in the courts? A judicial check on a chief executive seldom exists in parliamentary systems if the leader retains his party's backing. In what other countries do the legions of newspapers, pamphleteers, tabloids, talk radio, and bulletin-board chat groups flourish with such tolerance and even encouragement?

One of the more confusing aspects of presidential accountability is the way the American people find it convenient to blame presidents for a whole range of problems, regardless of whether the problems have been subject to presidential control. We generally withhold our applause when a president's work is good, but we seldom fail to hiss presidential blunders. As noted back in chapter 3, no matter what presidents do, their popularity is likely to decline. When news is good, a president's popularity goes down or stays about the same; when news is bad, it merely goes down faster and farther. The decline in approval of the president is in large part a function of the inability and unlikelihood of a president to live up to the buildup he received during the presidential honeymoon.

Ultimately, being paradoxical does not make the presidency incomprehensible. Can we rid the presidency of all paradoxes? We couldn't, even if we wanted to do so. And anyway, what is wrong with some ambiguity? It is in embracing the paradoxical nature of the American presidency that we may be able to arrive at understanding. And with understanding may come enlightened or constructive criticism. This is the basis for citizen democracy.

Plainly, the presidency is a dynamic, not a static institution. While there are standard role expectations and responsibilities faced by all occupants of the office, such uniformity must be seen as the other side of the coin of the rich variety each president brings to the office. Each president brings a unique set of skills, experiences, goals, and styles to the presidency, yet the office and the requirements of the times place certain demands on every president. Each new president has to both emulate the master performances of Washington, Lincoln, and FDR, yet find his (or her) own voice, acquire his own identity, and invent the kind of approach that will help build a better America.[45] It is this mix of the unique and the expected that makes the presidency such a fascinating institution. Since all presidents are expected to "lead," how individual differences (both inter- and intra-presidential) impact the institution often give us a clue of how well or poorly a president will perform.

In the end, presidents will be kept in line only if the people, according to their own personal views, exercise their rights and their political responsibilities. If the people insist both at and between elections that there be more respect for the doctrine of self-restraint, which all branches, including presidents, violate on occasion, it may happen. People can "vote" between elections in innumerable ways—by changing parties, by organizing protests, by their votes in mid-term elections, by civil suit and litigation; in short, by "sending them a message." Persistence and intensity have impact, especially if the political parties can be renewed as institutions for both education and accountability.

James Madison was right. A constitutional democracy must insist on the effectiveness of our "auxiliary precautions"—Congress, parties, the courts, the press, the Bill of Rights, concerned citizen groups, the independent counsel process—if we are to have a constitutionally acceptable presidency and preserve our constitutional democracy.

Notes

CHAPTER 1

1. Barbara Tuchman, *The Distant Mirror* (New York: Knopf, 1978), p. xvii.

2. F. Scott Fitzgerald, *The Crack-Up* (New Directions, 1956), p. 69.

3. Jeffrey Tulis, "The Two Constitutional Presidencies," in Michael Nelson, ed., *The Presidency and the Political System* (Washington, D.C.: Congressional Quarterly Press, 1984), p. 61.

4. See Harvey C. Mansfield, Jr., "The Ambivalence of Executive Power," in Joseph Bessette and Jeffrey Tulis, eds., *The Presidency in the Constitutional Order* (Baton Rouge: Louisiana State University Press, 1981), pp. 314–33; and Walter Berns, "The American Presidency: Statesmanship and Constitutionalism in Balance," *Imprimis* (January 1983).

5. See, for example, Godfrey Hodgson, *All Things to All Men: The False Promise of the American Presidency* (New York: Simon & Schuster, 1981); George Edwards, *The Public Presidency* (New York: St. Martin's Press, 1983); and Paul Brace and Barbara Hinckley, *Follow the Leader: Opinion Polls and the Modern Presidents* (New York: Basic Books, 1992).

6. Lance Morrow, "Keeping up the Presidential Style," *Time* Magazine, June 15, 1981, p. 52.

7. See Mervin Field, "Public Opinions and Presidential Response," in John Hoy and Melvin Bernstein, eds., *The Effective President* (Palisades Press, 1976), pp. 59–77.

8. William Davison Johnston, *TR, Champion of the Strenuous Life* (Theodore Roosevelt Association, 1958), p. 95.

9. Eugene Kennedy, "Political Power and American Ambivalence," *The New York Times Magazine,* March 19, 1978.

10. Richard M. Nixon, *Leaders* (New York: Warner Books, 1983), p. 341.

11. Charles de Gaulle, *The Edge of the Sword* (New York: Criterion, 1960).

12. Saul Alinsky, *Rules for Radicals* (New York: Random House, 1971).

13. Michael Maccoby, *The Gamesman: The New Corporate Leaders* (New York: Simon & Schuster, 1976).

14. Robert J. Morgan, *A Whig Embattled: The Presidency under John Tyler* (University of Nebraska Press, 1954), p. 183.

15. James MacGregor Burns, *Roosevelt: The Lion and the Fox* (San Diego: A Harvest/HBJ Book, 1984).

16. Quoted in *U.S. News & World Report,* January 28, 1985, p. 40.

17. Editorial, "Presidents and Politics," *The Washington Star,* January 10, 1981.

18. Ibid.

19. Harold J. Laski, *The American Presidency* (New York: Harper & Brothers, 1940), p. 34–35.

20. James MacGregor Burns, *Leadership* (New York: Harper & Row, 1978).

21. Bruce Miroff, *Pragmatic Illusions: The Presidential Politics of John F. Kennedy* (New York: David McKay, 1976), p. 31. For a somewhat similar treatment of several earlier presidents, see Richard Hofstadter, *The American Political Tradition* (New York: Vintage, 1948).

22. Arthur M. Schlesinger, Sr., *Paths to the Present* (Boston: Houghton Mifflin, 1964), p. 93.

23. Meg Greenfield "The Anxiety of Choosing Sides," *Newsweek,* October 20, 1980, p. 108.

24. For an examination of President Clinton and his flip-flops, see John Brummett, *High Wire: The Education of Bill Clinton* (New York: Hyperion Books, 1994).

25. See David H. Donald, *Lincoln* (New York: Simon & Schuster, 1995).

26. Laski, note 19, p. 93.

27. John Keegan, *The Mask of Command* (New York: Penguin Books, 1987).

28. *Boston Globe,* January 7, 1976, p. 19.

29. Bob Woodward, *The Agenda: Inside the Clinton White House* (New York: Simon & Schuster, 1994), p. 165.

30. This is how one of our students put it in a class discussion just a few years ago.

31. Henry Fairlie, *The Parties* (New York: St. Martin's Press, 1978), p. 213.

32. Former President Jimmy Carter, in a seminar at the U.S. Air Force Academy, August 26, 1985.

33. Aaron Wildavsky,"The Two Presidencies," *Trans-Action* (December 1966). See also Steven A. Shull, ed., *The Two Presidencies: A Quarter Century Assessment* (Chicago: Nelson-Hall, 1991).

34. Nelson Mandela, *Long Walk to Freedom* (Back Bay Books, 1993), p. 625.

CHAPTER 2

1. Elizabeth Drew, "Letter from Washington," *The New Yorker,* 1993. See also Thomas E. Patterson, *Out of Order* (New York: Vintage Books, 1994).

2. Karlyn Bowman, "Do You Want to Be President?," *The Public Perspective* (February–March 1997), pp. 39–41.

3. David Lauter, "Clinton's Image Hurts Him More Than His Ideas," *Los Angeles Times,* January 28, 1995, p. A1.

4. R. W. Apple, Jr., "Post-Cold War Candidates Find No Place Like Home," *The New York Times,* February 20, 1996, p. A12.

5. William R. Keech, "Selecting and Electing Presidents," in Thomas E. Cronin and Rexford G. Tugwell, eds., *The Presidency Reappraised* (New York: Praeger, 1977), p. 98.

6. Erwin Hargrove, "What Manner of Man?," in James David Barber, ed., *Choosing the President* (New York: Prentice-Hall, 1974), p. 19.

7. James David Barber, *The Presidential Character,* rev. ed. (New York: Prentice-Hall, 1992).

8. Stephen Ansolabehere and Shanto Iyengar, *Going Negative: How Political Advertisements Shrink and Polarize the Electorate* (New York: Free Press, 1996).

9. Tom Rosenstiel, *Strange Bedfellows: How Television and the Presidential Candidates Changed American Politics, 1992* (New York: Hyperion Books, 1993).

10. David S. Broder, "Snooping Hurts Candidates and the Press," *Los Angeles Times,* November 16, 1987. Editorials.

11. Bruce Buchanan, "Sizing up Candidates," *PS: Political Science and Politics,* (Spring 1988), p. 250.

12. Martin Walker, book review of Robert Wilson's *Character Above All, The Washington Monthly* (April 1996), p. 54.

13. Robert A. Wilson, ed., *Character Above All* (New York: Simon & Schuster, 1995).

14. Rexford G. Tugwell, *The Art of Politics* (New York: Doubleday, 1958), pp. 242–43.

15. On this point see Rexford G. Tugwell, *The Brains Trust* (New York: Viking, 1968), pp. 410, 423–24.

16. Emmett H. Buell, Jr., "The Invisible Primary," in William G. Mayer, ed., *In Pursuit of the Presidency* (Chatham N.J.: Chatham House, 1995).

17. Charles Lewis and the Center for Public Integrity, *The Buying of the President* (New York: Avon Books, 1996).

18. Quoted in *The Los Angeles Times,* February 2, 1995, p. E4.

19. Quoted in *The New York Times,* October 19, 1974, p. E18.

20. Jill Abramson and Thomas Petzinger, Jr., "Deja Vu: Big Political Donors Find Ways around Watergate Reforms," *Wall Street Journal,* June 11, 1992.

21. Federal Election Commission, *Record* (February 1993), p. 5.

22. Leslie Wayne, "Loopholes Allow Presidential Race to Set a Record," *The New York Times,* September 8, 1996, p. A1.

23. Herbert E. Alexander, *Financing Politics: Money, Elections and Political Reform,* 4th ed. (Washington, D.C. Congressional Quarterly Press, 1992).

24. Nelson Polsby and Aaron Wildowsky, *Presidential Elections* (Chatham, N.J.: Chatham House, 1995) p. 69.

25. Michael Duffy, "The Money Chase," *Time* Magazine, March 13, 1995, p. 93.

26. Frank J. Sorauf, *Money in American Politics* (Glenview, Ill.: Scott, Foresman, 1988); and Herbert Alexander's series of books on *Financing Presidential Campaigns.*

27. Quoted in Jasper B. Shannon, *Money and Politics* (New York: Random House, 1959), p. 35.

28. Jane Mayer, "Inside the Money Machine," *The New Yorker,* February 3, 1997, pp. 32–37; and Anthony Corrado, "Financing the 1996 Elections," in Gerald Pomper, ed., *The Election of 1996* (Chatham, N.J.: Chatham House, 1997).

29. Arthur T. Hadley, *The Invisible Primary* (New York: Prentice-Hall, 1976), pp. 14–15.

30. Elaine Ciulla Kamarck, "Structure as Strategy: Presidential Nominating Politics in the Post-Reform Era," in Sandy Maisel, ed., *The Parties Respond* (Boulder, Colo.: Westview Press, 1990).

31. James A. Barnes, "Rush to Judgment: Picking Presidents," *National Journal,* June 18, 1994, pp. 1412–15. See also Robert D. Loevy, *The Flawed Path to the Presidency* (Albany, N.Y.: SUNY Press, 1995).

32. See, for example, David S. Broder, "One Vote against the Primaries," *The New York Times Magazine,* January 31, 1960, p. 9H; and Jeane J. Kirkpatrick et al., *The New Presidential Elite* (Russell Sage Foundation and Twentieth Century Fund, 1976). See also James W. Ceasar, *Presidential Selection: Theory and Development* (Princeton, N.J.: Princeton University Press, 1979).

33. W. Wayne Shannon, "Evaluating the New Nominating System: Thoughts after 1988 from a Governance Perspective," in Emmett H. Buell, Jr. and Lee Sigelman, eds., *Nominating the President* (Knoxville: University of Tennessee Press, 1991), pp. 250–77.

34. One of the most useful and most balanced studies of the presidential primaries is found in William R. Keech and Donald R. Matthews, *The Party's Choice* (Washington, D.C.: Brookings, 1976). See also William J. Crotty, *Political Reform and the American Experiment* (Crowell, 1977), chap. 7.

35. See the provocative analysis in Robert D. Loevy, *The Flawed Path to the Presidency* (Albany, N.Y.: SUNY Press, 1995).

36. Quoted in Michael Kelly, "Glasshouse Conventions," *The New Yorker,* September 9, 1996, p. 39.

37. Robert T. Nakamura and Denis G. Sullivan, "Party Democracy and Democratic Control," in Walter Dean Burnham and Martha W. Weinberg, eds., *American Politics and Public Policy* (MIT Press, 1978), p. 33.

38. Ibid., p. 38. See also Denis G. Sullivan's review of the Kirkpatrick volume, *American Political Science Review* (September 1978), pp. 1068–69.

39. New York Times/CBS News poll, see Kelly, note 36, p. 37.

40. Alexander Heard and Michael Nelson, "Change and Stability in Choosing Presidents," in Heard and Nelson, eds., *Presidential Selection* (Durham, N.C.: Duke University Press, 1987).

41. Edward R. Tufte, *Political Control of the Economy* (Princeton, N.J.: Princeton University Press, 1978), pp. 2–3.

42. Myron A. Levine, *Presidential Campaigns and Elections,* 2d ed. (Itasca, Ill.: F.E. Peacock, 1995).

43. Harry A. Bailey, Jr., "Presidential Tenure and the Two-Term Tradition," *Publius* (Fall 1972), p. 106.

44. This point is also made by Nicholas von Hoffman, *Make Believe Presidents* (Pantheon, 1978), p. 219.

45. Robert A. Dahl, "Myth of the Presidential Mandate," *Political Science Quarterly* 105 (Fall 1990), pp. 355–72.

46. Nelson W. Polsby and Aaron Wildavsky, *Presidential Elections,* 9th ed. (Chatham, N.J.: Chatham House, 1995), p. 3.

47. David W. Abbott and James P. Levine, *Wrong Winner: The Coming Debacle in the Electoral College* (New York: Praeger, 1991).

48. Lawrence D. Longley, "Yes, the Electoral College Should Be Abolished," in Gary L. Rose, ed., *Controversial Issues in Presidential Selection,* 2nd ed. (Albany, N.Y.: SUNY Press, 1994), 206. But see also Judith A. Best, et al., *The Choice of the People? Debating the Electoral College,* (Lanham, Md.: Rowland & Littlefield, 1996).

49. Polsby and Aaron Wildavsky, note 46, p. 292.

50. Stephen J. Wayne, "Let the People Vote Directly for President," in Stephen J. Wayne and Clyde Wilcox, eds., *The Quest for the White House* (New York: St. Martin's Press, 1992), pp. 313–14.

51. *Electing the President: A Report of the Commission on the Electoral College Reform* (Washington, D.C.: American Bar Association, January 1967), p. 3.

52. Arthur Schlessinger, Jr., "The Electoral College Conundrum," *Wall Street Journal,* April 4, 1977.

53. Tom Wicker, *The New York Times,* March 27, 1977, p. E17.

54. William Keech, *Winner Take All,.* Report of the Twentieth Century Fund Task Force on Reform of the Presidential Election Process (Holmes and Meier, 1978), p. 67.

55. Ibid.

56. Jonathan B. Bingham, *Congressional Record,* February 26, 1979, H882–H883. The amendment carried the title, J.H.Res.223.

57. Peirce, *Winner Take All,* p. 15.

58. See, for example, Douglas Amy, *Real Choices/New Voices: The Case for Proportional Representation Elections in the United States* (New York: Columbia University Press, 1993) and Stephen J. Brams, *The Presidential Election Game* (New Haven, Conn.: Yale University Press, 1978).

CHAPTER 3

1. Arthur M. Schlesinger, Jr., "The Ultimate Approval Rating," *The New York Times Magazine,* December 15, 1996, pp. 46–51. See also Steve Neal, "Putting Presidents in Their Place," *Chicago Sun-Times,* November 19, 1995, pp. 30–31.

2. See, for example, Marjorie Connelly, "Americans Are Still Voting for JFK," *The New York Times,* Sunday, August 18, 1996; and New York Times/CBS News Poll, July 11–13, 1996, "Presidential Favorites," from *The New York Times* poll watcher service.

3. See, for example, Forrest McDonald, *The American Presidency* (Lawrence: University Press of Kansas, 1994); and Sidney Milkis and Michael Nelson, *The American Presidency: Origins and Development,* 2d ed. (Washington, D.C.: Congressional Quarterly Press, 1994).

4. Roberta S. Sigel and David J. Butler, "The Public and the No Third Term Tradition: Inquiry into Attitudes toward Power," *Midwest Journal of Political Science* 8 (1964), p. 54.

5. Walter Berns, "The American Presidency: Statesmanship and Constitutionalism in Balance," *Imprimis* (January 1983), p. 3.

6. See David H. Donald, *Lincoln* (New York: Simon & Schuster, 1995), pp. 128–29.

7. See, for example, Nathan Miller, *Theodore Roosevelt: A Life* (New York: Quill, 1992).

8. See Woodrow Wilson, *Constitutional Government in the United States* (New York: Columbia University Press, 1908). See also James Ceasar et al., "The Rise of the Rhetorical Presidency," in Thomas E. Cronin, ed., *Rethinking the Presidency* (Boston: Little, Brown, 1982).

9. James L. Sundquist, *The Decline and Resurgence of Congress* (Washington, D.C.: Brookings, 1981), p. 33.

10. Jack Citrin and Donald Philip Green, "Presidential Leadership and the Resurgence of Trust in Government," *British Journal of Political Science* (October 1986), p. 450.

11. John Steinbeck, *America and Americans* (Bonanza Books, 1966), p. 46.

12. Roberta S. Sigel, "Image of the American Presidency: Part II of an Exploration into Popular Views of Presidential Power," *Midwest Journal of Political Science* (February 1966), pp. 123–37.

13. Kenneth E. Clark and Miriam B. Clark, *Choosing to Lead* (Greensboro, N.C.: Leadership Press, 1994), p. 173.

14. William W. Welch, "Tax Cut Supporters Walking a Tightrope," *USA Today,* April 3, 1995, p. 10A.

15. Samuel Kernell, *Going Public: New Strategies of Presidential Leadership* (Washington, D.C.: Congressional Quarterly Press, 1986), p. 174.

16. Anthony Downs, "Up and Down with Ecology—the Issue Attention Cycles," *Public Interest* (September 1972), p. 38.

17. Paul Brace and Barbara Hinckley, *Follow the Leader: Opinion Polls and the Modern Presidents* (New York: Basic Books, 1992).

18. Richard E. Neustadt, *Presidential Power and the Modern Presidents* (New York: Free Press, 1990).

19. George C. Edwards, *Presidential Influence in Congress* (San Francisco: W.H. Freeman, 1980), pp. 86–100.

20. Neustadt, note 18, p. 76.

21. Quoted in George C. Edwards and Stephen Wayne, *Presidential Leadership,* 3rd ed. (New York: St. Martin's Press, 1994), p. 90.

22. Arthur M. Schlesinger, Jr., "The Ultimate Approval Rating," *The New York Times Magazine,* December 15, 1996, p. 51.

23. Richard A. Brody, *Assessing the President: The Media, Elite Opinion, and Public Support* (Stanford, Calif.: Stanford University Press, 1991).

24. Robert H. Ferrell, ed., *Off the Record: The Private Papers of Harry S. Truman* (New York: Harper & Row, 1980), p. 310.

25. Brace and Hinckley, note 17, p. 45.

26. Roderick Hart, *The Sound of Leadership* (Chicago: University of Chicago Press, 1987).

27. George Edwards, *The Public Presidency* (New York: St. Martin's Press, 1983), p. 233.

28. Ibid.

29. Kristen R. Monroe, *Presidential Popularity and the Economy* (New York: Praeger, 1984), pp. 145–46. But see also Samuel Kernell, "Explaining Presidential Popularity," *American Political Science Review* 72 (June 1978), pp. 506–22; and James A. Stimson, "Public Support for American Presidents," *Public Opinion Quarterly* 40 (1976), pp. 1–21.

30. Summaries of the literature on the impact of the economy can be found in Kernell, note 15, chap. 7; and Edwards, note 27, chap. 6.

31. Thomas E. Yantek, "Public Support for Presidential Performance: A Study of Macroeconomic Effects," *Polity* (Winter 1982), p. 278. For somewhat differing but parallel studies see Samuel Kernell, "The Presidency and The People" in Michael Nelson, ed., *The Presidency and The Political System* (Washington, D.C.: Congressional Quarterly Press, 1984), pp. 233–63. See Richard Brody, "Public Evaluations and Expectations and the Future of the Presidency," in James Young, ed., *Problems and Prospects of Presidential Leadership in the Nineteen Eighties,* vol. 1 (University Press of America, 1982), pp. 37–55; and Charles W. Ostrom, Jr. and Dennis M. Simon, "Promise and Performance: A

Dynamic Model of Presidential Popularity," *American Political Science Review* (June 1985), pp. 334–58.

32. James W. Ceaser, "The Reagan Presidency and American Public Opinion," in Charles O. Jones, ed., *The Reagan Legacy: Promise and Performance* (Chatham, N.J.: Chatham House, 1988), p. 220.

33. But see the valuable analysis provided in Stanley A. Renshon, *The Psychological Assessment of Presidential Candidates* (Albany, N.Y.: New York University Press, 1996).

34. See Gary M. Maranell and Richard D. Dodder, "Political Orientation and the Evaluation of Presidential Prestige: A Study of American Historians," *Social Science Quarterly* (September 1970), p. 418. These authors found that conservatives rated Lincoln, Washington, and Jefferson slightly higher in prestige than did liberals, while the liberal "experts" not surprisingly rated FDR, Truman, Wilson, Jackson, and Kennedy somewhat more favorably than did the conservative experts. Authors who suggest the problem of the liberal bias in these polls are Thomas A. Bailey, *Presidential Greatness* (Appleton-Century, 1966), pp. 25–28; and Daniel Bratton, "The Rating of Presidents," *Presidential Studies Quarterly* (Summer 1983), pp. 400–4.

35. Henry Steele Commager, *Parade Magazine,* May 8, 1977, p. 17. For an innovative effort by a political psychologist to examine this, see Dean Keith Simonton, *Why Presidents Succeed* (New Haven, Conn.: Yale University Press, 1987).

36. Bailey, note 34, p. 292.

37. Bruce Miroff, *Icons of Democracy: Heroes, Aristocrats, Dissenters, and Democrats* (New York: Basic Books, 1993).

38. Schlesinger, note 22, p. 46.

39. See Steve Neal, "Putting Presidents in Their Places," *Chicago Sun-Times,* November 19, 1995, pp. 30–31.

40. Bush is viewed as inadequate and not a strong president even by conservatives and former associates. For example, see Terry Eastland's *Energy in the Executive: The Case for the Strong Presidency* (New York: Free Press, 1992).

41. Bailey, note 34, p. 259. Also see Simonton, note 35, chap. 6.

42. George F. Will "The Presidency in the American Political System," *Presidential Studies Quarterly* (Summer 1984), pp. 328–29. See also Garry Wills, "What Is Political Leadership?" *Atlantic Monthly,* (April 1994), pp. 63–80.

43. Fred Greenstein, "What the President Means to Americans," in James D. Barber, ed., *Choosing the President* (New York: American Assembly, 1974), pp. 130–31.

44. Theodore J. Lowi, *The Personal President: Power Invested, Promise Unfulfilled* (Ithaca, N.Y.: Cornell University, 1985).

45. Brace and Hinckley, note 17, p. 1.

46. Michael B. Grossman and Martha Joynt Kumar, *Portraying the President* (Baltimore, Md.: Johns Hopkins University Press, 1981), pp. 255–63.

47. Michael J. Robinson et al., "With Friends Like These . . . ," *Public Opinion* (June/July 1983), pp. 2–3.

48. Kenneth T. Walsh, *Feeding the Beast, the White House vs. Press* (New York: Random House, 1996).

49. Larry Sabato, *Feeding Frenzy* (New York: Free Press, 1993), p. 1.

50. Daniel Hallin, "Sound Bite News: Television Coverage of Elections," *Journal of Communication* (Spring 1992).

51. See Howard Kurtz, "There's Anger in the Air," *The Washington Post National Weekly Edition,* October 31–November 6, 1994, pp. 8–10.

52. John Orman, "Covering the American Presidency: Valenced Reporting in the Presidential Press," *Presidential Studies Quarterly* 14, (1984), pp. 381–82.

53. Norman Ornstein and Michael Robinson, "The Case of Our Disappearing Congress," *TV Guide,* January 11, 1986, pp. 4–10.

54. John Maltese, *Spin Control: The White House Office of Communications and the Management of Presidential News* (Chapel Hill: University of North Carolina Press, 1992). See also, for an inside view, Michael K. Deaver, *Behind the Scenes* (New York: William Morrow, 1987).

55. Hedrick Smith, *The Power Game: How Washington Works* (New York: Ballantine, 1988), p. 407.

56. Kernell, note 15.

57. See Walsh, note 48.

58. Quoted in Michael J. Robinson and Margaret A. Sheehan, *Over the Wire and on TV* (New York: Russell Sage Foundation, 1983), p. 103.

59. Mark Hertsgaard, *On Bended Knee: The Press and the Reagan Presidency* (New York: Schocken Brooks, 1988).

60. See series by David Shaw, *Los Angeles Times,* September 15, 16, and 17, 1993, for an examination of the media's treatment of the Clinton presidency.

61. Sabato, note 49.

CHAPTER 4

1. Bruce Miroff, *Icons of Democracy: American Leaders as Heroes, Aristocrats, Dissenters, and Democrats* (New York: Basic Books, 1993).

2. Charles O. Jones, *The Presidency in a Separated System* (Washington, D.C.: Brookings, 1994).

3. Michael A. Genovese, *The Presidential Dilemma: Leadership in the American System* (New York: HarperCollins, 1995), chap. 2.

4. Samuel P. Huntington, *American Politics: The Promise of Disharmony* (Cambridge, Mass.: Harvard University Press, 1981), pp. 4, 33.

5. Clinton Rossiter, *Conservatism in America* (New York: Vintage, 1962), p. 72.

6. Max Lerner, *America as a Civilization* (New York: Knopf, 1957), p. 718.

7. Alexis de Tocqueville, *Democracy in America* (New York: Knopf, 1969), p. 430.

8. Michel Crozier, *The Trouble With America* (Berkeley: University of California Press, 1984), p. 71.

9. See Robert Bellah, Richard Masden, William Sullivan, Ann Swidel, and Steven Tipton, *Habits of the Heart,* (New York: Harper & Row, 1985). Also see Arthur M. Schlesinger, Jr., *The Disuniting of America* (New York: Norton, 1991).

10. Warren Bennis, *Why Leaders Can't Lead: The Unconscious Conspiracy Continues* (San Francisco: Jossey-Bass Publishers, 1989), p. 46.

11. See Stephen Skowronek, *The Politics Presidents Make* (Cambridge, Mass.: Belknap Press, 1993).

12. Bert Rockman: *The Leadership Question: The Presidency and the American System* (New York: Praeger, 1984), p. 84.

13. Stephen Skowronek, "Presidential Leadership in Political Time," in Michael J. Nelson, ed., *The President and the Political System,* 3rd ed. (Washington: *Congressional Quarterly Press,* 1990), pp. 117–62.

14. Valerie Bunce, *Do New Leaders Make a Difference?* (Princeton, N.J.: Princeton University Press, 1981).

15. Paul C. Light, *The President's Agenda* (Baltimore, Md.: Johns Hopkins University Press, 1982).

16. Arthur M. Schlesinger, Jr., *The Cycles of American History* (Boston: Houghton Mifflin, 1986), pp. 22–27.

17. Ibid., p. 34.

18. Many constitutional scholars contend that the framers were explicit in limiting and constraining governmental powers. See the useful collection of essays in David G. Adler and Larry George, eds., *The Constitution and the Conduct of Foreign Policy* (Lawrence: University Press of Kansas, 1996).

19. Edward S. Corwin, *The President: Office and Powers,* rev. ed. (New York: New York University Press, 1957).

20. de Tocqueville, note 7, p. 126.

21. Lyndon Johnson to Richard Nixon, quoted in Bobby Baker with Larry Nixon, *Wheeling and Dealing* (Norton, 1978), p. 265.

22. This work has gone through several revisions: Richard Neustadt, *Presidential Power: The Politics of Leadership with Reflections on Johnson and Nixon* (New York: Wiley & Sons, 1976). The latest edition (Free Press, 1990) includes commentary up to and including the Reagan presidency. All quotes are drawn from the 1976 edition.

23. Ibid., p. 229.

24. Ibid., 230.

25. Ibid., p. 239.

26. Ibid., p. 33.

27. Ibid., pp. 1–2.

28. Ibid., p. 41.

29. Ibid., p. 50.

30. Ibid., pp. 21–22.

31. John Hart, "Presidential Power Revisited," *Political Studies* (March 1977), p. 56.

32. Charles E. Lindblom, "The Market as Prison," *The Journal of Politics* 44 (1982), pp. 324–36; and William F. Grover, *The President as Prisoner* (Albany, N.Y.: SUNY Press, 1989).

33. Bruce Miroff, *Pragmatic Illusions* (New York: David McKay, 1976).

34. Others make the case for strong presidents and a strong presidency because they see the possibilities of principled conservative leadership. One example is Terry Eastland's *Energy in the Executive: The Case for the Strong Presidency* (New York: Free Press, 1992).

35. Quoted in Henry Etzkowitz and Peter Schwab, eds., *Is America Necessary?* (West Publishing, 1976), p. 579.

36. Eastland, note 34; and L. Gordon Corvitz and Jeremy A. Rabkin, eds., *The Fettered Presidency: Legal Constraints on the Executive Branch* (Washington, D.C.: America Enterprise Institute, 1989). And Republican contenders for the 1996 Republican nomination also saw the White House and the presidency as important tools to advance a conservative political agenda.

37. Hedrick Smith, *The Power Game: How Washington Works* (New York: Ballantine, 1988).

38. Robert Shogan, *The Riddle of Power* (New York: Dutton, 1991), p. 5.

39. Robert A. Dahl, *Pluralist Democracy in the United States* (New York: Rand McNally, 1967), p. 90.

40. James P. Pfiffner, *The Strategic Presidency: Hitting the Ground Running,* 2d ed. (Lawrence: University Press of Kansas, 1996).

41. James MacGregor Burns, *Roosevelt: The Lion and the Fox* (San Diego: Harcourt Brace Jovanovich, 1956), p. 197. See also Ronald Heifetz, *Leadership without Easy Answers* (Cambridge, Mass.: Belknap Press, 1994).

42. Corwin, note 19.

43. John F. Kennedy, "Foreword" to Theodore C. Sorensen, *Decision-Making in the White House* (New York: Columbia University Press, 1963), p. xi. Kennedy's noted speechwriter maintained that vigorous presidential leadership with a strong Democratic party is the best prescription for solving our nation's problems. See Theodore C. Sorensen, *Why I Am a Democrat* (New York: Henry Holt, 1996).

44. Warren Bennis and Burt Nanus, *Leaders* (New York: Harper & Row, 1985), pp. 39 and 42. See also Warren Bennis and Patricia Ward Biederman, *Organizing Genius* (Reading, Mass.: Addison Wesley, 1997).

45. Lance Blakesley, *Presidential Leadership: From Eisenhower to Clinton* (Chicago: Nelson-Hall, 1995).

46. Quoted in Sidney Blumenthal, "The Education of a President," *The New Yorker,* January 24, 1994, p. 33.

47. Erwin C. Hargrove, "Presidential Personality and Leadership Style," in George C. Edwards, John H. Kessel, and Bert A. Rockman, eds., *Researching the Presidency: Vital Questions, New Approaches* (Pittsburgh: University of Pittsburgh Press, 1993), p. 70.

48. Ibid., p. 82.

49. Ibid., p. 69.

50. For a view that suggests we should spend less time studying skill and more studying institutional pressures, see Terry M. Moe, "Presidents, Institutions, and Theory," in Edwards, Kessel, and Rockman, note 47, pp. 337–85.

51. Robert K. Murray and Tim H. Blessing, *Greatness in the White House: Rating the Presidents, Washington through Carter* (University Park: Pennsylvania State University Press, 1988).

52. Burns, note 41, p. 264.

53. See James David Barber's oft criticized but influential work, *The Presidential Character: Predicting Performance in the White House* 4th ed. (Englewood Cliffs, N.J.: Prentice-Hall, 1992).

54. James P. Pfiffner, *The Managerial Presidency* (Pacific Grove, Calif.: Brooks/ Cole, 1991).

55. Pfiffner, note 40, p. 4.

56. See Smith, note 37, pp. 340–42.

57. See James Perry, "For President Bush, the First 100 Days Offer Little to Provide Strength for Tough Times Ahead, Political Experts Contend," *Wall Street Journal,* April 28, 1989. But see retired President George Bush's retrospective and, not surprisingly, more positive view, "An Interview with Former President George Bush," *Parade Magazine,* December 1, 1996, pp. 4–6.

58. See, for example, Eastland, note 34; and Forrest McDonald, *The American Presidency: An Intellectual History* (Lawrence: University Press of Kansas, 1994).

59. Samuel P. Huntington, "The Democratic Distemper," in Nathan Glazer and Irving Kristol, eds., *The American Commonwealth* (New York: Basic Books, 1975), p. 24.

60. Arthur M. Schlesinger, Jr., *The Imperial Presidency* (Boston: Houghton Mifflin, 1973), p. 404. See also Arthur M. Schlesinger, Jr., "The Ultimate Approval Rating," *The New York Times Magazine,* December 15, 1996, pp. 49–51. For a similar post-Watergate reaffirmation of the presidency see Theodore C. Sorensen, *Watchmen in the Night: Presidential Accountability after Watergate* (MIT Press, 1975), and *Why I Am A Democrat,* note 43.

61. James Bryce, *The American Commonwealth,* vol. 2 (New York: MacMillan & Co., 1888), p. 460.

62. Benjamin R. Barber, "Neither Leaders nor Followers: Citizenship under Strong Democracy," in Michael R. Beschloss and Thomas E. Cronin, eds., *Essays in Honor of James MacGregor Burns* (Englewood Cliffs, N.J.: Prentice-Hall, 1989), pp. 117–32.

63. Miroff, note 1, p. 1.

64. Arthur M. Schlesinger, Jr., *The Cycles of American History* (Boston: Houghton Mifflin, 1986), p. 430. See also how several authors wrestle with this enduring tension in David G. Adler and Larry George, eds., *The Constitution and the Conduct of Foreign Policy* (Lawrence: University Press of Kansas, 1996).

65. Willard Sterne Randall, *Thomas Jefferson: A Life* (New York: Henry Holt, 1993). See also Merrill D. Peterson, ed., *The Portable Thomas Jefferson* (New York: Viking Press, 1975).

66. Jimmy Carter, "Prayer and Civic Religion," *The New York Times,* December 29, 1996, p. A13.

CHAPTER 5

1. For an examination of these conflicting views, see Charles Beard, *An Economic Interpretation of the Constitution* (New York: Macmillan, 1913); and Robert E. Brown, *Charles Beard and the Constitution* (Princeton, N.J.: Princeton University Press, 1956).

2. Alexander Hamilton, James Madison, and John Jay, *The Federalist Papers, No. 51* (New York: New American Library, 1961), p. 322.

3. Madison, *The Federalist Papers, No. 45,* Hamilton, *The Federalist Papers, No. 70,* respectively.

4. Edward S. Corwin, *The President: Office and Powers, 1978–1984* 5th ed. (New York: New York University Press, 1984; originally published in 1940). See also Joseph M. Bessette and Jeffrey Tulis, *The Presidency in the Constitutional Order* (Baton Rouge: Louisiana State University Press, 1981); and Louis Fisher, *The Constitution between Friends* (New York: St. Martin's Press, 1978).

5. James P. Pfiffner, *The Modern Presidency* (New York: St. Martin's Press, 1993), p. 13.

6. Glenn A. Phelps, *George Washington and American Constitutionalism* (Lawrence: University Press of Kansas, 1993).

7. Leonard White, *The Federalists: A Study in Administrative History* (New York: Macmillan, 1967), p. 99.

8. Stephen Showronek, *The Politics Presidents Make: Leadership from John Adams to George Bush* (Cambridge, Mass.: Belknap Press of Harvard University Press, 1993).

9. Willard Sterne Randall, *Thomas Jefferson: A Life* (New York: Henry Holt, 1993).

10. Robert V. Remini, *Andrew Jackson* (New York: Harper & Row, 1966).

11. Herman Belz, *Lincoln and the Constitution: The Dictatorship Question Reconsidered* (Fort Wayne, Ind.: Louis A. Warren Lincoln Library and Museum, 1984); James G. Randall, *Constitutional Problems under Lincoln,* rev. ed. (Urbana: University of Illinois Press, 1951); and David Donald, *Lincoln* (New York: Simon & Schuster, 1995).

12. Jeffrey K. Tulis, *The Rhetorical Presidency* (Princeton, N.J.: Princeton University Press, 1987).

13. Theodore Roosevelt, *The Autobiography of Theodore Roosevelt* (New York: Charles Scribner's Sons, 1941), pp. 197–98.

14. Arthur S. Link, *Wilson and the New Freedom* (Princeton N.J.: Princeton University Press, 1956); and Hendrick A. Clements, *The Presidency of Woodrow Wilson* (Lawrence: University Press of Kansas, 1992).

15. William E. Leuchtenburg, *In the Shadow of FDR: From Harry Truman to Ronald Reagan,* (Ithaca, N.Y.: Cornell University Press, 1983). See also Philip Abbott, *The Exemplary Presidency* (Amherst: University of Massachusetts Press, 1990).

16. Fred I. Greenstein, *The Hidden-Hand Presidency: Eisenhower as Leader* (New York: Basic Books, 1982).

17. Clinton Rossiter, *The American Presidency* (New York: Harcourt, Brace & World, 1956).

18. Ibid., p. 251.

19. Alfred DeGrazia, *Congress and the Presidency* (Washington, D.C.: American Enterprise Institute, 1967) and *Republic in Crisis: Congress against the Executive Force* (New York: Federal Legal Publications, 1965).

20. Albert Cantrill, *The American People, Vietnam, and the Presidency* (Princeton, N.J.: Princeton University Press, 1976). David Halberstam, *The Best and the Brightest* (New York: Random House, 1972); and Stanley Karnow, *Vietnam: A History* (New York: Viking, 1983).

21. Larry Berman, *Lyndon Johnson's War* (New York: Norton, 1989).

22. William Carey, "Presidential Staffing in the Sixties and Seventies," *Public Administration Review* (September–October 1969), p. 453.

23. Lyndon Johnson, quoted in Hugh Sidey, *A Very Personal Presidency* (Atheneum, 1978), p. 260.

24. Aaron Wildavsky, "The Two Presidencies," *Trans-Action* (December 1966), in Wildavsky, ed., *Perspectives on the Presidency* (Boston: Little, Brown, 1975), p. 418.

25. John F. Kennedy, quoted in Richard Nixon, *RN: Memoirs of Richard Nixon* (Grosset & Dunlap, 1978), p. 235.

26. See the discussion in Michael R. Beschloss and Strobe Talbott, *At the Highest Levels: The Inside Story of the End of the Cold War* (Boston: Little, Brown, 1993), pp. 463–74.

27. Richard Nixon, quoted in Theodore White, *The Making of the President 1968* (Atheneum, 1969), p. 147.

28. William Manchester, based on an interview with John F. Kennedy, quoted in Manchester, *Portrait of a President* (Macfadden Books, 1962), p. 33. Manchester also quoted a Kennedy family member as saying, "Of course he's preoccupied [with foreign policy]. It would be a miracle if he weren't. Saigon, Germany, fifty-megaton bombs—that's why he can't get to sleep until two or three in the morning" (p. 33).

29. John F. Kennedy, State of the Union Address, January 30, 1962, *Public Papers of the Presidents of the United States* (Washington, D.C.: U.S. Government Printing Office, 1962), pp. 22–23.

30. John H. Kessel, "The Parameters of Presidential Politics," paper presented at the 1972 Annual Meeting of the American Political Science Association: processed, pp. 8–9. Later published in *Social Science Quarterly* (June 1974), pp. 8–21.

31. Leonard Garment, quoted in White, note 27, p. 52.

32. Paul Brace and Barbara Hinckley, *Follow the Leader: Opinion Polls and the Modern Presidents* (New York: Basic Books, 1992).

33. Wildavsky, note 24, p. 7. See also Barbara Kellerman and Ryan J. Barilleaux, *The President as World Leader* (New York: St. Martin's Press, 1991); and Cecil V. Crabb and Kevin V. Mulcahy, *American National Security* (Pacific Grove, Calif.: Brooks/Cole, 1991).

34. See David G. Adler and Larry George, eds., *The Constitution and the Conduct of Foreign Policy* (Lawrence: University Press of Kansas, 1996).

35. Louis Fisher, *Presidential War Power* (Lawrence: University Press of Kansas, 1995).

36. Jon R. Bond and Richard Fleisher, *The President in the Legislative Arena* (Chicago: University of Chicago Press, 1990), p. 171. See also David P. Forsythe and Ryan C. Henderson, "U.S. Use of Force Abroad: What Law for the President?" *Presidential Studies Quarterly* (Fall 1996), pp. 950–61.

37. Marcia Lynn Whicker, James P. Pfiffner, and Raymond A. Moore, eds., *The Presidency and the Persian Gulf War* (Westport, Conn.: Praeger, 1993).

38. See Allen Schick, *The Federal Budget Process: Politics, Policy and Process* (Washington, D.C.: Brookings Institution, 1995); and Calvin MacKenzie and Saranna Thornton, *Bucking the Deficit: Economic Policymaking in America* (Boulder, Colo.: Westview, 1996).

39. See Roger H. Davidson and Walter J. Oleszek, *Congress and Its Members,* 5th ed. (Washington, D.C.: Congressional Quarterly Press, 1995), chap. 13.

40. See Haynes Johnson and David Broder, *The System* (Boston: Little Brown, 1996); and Shirley Ann Warshaw, *The Domestic Presidency* (Boston: Allyn and Bacon, 1997).

41. John F. Kennedy, quoted in Richard Rovere, "Letter from Washington," *New Yorker,* November 30, 1963, p. 53.

42. For a full examination of the presidency and crisis management, see Graham T. Allison, *Essence of Decision: Explaining the Cuban Missile Crisis* (Boston: Little Brown, 1971); Alexander L. George, *Presidential Decision-making in Foreign Policy: The Effective Use of Information and Advice,* (Boulder, Colo.: Westview, 1980); and Alexander L. George, *Avoiding War: Problems of Crisis Management,* (Boulder, Colo.: Westview, 1991).

43. Richard M. Pious, *The Presidency* (Boston: Allyn and Bacon, 1995), p. 8.

44. Thomas S. Langston, *With Reverence and Contempt: How Americans Think about Their President* (Baltimore: Johns Hopkins University Press, 1995).

45. Richard Brookhiser, *Founding Father: Rediscovering Washington* (New York: Free Press, 1996).

46. Barry Schwartz, *George Washington: The Making of an American Symbol* (New York: Free Press, 1987).

47. Clinton Rossiter, *The American Presidency* (New York: Mentor Books, 1960), pp. 102–3.

48. Bruce Miroff, "Abraham Lincoln: Democratic Leadership and the Tribe of the Eagle," in Miroff, *Icons of Democracy: American Leaders as Heroes, Aristocrats, Dissenters, and Democrats* (New York: Basic Books, 1993).

49. See David Donald's account in *Lincoln* (New York: Simon and Schuster, 1995).

50. Abraham Lincoln, *Collected Works,* Roy P. Basler, ed., 9 vols. (New Brunswick, N.J.: Rutgers University Press, 1953–55), pp. 530–37.

51. On the impact and significance of Lincoln's celebrated Gettysburg address, see Garry Wills, *Lincoln at Gettysburg: The Words That Remade America* (New York: Simon & Schuster, 1992).

52. See the discussion of "the politics of meaning" and a critical evaluation of the Clintons' attempts in Michael Lerner, *The Politics of Meaning* (Reading, Mass.: Addison Wesley, 1996); see especially the epilogue, pp. 309–39.

53. Bruce Miroff, "The Presidency and the Public: Leadership as Spectacle," in Michael Nelson, ed., *The Presidency and the Political System* (Washington, D.C.: *Congressional Quarterly Press,* 1995).

54. Ben Wattenberg, *Values Matter Most* (New York: Free Press, 1995).

55. Quoted in Arthur M. Schlesinger, Jr., *A Thousand Days* (Boston: Houghton Mifflin, 1965), p. 127.

56. See Irving I. Janis, *Victims of Groupthink* (Boston: Houghton Mifflin, 1972).

57. See Robert Scigliano, *The Supreme Court and the Presidency* (New York: Free Press, 1971), pp. 96–99, 146–58. The author shows persuasively that both partisanship and the timing of appointments are critical factors, especially in combination. See also Harold Chase, *Federal Judges: The Appointing Process* (Minneapolis: University of Minnesota Press, 1972); Henry Abraham, *Justices and*

Presidents (New York: Oxford University Press, 1974); and Michael A. Genovese, *The Supreme Court, the Constitution and Presidential Power* (Landham, Md.: University Press of America, 1980).

58. G. Calvin Mackenzie, "The Politics of the Appointment Process," mimeo, 1977, pp. 66–67. See also F. V. Malek, *Washington's Hidden Tragedy* (New York: Free Press, 1978), chap. 4; and Kenneth Prewitt and William McAllister, "Changes in the American Executive Elite, 1930–1970," in Heinz Eulau and Moshe Czudnowski, eds., *Elite Recruitment in Democratic Politics* (Newberry Park, Calif.: Sage, 1976), pp. 105–32.

59. Benjamin Ginsberg and Martin Shefter, *Politics by Other Means* (New York: Basic Books, 1990), pp. 164–65.

60. James MacGregor Burns, quoted in Hedrick Smith, *The Power Game: How Washington Works* (New York: Ballantine Books, 1988), p. 679.

61. John F. Kennedy, television interview, *Public Papers of the President of the United States* (U.S. Government Printing Office, 1963), pp. 892–94.

62. See Charles O. Jones, *The Presidency in a Separated System* (Washington, D.C.: Brookings, 1994).

63. Two classic studies of implementation are Jeffrey Pressman and Aaron Wildavsky, *Implementation* (University of California Press, 1973); and Eugene Bavdach, *The Implementation Game* (Cambridge: MIT Press, 1977).

64. An excellent review of the problems a president encounters in implementing policy decisions is found in George C. Edwards III, "Presidential Policy Implementation: The Critical Variable," paper prepared for the Annual Meeting of the Midwest Political Science Association, Chicago, April 20–22, 1978, p. 1. See also George C. Edwards III and Ira Sharkansky, *The Policy Predicament* (San Francisco: W.H. Freeman, 1978).

65. James P. Pfiffner, *The Managerial Presidency* (Pacific Grove, Calif.: Brooks/Cole, 1991).

66. Jonathan Daniels, *Frontiers on the Potomac* (New York: Macmillan, 1946), pp. 31–32.

67. James P. Pfiffner, "Can the President Manage the Government? Should He?" in Pfiffner, note 65, pp. 1–16.

68. Marver Bernstein, "The Presidency and Management Improvement," *Law and Contemporary Problems* (Summer 1970), p. 516. See also Otis I. Graham, Jr., *Toward a Planned Society: From Roosevelt to Nixon* (Oxford University Press, 1976).

69. Hugh Heclo, *A Government of Strangers* (Washington, D.C.: Brookings, 1977).

70. Richard Rose, *Managing Presidential Objectives* (New York: Free Press, 1976), p. 147.

71. Colin Campbell, *Managing the Presidency* (Pittsburgh: University of Pittsburgh Press, 1986); and Robert F. Durant, *The Administrative Presidency Revisited* (Albany, N.Y.: SUNY, 1992).

CHAPTER 6

1. James S. Young, *The Washington Community, 1800–1828* (New York: Columbia University Press, 1966), pp. 75–76.

2. James A. Thurber, ed., *Rivals for Power: Presidential-Congressional Relations* (Washington, D.C.: Congressional Quarterly Press, 1996).

3. Emmet John Hughes, *The Living Presidency* (New York: Coward, McCann & Geoghegan, 1973), p. 208.

4. Jon R. Bond and Richard Fleisher, *The President in the Legislative Arena* (Chicago: University of Chicago Press, 1990).

5. Robert J. Spitzer, *President and Congress: Executive Hegemony at the Crossroads of American Government* (New York: McGraw-Hill, 1993), p. xx.

6. Lance T. LeLoup and Steven A. Shull, *Congress and the President: The Policy Connection* (Belmont, Calif.: Wadsworth, 1993).

7. Clinton Rossiter, *The American Presidency* (New York: Harcourt, Brace & World, 1980), p. 26.

8. Roger H. Davidson, "The Presidency and Congress," in Michael Nelson, ed., *The Presidency and the Political System* (Washington, D.C.: Congressional Quarterly Press, 1984), pp. 363–91.

9. George C. Edwards III, *At The Margins: Presidential Leadership of Congress* (New Haven, Conn.: Yale University Press, 1989).

10. *Youngstown Sheet and Tube Co. v. Sawyer,* 343 U.S. 579, 635 (1952).

11. Jean Edward Smith, *George Bush's War* (New York: Henry Holt, 1992). See also Bob Woodward, *The Commanders* (New York: Simon & Schuster, 1991).

12. Anthony Lewis, "'On His Word Alone,'" *The New York Times,* January 12, 1992, p. E19. See also several of the essays in David G. Adler and Larry George, *The Constitution and the Conduct of Foreign Policy* (Lawrence: University Press of Kansas, 1996).

13. Oliver L. North, *Under Fire: An American Story* (New York: HarperCollins, 1991), p. 323.

14. Jane Mayer and Jill Abramson, *Strange Justice: The Selling of Clarence Thomas* (New York: Plume, 1995); and John Danforth, *Resurrection: The Confirmation of Clarence Thomas* (New York: Free Press, 1994).

15. David Rogers and Jackie Calmes, "Thomas Is Confirmed by Senate," *The Wall Street Journal,* October 16, 1991, p. A18.

16. Lester G. Seligman and Cary R. Covington, *The Coalitional Presidency* (Chicago: Dorsey Press, 1989).

17. James A. Davis and David L. Nixon, "The President's Party," *Presidential Studies Quarterly* (Spring 1994), pp. 363–65.

18. Roger H. Davidson and Walter J. Oleszik, *Congress and Its Members,* 5th ed. (Washington, D.C.: Congressional Quarterly Press, 1996), p. 289. See also

Mark A. Peterson, *Legislating Together* (Cambridge, Mass.: Harvard University Press, 1990), and David Mayhew, *Divided We Govern* (New Haven, Conn.: Yale University Press, 1991).

19. Edwards, note 9; Bond and Fleisher, note 4.

20. Paul Brace and Barbara Hinckley, *Follow the Leader* (New York: Basic Books, 1992); see Douglas Rivers and Nancy Rose, "Passing the President's Program: Public Opinion and Presidential Influence in Congress," *American Journal of Political Science* 29 (May 1985), 183–96; and Richard Brody, *Assessing the President: The Media, Elite Opinion, and Public Support* (Stanford, Calif.: Stanford University Press, 1991), p. 21.

21. Brody, note 20, p. 22.

22. Bond and Fleisher, note 4, p. 218; Edwards, note 9; and Aage R. Clausen, *How Congressmen Decide: A Policy Focus* (New York: St. Martin's Press, 1973).

23. Terry Sullivan, "Headcounts, Expectations, and Presidential Coalitions in Congress," *American Journal of Political Science* 32:(3) (August 1988), pp. 567–89.

24. Edwards, note 9, p. 216.

25. Rowland Evans and Robert Novak, *Lyndon B. Johnson: The Exercise of Power* (New York: New American Library, 1966), p. 104.

26. Douglas Jehl, "Clinton Assails G.O.P. Delays in Bitter Tones," *The New York Times,* July 27, 1993, p. 1.

27. See Mark A. Peterson, *Legislating Together* (Cambridge, Mass.: Harvard University Press, 1990).

28. On the veto power, see Robert Spitzer, *The Presidential Veto* (Albany, N.Y.: SUNY Press, 1988).

29. Barbara Hinkson Craig, *The Legislative Veto* (Boulder, Colo.: Westview Press, 1983).

30. 77 I. Ed 2d 317 (1983).

31. Daniel P. Franklin, "Why the Legislative Veto Isn't Dead," *Presidential Studies Quarterly* (Summer 1986), p. 499.

32. Louis Fisher, quoted in Martin Tolchin, "The Legislative Veto, an Accommodation That Goes On and On," *The New York Times,* March 31, 1989, p. A8. See also Louis Fisher, *Constitutional Dialogues: Interpretation as Political Process* (Princeton, N.J.: Princeton University Press, 1988).

33. David Gray Adler, "The President's War-making Power," in Thomas E. Cronin, ed., *Inventing the American Presidency* (Lawrence: University Press of Kansas, 1989), pp. 119–53; Louis Fisher, *Constitutional Conflicts between Congress and the President,* (Lawrence: University Press of Kansas, 1991), chaps. 8, 9.

34. Leonard W. Levy, *Original Intent and the Framers' Constitution* (New York: Macmillan, 1988), p. 30.

35. Stephen A. Ambrose, "The Presidency and Foreign Affairs," *Foreign Affairs* (Winter 1991–92), p. 137.

36. Ibid., p. 136.

37. Arthur M. Schlesinger, Jr., *The Imperial Presidency* (Boston: Houghton Mifflin, 1973).

38. Theodore Lowi, *The Personal President* (Cornell University Press, 1985), p. 179.

39. Leonard C. Meeker, "The Legality of U.S. Participation in the Defense of Vietnam," *Department of State Bulletin,* March 28, 1966, pp. 484–85.

40. See James Sundquist, *The Decline and Resurgence of Congress* (Washington, D.C.: Brookings Institution, 1981).

41. John Hart Ely, *War and Responsibility* (Princeton, N.J.: Princeton University Press, 1993); and Louis Fisher, *Presidential War Power* (Lawrence: University Press of Kansas, 1995).

42. Ronald Reagan, quoted in Christopher Madison, "Despite His Complaints Reagan Going Along with Spirit of War Powers Law," *National Journal,* May 19, 1984, p. 990.

43. President George Bush to Congress, January 11, 1991, reprinted in *Congressional Quarterly,* January 12, 1991, p. 70.

44. Senator George Mitchell, quoted in *Congressional Quarterly,* January 12, 1991, p. 70.

45. Michael Ross, "House Defeats GOP Effort to Kill War Powers Act," *Los Angeles Times,* June 8, 1995, p. A4.

46. See, too, Edward Keynes, "The War Powers Resolution and the Persian Gulf War," in Adler and George, note 12, pp. 241–56.

47. Senator Leverett Saltonstall (R-Mass.), quoted in Henry Howe Ransom, *The Intelligence Establishment* (Cambridge, Mass.: Harvard University Press, 1970), p. 169.

48. Stansfield Turner, *Secrecy and Democracy: The CIA in Transition* (Boston: Houghton Mifflin, 1985).

49. See, for example, Bob Woodward, *Veil: The Secret Wars of the CIA* (New York: Simon & Schuster, 1987).

50. Fisher, note 33. See, in general, Allen Schick, *The Federal Budget: Politics, Policy and Process* (Washington, D.C.: Brookings, 1995); and David A. Stockman, *The Triumph of Politics* (New York: Harper & Row, 1986).

51. Bill Bradley, "Congress at Its Worst," *The Washington Post National Weekly Edition,* October 28, 1985, p. 29. See Warren Rudman's vivid personal account of it in Rudman, *Combat: Twelve Years in the U.S. Senate* (New York: Random House, 1996), chap. 2.

52. Thomas E. Mann, quoted in "How to Get the Hill Humming Again," *Business Week,* April 16, 1990, p. 63.

53. James A. Thurber, "The Impact of Budget Reform on Governance," in Thurber, ed., *Divided Democracy,* (Washington, D.C.: Congressional Quarterly

Press, 1991), p. 167. See also John B. Gilmour, *Reconcilable Differences: Congress, the Budget Process and the Deficit* (Berkeley: University of California Press, 1990).

54. Christopher J. Deering, "Damned If You Do and Damned If You Don't: The Senate's Role in the Appointment Process," in G. Calvin MacKenzie, ed., *The In-and-Outers: Presidential Appointees and the Problems of Transient Government in Washington* (Baltimore, Md.: Johns Hopkins Press, 1987), chap. 5.

55. Stephen L. Carter, *The Confirmation Mess* (New York: Basic Books, 1994).

56. John Massaro, *Supremely Political: The Role of Ideology and Presidential Management in Unsuccessful Supreme Court Nominations* (Albany, N.Y.: SUNY, 1990).

57. Stephen L. Elkin, "Contempt of Congress: The Iran-Contra Affair, and the American Constitution," *Congress and the Presidency* 18 (1) (Spring 1991), p. 3.

58. Lloyd Cutler, "Party Government under the Constitution," in Donald L. Robinson, ed., *Reforming American Government: The Bicentennial Papers of the Committee on the Constitutional System* (Boulder, Colo.: Westview Press, 1985); and Donald L. Robinson, *To the Best of My Ability: The Presidency and the Constitution* (New York: Norton, 1987).

59. James L. Sundquist, *Constitutional Reform and Effective Government* (Washington, D.C.: Brookings, 1986), p. 15.

60. Ibid., p. 75.

61. Ibid., p. 135.

62. Ibid., p. 203–4.

63. Ibid., p. 205.

64. James M. Burns, *The Power to Lead: The Crisis of the American Presidency* (New York: Simon & Schuster, 1984), p. 117.

65. Ibid., p. 214.

66. Ibid., p. 228.

67. Ibid., p. 237.

68. Ibid., pp. 234–35.

69. See the analysis and critique of Burns's proposal in Thomas E. Cronin, "James MacGregor Burns and the Idea of a Vigorous, Programmatic, Partisan, Yet Accountable Presidency," in Michael R. Beschloss and Thomas E. Cronin, eds., *Essays in Honor of James MacGregor Burns* (Englewood Cliffs, N.J.: Prentice-Hall, 1989), pp. 38–65.

70. These views are presented in detail in L. Gordon Crovitz and Jeremy A. Rabkin, eds., *The Fettered Presidency: Legal Constraints on the Executive Branch* (Washington, D.C.: American Enterprise Institute, 1989); Gordon Jones and John Marini, eds., *The Imperial Congress* (New York: Pharos Books, 1988).

71. North, note 13, pp. 173–74. Reagan also shared some of these views. See Ronald Reagan, *An American Life: The Autobiography* (New York: Simon & Schuster, 1990).

CHAPTER 7

1. James W. Davis, *The President as Party Leader* (New York: Praeger, 1993), pp. ix–x.

2. Robert V. Remini, "The Emergence of Political Parties and Their Effect on the Presidency" in Philip C. Dolce and George H. Skau, eds., *Power and The Presidency* (New York: Scribners, 1976), p. 32.

3. Ibid., p. 33. See also Robert Remini, *Martin Van Buren and the Democratic Party* (New York: Columbia University Press, 1959).

4. Quoted in Arthur B. Tourtellot, *The Presidents on the Presidency* (New York: Doubleday, 1964), p. 387.

5. Ibid., p. 5.

6. Michael Duffy and Dan Goodgame, *Marching in Place: The Status Quo Presidency of George Bush* (New York: Simon & Schuster, 1992).

7. Roger G. Brown and David M. Welborn, "Presidents and their Parties: Performance and Prospects," *Presidential Studies Quarterly* 12 (3) (Summer 1982).

8. Robert Harmel, "President-Party Relations in the Modern Era: Past, Problems, and Prognosis," in Harmel, ed., *Presidents and Their Parties: Leadership or Neglect* (New York: Praeger, 1984), p. 250.

9. Arthur Schlesinger, Jr., "Crisis of the Party System, I," *Wall Street Journal,* May 10, 1979, editorial page.

10. Sidney M. Milkis, *The President and the Parties* (New York: Oxford University Press, 1993).

11. See also Kay Lawson, "Party Renewal and American Democracy," in Michael R. Beschloss and Thomas E. Cronin, eds., *Essays in Honor of James MacGregor Burns* (Englewood Cliffs, N.J.: Prentice-Hall, 1989); and James MacGregor Burns, *The Power to Lead: The Crisis of the American Presidency* (New York: Simon & Schuster, 1984).

12. See Gerald Ford's response to this problem in his memoir, *A Time to Heal* (New York: Harper & Row/Reader's Digest, 1979), p. 295.

13. See Michael A. Genovese, *Watergate: Opening the Age of Cynicism* (Westport, Conn.: Greenwood, 1998).

14. Austin Ranney, *The Federalization of Presidential Primaries* (Washington, D.C.: American Enterprise Institute, 1978), p. 39.

15. See Martin Wattenberg, *The Decline of American Parties, 1952–1980* (Cambridge, Mass.: Harvard University Press, 1984).

16. Quoted in Donald Robinson, "Cleaning Up Campaigns: How to Reform Election Finances Without Hurting Political Parties," *Chronicle of Higher Education,* December 20, 1996, p. 134.

CHAPTER 8

1. Terry Eastland, *Energy in the Executive: The Case for the Strong Presidency* (New York: Free Press, 1992), p. 235.

2. Selections from the *Correspondence of Theodore Roosevelt and Henry Cabot Lodge, 1884–1918* (New York: Charles Scribner's Sons, 1925), vol. 2, p. 519. For a good study of President Harding's appointment of a Supreme Court justice, see David J. Danelski, *A Supreme Court Justice Is Appointed* (New York: Random House, 1964).

3. Oral history of Leonard Hall, Dwight D. Eisenhower Library, Abilene, Kansas, May 19, 1975, p. 26. For a general study of the appointment process in the 1950s and 1960s, see Harold W. Chase, *Federal Judges: The Appointing Process* (Minneapolis: University of Minnesota Press, 1972).

4. Henry J. Abraham, *Justices and Presidents: A Political History of Appointments to the Supreme Court,* 2d ed. (New York: Oxford University Press, 1985), p. 65. See also Robert J. Steamer, *Chief Justice: Leadership and the Supreme Court* (Columbia: University of South Carolina Press, 1986). For a useful study of Lyndon B. Johnson's criteria and political decision making in the appointment of judges, see Neil D. McFeeley, *Appointment of Judges: The Johnson Years* (Austin: University of Texas, 1987), esp. chaps. 5 and 6.

5. Stephen L. Carter, *The Confirmation Mess* (New York: Basic Books, 1994).

6. See Warren Rudman, *Combat: Twelve Years in the U.S. Senate,* chap. 7, "The Nomination of David Hackett Souter" (New York: Random House, 1996,) pp. 152–94.

7. Jane Mayer and Jill Abramson, *Strange Justice: The Selling of Clarence Thomas* (New York: Houghton Mifflin, 1994); and John C. Danforth, *Resurrection: The Confirmation of Clarence Thomas* (New York: Viking, 1994). For a reaction to both the Bork and Thomas nominations, see Carter, note 5.

8. William H. Rehnquist, "Rehnquist on Justices," *The New York Times,* October 20, 1984, p. 9. For a somewhat different point of view and one that emphasizes that presidents can pack the court unless the Senate aggressively asserts itself in the confirmation process, see Laurence Tribe, *God Save This Honorable Court: How the Choice of Supreme Court Justices Shapes Our History* (New York: Random House, 1985).

9. *Brown v. Board of Education,* 347 U.S. 483 (1954).

10. Stephen E. Ambrose, *Eisenhower: The President* (New York: Simon & Schuster, 1984), p. 90. See also the oral history of top Eisenhower aide Sherman Adams, conducted for Columbia University, April 11, 1967, vol. 2 of 4, p. 121. Eisenhower Library, Abilene, Kansas.

11. Quoted in Anthony Lewis, "A Talk With Warren on Crime, the Court, the Country," *The New York Times Magazine,* October 19, 1969, pp. 128–29. See also Earl Warren, *The Memoirs of Earl Warren* (Garden City, N.Y.: Doubleday, 1977).

12. Robert Scigliano, *The Supreme Court and the Presidency* (New York: Free Press, 1971), pp. 146–47. Another study concludes that most presidents by and large do rather well in appointing individuals who will be supportive of the administration that supports them, at least during their first few years on the Court. "After that initial period, there is an apparent falling off." See Roger Handberg and Harold F. Hill, Jr., "Predicting the Judicial Performance of Presidential Appointments to the United States Supreme Court," *Presidential Studies Quarterly* (Fall 1984), pp. 538–47.

13. Personal interview with Justice Harry Blackmun, interviewed in Aspen, Colorado, by Thomas E. Cronin.

14. *Roe v. Wade,* 410 U.S. 113 (1973).

15. *Marbury v. Madison,* 5 U.S. (1 Crouch) 137 (1803); *Ex-parte Milligan,* 4 Wall. 2 (1966); *Youngstown Sheet and Tube Co. v. Sawyer,* 343 U.S. 579 (1952); and *U.S. v. Nixon,* 418 U.S. 683 (1974).

16. Robert H. Jackson, *The Struggle for Judicial Supremacy* (New York: Vintage, 1941), p. 314.

17. Alan Dershowitz, "U.S. Supreme Court Isn't Supposed to Be A President's Tool," *The Denver Post,* November 11, 1984, p. 3–1. Also see Laurence Tribe, *God Save This Honorable Court.*

18. Scigliano, note 12, p. 159. This view is echoed in David M. O'Brien, *Storm Center: The Supreme Court in American Politics* (New York: Norton, 1986), chap. 2.

19. Quoted in Linda Greenhouse, "Rehnquist Asserts Most Attempts by Presidents to Pack Court Fail," *The New York Times,* October 10, 1984, p. 9.

20. Quoted in *Time* magazine, May 23, 1969, p. 24.

21. *Dred Scott v. Sandford,* 60 U.S. (19 How.) 393 (1856).

22. *Ex-parte Merryman,* 17 Fed. Case No. 9487 (1861). See also on this clash, David M. Silver, *Lincoln's Supreme Court* (Urbana: University of Illinois Press, 1957).

23. Clinton Rossiter, *The Supreme Court and the Commander in Chief* (Ithaca, N.Y.: Cornell University Press, 1951), p. 25.

24. Richard N. Current, "The Lincoln Presidents," *Presidential Studies Quarterly* (Winter 1979), p. 32. See also Arthur M. Schlesinger, Jr., *The Imperial Presidency* (Boston: Houghton Mifflin, 1973), pp. 64–67; and J. G. Randall, *Constitutional Problems under Lincoln,* rev. ed. (Urbana: University of Illinois Press, 1951), pp. 513–22. For a broader contrast, see David Donald, *Lincoln* (New York: Simon & Schuster, 1995).

25. Robert H. Jackson, *The Struggle For Judicial Supremacy* (New York: Vintage, 1941).

26. Philip Abbott, *The Exemplary Presidency: Franklin D. Roosevelt and the American Political Tradition,* chap. 7, "They Have Retired into the Judiciary" (Amherst: University of Massachusetts Press, 1990).

27. William E. Leuchtenburg, "Court-Packing Plan," in Otis L. Graham and Meghan Robinson Wander, *Franklin D. Roosevelt: His Life and Times* (Boston: G.K. Hall, 1985), p. 86. For related information and a somewhat different interpretation, see Theresa A. Niedziela, "Franklin D. Roosevelt and the Supreme Court," *Presidential Studies Quarterly* (Fall 1976), p. 51–57.

28. David Gray Adler, "Court, Constitution, and Foreign Affairs," in Adler and Larry George, eds., *The Constitution and the Conduct of American Foreign Policy* (Lawrence: University Press of Kansas, 1996), p. 25.

29. David J. Danelski, "The Saboteurs' Case," *Journal of Supreme Court History* 1 (1996), p. 80. (Complete essay pp. 61–82.)

30. *Korematsu v. United States,* 323 U.S. 214 (1944).

31. See the discussion of Truman's plight in this case in Robert J. Donovan, *Tumultuous Years: The Presidency of Harry S Truman, 1949–1953* (New York: Norton, 1982), pp. 382–91. Also see Maeva Marcus, *Truman and the Steel Seizure Case: The Limits of Presidential Power* (New York: Columbia University Press, 1977); and Alan Westin, *The Anatomy of a Constitutional Law Case* (New York: Macmillan, 1958).

32. Stanley K. Kutler, *The Wars of Watergate* (New York: Knopf, 1990); and Fred Emery, *Watergate* (New York: Touchstone, 1995).

33. Theodore Draper, *A Very Thin Line: The Iran-Contra Affair* (New York: Touchstone Books, 1991); Jane Mayer and Doyle McManus, *Landslide: The Unmaking of the President, 1984–1988* (Boston: Houghton Mifflin, 1988); and Lawrence E. Walsh, *Firewall: The Iran-Contra Conspiracy and Cover-up* (New York: Norton, 1997).

34. Nancy V. Baker, *Conflicting Loyalties: Law and Politics in the Attorney General's Office, 1789–1990* (Lawrence: University Press of Kansas, 1992); Katy J. Harriger, *Independent Justice: The Federal Special Prosecutor in American Politics* (Lawrence: University Press of Kansas, 1992); and Rebecca Mae Salokar, *The Solicitor General: The Politics of Law* (Philadelphia: Temple University Press, 1992).

35. For a discussion of this issue, see: Christine E. Burgess, "When May a President Refuse to Enforce the Law?," *Texas Law Review,* 72 (3) (1994).

36. Harold Hongju Koh, *The National Security Constitution: Sharing Power after the Iran-Contra Affair* (New Haven, Conn.: Yale University Press, 1990).

37. Daniel P. Franklin, *Extraordinary Measures: The Exercise of Prerogative Powers in the United States* (Pittsburgh: University of Pittsburgh Press, 1991).

38. Michael A. Genovese, "Presidential Leadership and Crisis Management," *Presidential Studies Quarterly* (Spring 1986); and Michael A. Genovese, "Presidents and Crisis: Developing a Crisis Management System in the Executive Branch," *International Journal on World Peace* (Spring 1987).

39. John Locke, *Treatise of Civil Government and a Letter Concerning Toleration* (New York: Appleton-Century-Crofts, 1937), chap. 14, p. 109.

40. Ibid., p. 113.

41. Jean Jacques Rousseau, *The Social Contract and Discourses* (New York: E.P. Dutton, 1950), Book IV, chap. 6, p. 123.

42. John Stuart Mill, *Considerations on Representative Government* (New York: Liberal Arts Press, 1958), p. 274.

43. J. Malcolm Smith and Cornelius P. Cotter, *Powers of the President during Crises* (Washington, D.C.: Public Affairs Press, 1960), p. 7.

44. Niccolo Machiavelli, *The Prince and the Discourses* (New York: Modern Library, 1950), Book 1, chap. 34, p. 203.

45. Clinton Rossiter, *Constitutional Dictatorship: Crisis Government in the Modern Democracy* (Princeton, N.J.: Princeton University Press, 1948).

46. Abe Fortas, "The Constitution and the Presidency," *Washington Law Review* 49 (August 1974), p. 100.

47. Edward S. Corwin, *Total War and the Constitution* (Westminister, Md: Knopf, 1947), p. 80.

48. Rossiter, note 45, p. 12.

49. *The New York Times,* September 8, 1942, p. 1.

50. See Harold Hongju Koh, "Why the President Almost Always Wins in Foreign Affairs," in Adler and George, note 28, pp. 158–80.

51. Richard Longaker, Introduction to Rossiter, *The Supreme Court and the Commander in Chief,* exp. ed. (Ithaca, N.Y.: Cornell University Press, 1976), p. xii.

52. Glendon Schubert, *The Presidency in the Courts* (Minneapolis: University of Minnesota Press, 1957), pp. 4, 354.

53. See, for example, David G. Adler, "Foreign Policy and the Separation of Powers under the Constitution: The Influence of the Judiciary," paper delivered at the 1987 Annual Meeting of the Western Political Science Association, Anaheim, California, March 26–28, 1987; and his more specialized book *The Constitution and the Termination of Treaties* (New York: Garland Publishing, 1986). See also Francis D. Wormuth and Edwin B. Firmage, *To Chain the Dog of War* (Dallas: Southern Methodist University Press, 1986). For a more modified verdict see Louis Fisher, *The Politics of Shared Power,* 2d ed. (Washington, D.C.: Congressional Quarterly Press, 1987), pp. 105–7. For a much more modified view, see Robert Scigliano, "The Presidency and the Judiciary," in Michael Nelson, ed., *The Presidency and the Political System* (Washington, D.C.: Congressional Quarterly Press, 1984), p. 414. See also a helpful quantitative study by Craig R. Ducat and Robert L. Dudley, "Presidential Power in the Federal Courts during the Post War Era," paper delivered at the 1985 Annual Meeting of the American Political Science Association, New Orleans, August, 1985.

54. *The Prize Cases,* 67 U.S. (2 Black) 635 (1863).

55. *In re Neagle,* 135 U.S. 1 (1890).

56. *In re Debs,* 158 U.S. 564 (1895).

57. *Myers v. United States,* 272 U.S. 52 (1926).

58. *United States v. Curtiss-Wright Export Corporation,* 299 U.S. 304 (1936).

59. *United States v. Belmont,* 301 U.S. 324 (1937).

60. Hugo Black, *Korematsu v. United States,* 323 U.S. 214 (1944).

61. Eugene V. Rostow, book review, *The Washington Post National Weekly Edition,* January 2, 1984, p. 34. See also Peter Irons, *Justice At War: The Story of the Japanese American Internment Cases* (New York: Oxford University Press, 1983).

62. Quoted in Robert Pear, "A Japanese Relocation Case Tests the Verdict of History," *The New York Times,* June 30, 1985, p. 8E.

63. David G. Adler, *The Constitution and the Termination of Treaties* (New York: Garland Publishing, 1986), p. 306. For other views see Louis W. Koenig et al., eds., *Congress, the Presidency and the Taiwan Relations Act* (New York: Praeger, 1985).

64. *Dames and Moore v. Reagan,* 453 U.S. 654 (1981).

65. *Nixon v. Fitzgerald,* 457 U.S. 731 (1982).

66. *Nixon v. Fitzgerald* (1982), excerpts from *The New York Times,* June 25, 1982, p. Y9.

67. *INS v. Chadha,* 462 U.S. 919 (1983).

68. Ibid.

69. See Joseph Cooper, "The Legislative Veto in the 1980's," in Larry C. Dodd and Bruce J. Oppenheimer, eds., *Congress Reconsidered,* 3rd ed., (Washington, D.C.: Congressional Quarterly Press, 1985), pp. 364–89; Louis Fisher, "Judicial Misjudgments about the Lawmaking Process: The Legislative Veto Case," *Public Administration Review* (November 1985), pp. 705–11; Louis Fisher "Legislative Vetoes Enacted after Chadha," mimeo report, Congressional Research Service, March 25, 1985; and Stephan Labaton, "Wrong Again, Supreme Court," *The Washington Post National Weekly Edition,* August 19, 1985, p. 10.

70. *Bowsher v. Synar,* 106 U.S. 3181 (1986).

71. Michael A. Genovese, *The Supreme Court, the Constitution, and Presidential Power* (Lanham, Md.: University Press of America, 1980), p. 121.

72. *Marbury v. Madison,* 1 Cranch 137, 2 L. Ed. 60 (1803).

73. *Little v. Bareme,* 2 Cranch 170 (1804).

74. *Humphrey's Executor v. United States,* 295 U.S. 602, 628 (1935).

75. See Louis Fisher, "Congress and the Removal Power," *Congress and the Presidency* (Spring 1983), pp. 64–65.

76. Robert J. Donovan, *Tumultuous Years* (New York: Norton, 1982), p. 387.

77. Ibid., p. 391.

78. *New York Times Co. v. United States,* 403 U.S. 713 (1971).

79. Ibid., 712.

80. *United States v. U.S. District Court,* 407 U.S. 297 (1972).

81. *Morrison v. Olsen,* 487 U.S. 654 (1988).

82. See Charles Fried, *Order and Law: Arguing the Reagan Revolution—A Firsthand Account* (New York: Simon & Schuster, 1991).

83. Charles F. C. Ruff, White House counsel, questioned in Stephen Labaton, "Clintons Lose before Justices on Notes Fight," *The New York Times,* June 24, 1997, p. A12.

84. This is nicely shown in Ducat and Dudley, note 53.

85. Justice Robert Jackson, *Youngstown Case,* 343 U.S. 579, 654 (1952).

CHAPTER 9

1. Charles O. Jones, *The Presidency in a Separated System* (Washington, D.C.: Brookings, 1994), p. 103. See also Anthony J. Bennett, *The American President's Cabinet* (New York: St. Martin's, 1996).

2. Jimmy Carter, *Keeping Faith: Memoirs of a President* (New York: Bantam, 1982), p. 60.

3. Peri E. Arnold, *Making the Managerial Presidency* (Princeton, N.J.: Princeton University Press, 1986), p. 361.

4. Henry Barret Learned, *The President's Cabinet: Studies in the Origin, Formation and Structure of an American Institution* (Burt Franklin, 1912, reissued in 1972), p. 119. This is a useful study of the origins of the cabinet. The best study of the cabinet in the mid-twentieth century was that of Richard Fenno, *The President's Cabinet* (New York: Vintage, 1959).

5. Lyndon B. Johnson, quoted in Charles Maguire, oral history, August 19, 1969, Lyndon B. Johnson Presidential Library, Austin, Texas, p. 28.

6. Richard Nixon, *RN: The Memoirs of Richard Nixon* (Grosset & Dunlap, 1978), p. 338.

7. Keith Nicholls, "Presidential Cabinets: The Politics of Selection from Washington to Reagan," *Congress and the Presidency* (Autumn 1989).

8. Robert B. Reich, *Locked in the Cabinet* (New York: Knopf, 1997), pp. 51–52.

9. Shirley Anne Warshaw, *Powersharing: White House–Cabinet Relations in the Modern Presidency* (Albany, N.Y.: SUNY Press, 1996).

10. Reich, note 8, pp. 43–44.

11. Carter White House aide, interview with Thomas E. Cronin.

12. McGeorge Bundy, *The Strength of Government* (Harvard University Press, 1968), p. 39.

13. Bradley H. Patterson, Jr., *The President's Cabinet: Issues and Questions* (American Society for Public Administration, 1976), pp. 17–18.

14. Shirley M. Hufstedler, "Open Letter to a Cabinet Member," *The New York Times Magazine,* January 11, 1981, p. 38. Similar themes are echoed in Terrel M. Bell, *The Thirteenth Man: A Reagan Cabinet Memoir* (New York: Free Press, 1987).

15. Quoted in Kermit Gordon, "Reflections on Spending," in J. D. Montgomery and Arthur Smithies, eds., *Public Policy* (Cambridge, Mass.: Harvard University Press, 1966), p. 15).

16. Richard M. Nixon, November 27, 1972, *The New York Times,* November 28, 1972, p. C40.

17. Lyndon B. Johnson, quoted in Bobby Baker, with Larry King, *Wheeling and Dealing* (Norton, 1978), p. 265.

18. Walter J. Hickel, *The New York Times,* May 7, 1970, p. C18.

19. Lyndon B. Johnson, note 17, pp. 265–66. It should be noted that this statement was made some four years after LBJ had left the White House.

20. "Secretary of Collision," *The New York Times,* October 3, 1985, p. A26. For similar struggles between White House staffers, especially Ed Meese and a secretary of education in the Reagan administration, see Bell, note 14.

21. Walt Williams, "George Bush and Executive Branch Domestic Policy Making Competence," *Policy Studies Journal* (Winter 1993).

22. Harold Gosnell, *Harry S Truman* (New York: Greenwood Press, 1980).

23. Quoted in *U.S. News & World Report,* March 29, 1982, p. 28.

24. Interview at the White House with Thomas E. Cronin.

25. Jimmy Carter, *Keeping Faith: Memoirs of a President* (New York: Bantam, 1982), p. 59.

26. Ibid., p. 60.

27. Theodore C. Sorensen, transcript of a panel discussion at the Annual Meetings of the American Society for Public Administration, New York City, March 22, 1972, p. 22. Reprinted as "Advising the President: A Panel," *Bureaucrat* (April 1974), p. 33.

28. Nixon, note 6.

29. Elliot Richardson, *The Creative Balance* (Holt Rinehart and Winston, 1976), p. 74.

30. Michael K. Deaver, *Behind the Scenes* (New York: Morrow, 1988), p. 127.

31. Harold L. Ickes, *The Secret Diaries of Harold L. Ickes,* vol. 1 (New York: Simon & Schuster, 1953), p. 308.

32. Griffin B. Bell and Ronald J. Ostrow, *Taking Care of the Law* (New York: Morrow, 1982), p. 47.

33. Patterson, note 13, p. 113.

34. See, for example, David A. Stockman, *The Triumph of Politics: Why the Reagan Revolution Failed* (New York: Harper & Row, 1986), chap. 4.

35. Terry M. Moe, "The Politicized Presidency," in John E. Chubb and Paul E. Peterson, eds., *The New Direction in American Politics* (Washington, D.C.: Brookings, 1985), pp. 260–61.

36. Edwin Meese, "The Institutional Presidency: A View from the White House," *Presidential Studies Quarterly* (Spring 1983), p. 196.

37. Dick Kirschten, "Reagan's Cabinet Councils May Have Less Influence Than Meets the Eye" *National Journal,* July 11, 1981, p. 1242.

38. James P. Pfiffner, *The Strategic Presidency* (Chicago: Dorsey Press, 1988), p. 63.

39. Moe, note 35, pp. 261–62.

40. Keith Nicholls, "The Dynamics of National Executive Service: Ambition Theory and the Careers of Presidential Cabinet Members," *Western Political Quarterly* (March 1991).

41. John Leacacos, *Fires in the In-Basket* (World, 1968), p. 110.

42. Quoted in Richard H. Rovere, "Eisenhower: A Trial Balance," *The Reporter,* April 21, 1955, pp. 19–20.

43. Leonard Silk, *Nixonomics* (New York: Praeger, 1972), p. 81.

44. Myra Gutin, *The President's Partner* (New York: Greenwood Press, 1989); and Edith Mayo, "The Influence and Power of First Ladies," *The Chronicle of Higher Education,* September 15, 1993, sec. A; Alice E. Anderson and Hadley V. Baxendale, *Behind Every Successful President: The Hidden Power and Influence of America's First Ladies,* (Shapolsky Publishers, 1992); Carl S. Anthony, *First Ladies: The Saga of the President's Wives and Their Power* (William Morrow, 1990); Lewis L. Gould, "Modern First Ladies and the Presidency," *Presidential Studies Quarterly* 20 (1990), pp. 677–83; Lewis L. Gould, "First Ladies," *The American Scholar* 55 (1986), pp. 528–35; and Karen O'Connor, Bernadette Nye, and Laura van Assendalft, "Wives in the White House: The Political Influence of the First Ladies," *Presidential Studies Quarterly* 26 (1996), pp. 835–53.

45. Betty Boyd Caroli, *First Ladies* (New York: Oxford University Press, 1987).

46. Margaret Mead, quoted in *The New York Times,* December 12, 1974, sec. VI, p. 36.

47. Donald Young, *American Roulette: The History and Dilemma of the Vice Presidency* (New York: Viking, 1974), p. 134.

48. Mayo, note 44, p. 2; and Edith Mayo and Denise Meringolo, *First Ladies: Political Role and Public Image* (Washington, D.C.: Smithsonian Institute, 1994).

49. Caroli, note 45, p. xvii.

50. *The New York Times,* July 16, 1985, p. 1.

51. James P. Pfiffner, "The President's Chief of Staff: Lessons Learned," Working Paper 92, Institute of Public Policy, George Mason University, October 1992, p. 19.

52. See Warshaw, note 9; see also Richard T. Johnson, *Managing the White House* (New York: Harper & Row, 1974).

53. Two of the useful surveys of White House–cabinet relations can be found in Anthony J. Bennett, *The American President's Cabinet* (New York: St. Martin's Press, 1996); and Warshaw, note 9.

54. Personal interview.

55. Jesse Jones, quoted in Richard Fenno, "President-Cabinet Relations and Pattern and a Case Study," *American Political Science Review* (March 1958), p. 394.

56. Zbigniew Brzezinski, *Power and Principle: Memoirs of the National Security Adviser, 1977–1981* (New York: Farrar, Strauss, Giroux, 1983), p. 67.

57. Joseph A. Califano, Jr., *Governing America: An Insider's Report from the White House and the Cabinet* (New York: Simon & Schuster, 1981), p. 407.

58. Ibid., p. 434–35.

59. John Ehrlichman, *Witness to Power: The Nixon Years* (New York: Simon & Schuster, 1982), pp. 110–11.

60. Reich, note 8, p. 179.

61. Bill Clinton, quoted in Dick Morris, *Behind the Oval Office: Winning the Presidency in the Nineties* (New York: Random House, 1996), pp. 97–98.

62. Harry S Truman, quoted in Richard Neustadt, *Presidential Power* (New York: Mentor, 1960), p. 22.

63. Arthur M. Schlesinger, Jr., *The Crisis of Confidence* (New York: Houghton Mifflin, 1969), p. 291.

64. Kermit Gordon, "Reflections on Spending," in J. D. Montgomery and A. Smithies, eds., *Public Policy* (Cambridge, Mass.: Harvard University Press, 1966), pp. 13–15.

65. John W. Gardner, testimony before the Senate Government Operations Committee, *Congressional Record,* 92nd Congress, 1st session, June 3, 1971, S.8140.

CHAPTER 10

1. Paul C. Light, *Vice Presidential Power: Advice and Influence in the White House* (Baltimore, Md.: Johns Hopkins University Press, 1984), p. 1.

2. Jules Whitcover, *Crapshoot: Rolling the Dice on the Vice Presidency* (New York: Crown, 1992).

3. Allan P. Sindler, *Unchosen Presidents* (Berkeley: University of California Press, 1976), p. 41.

4. Henry Kissinger, *White House Years* (Boston: Little, Brown, 1979), p. 713.

5. Doris Kearns, *Lyndon Johnson and the American Dream* (New York: Harper & Row, 1976), p. 164.

6. George S. Sirgiovanni, "Dumping the Vice President," *Presidential Studies Quarterly* (Fall 1994), pp. 765–82.

7. The most comprehensive account of the Twenty-fifth Amendment is by John D. Feerick, *The Twenty-fifth Amendment* (New York: Fordham University Press, 1976).

8. *Time,* January 20, 1975, p. 23.

9. Alben W. Barkley, *That Reminds Me: The Autobiography of the Veep* (Garden City, N.J.: Doubleday, 1954), p. 209.

10. James M. Naughton, "Above the Battle," *The New York Times Magazine,* June 24, 1973, p. 49.

11. Senator Clinton Anderson, quoted in Roland Evans and Robert Novak, *Lyndon B. Johnson: The Exercise of Power* (New York: New American Library, 1966), p. 307.

12. Donald Young, *American Roulette: The History and Dilemma of the Vice Presidency* (New York: Holt, Rinehart & Winston, 1972), pp. 353–54.

13. Paul T. David, "The Vice Presidency: Its Institutional Evolution and Contemporary Status," *Journal of Politics* (November 1967), p. 733. President Eisenhower's dealings with his own vice president, Richard Nixon, were also affected by his ambivalence toward Nixon. According to his leading biographer, Ike often doubted Nixon's maturity and thought him overly ambitious. "The more Eisenhower tried to praise him [Nixon], it somehow seemed, the more tongue-tied he got; the more he tried to endorse Nixon's leadership qualities, the more doubtful he sounded." Stephen E. Ambrose, *Eisenhower: The President* (New York: Simon & Schuster, 1984), p. 297.

14. James F. Byrnes, *All in One Lifetime* (New York: Harper & Brothers, 1958), p. 233.

15. Joseph E. Kallenbach, *The American Chief Executive* (New York: Harper & Row, 1966), pp. 234–35.

16. William A. Cohen, *The Art of the Leader* (Englewood Cliffs, N.J.: Prentice-Hall, 1990).

17. Dan Quayle, *Standing Firm: A Vice Presidential Memoir* (New York: Harper Collins, 1994).

18. Milton Eisenhower, *The President is Calling* (New York: Doubleday, 1974), pp. 540–41.

19. Michael Novak, *Choosing our King* (New York: Macmillan, 1974), pp. 263–64.

20. James MacGregor Burns, *The Lion and the Fox* (New York: Harcourt, Brace, 1956), p. 414.

21. Sam Houston Johnson, *My Brother Lyndon* (New York: Cowles, 1970), p. 109.

22. Arthur M. Schlesinger, Jr., "Is the Vice Presidency Necessary?" *Atlantic* (May 1974), p. 37.

23. Hubert H. Humphrey, "A Conversation with Vice President Hubert H. Humphrey," National Educational Television, April 1965.

24. Gerald R. Ford, *A Time To Heal* (New York: Harper & Row/Reader's Digest, 1979), p. 327.

25. Ibid., p. 328. For the Rockefeller side of this story see Joseph E. Persico, *The Imperial Rockefeller: A Biography of Nelson A. Rockefeller* (New York: Simon & Schuster, 1982); and Michael Turner, *The Vice President as Policy Maker: Rockefeller in the Ford White House* (Westport, Conn.: Greenwood Press, 1982).

26. Light, note 1, p. 63.

27. Walter F. Mondale, quoted in *National Journal,* December 1, 1979, p. 2016.

28. Quoted in *National Journal,* March 11, 1978, p. 379. See also the comparative analysis of Paul Light, "Vice Presidential Influence under Rockefeller and Mondale," *Political Science Quarterly,* Winter 1983–84, pp. 617–40.

29. President Jimmy Carter news conference, *The New York Times,* September 30, 1977, p. A18.

30. Jimmy Carter, *Keeping Faith: Memoirs of a President* (New York: Bantam, 1982), p. 40. See also *The New York Times,* November 13, 1980, p. 14. Two recent studies of the vice presidency also document the unusually close and sensible relationships between Carter and Mondale and their staffs—Joel K. Goldstein *The Modern American Vice Presidency* (Princeton, N.J.: Princeton University Press, 1982); and Paul C. Light, *Vice-Presidential Power* (Baltimore: Johns Hopkins Press, 1984).

31. Walter F. Mondale, lecture on "The American Vice Presidency" at the University of Minnesota, Minneapolis, February 18, 1981, mimeo, p. 3.

32. Walter F. Mondale, interview, *National Journal,* December 1, 1979, p. 2016.

33. James Reston, "Fritz Mondale's Blitz," *The New York Times,* November 2, 1979, p. A31.

34. See Mondale interview, *Politics Today,* March/April, 1980, p. 48.

35. Adapted and condensed from Mondale, note 31, p. 7.

36. Dick Kirschten, "George Bush: Keeping His Profile Low so He Can Keep His Influence High" *National Journal,* June 20, 1981, p. 1096.

37. *The Houston Post,* May 1, 1983, p. 1.

38. *Time,* February 22, 1982, p. 22.

39. Ibid. A later criticism along these lines is found in Dale Russakoff, "Bush: The Loyal Subordinate. . . . ," *The Washington Post,* January 21, 1985, pp. G4–5.

40. *The Houston Post,* note 37.

41. Ronald Reagan, interviewed by reporters from *The New York Times,* February 12, 1985, p. 6.

42. E. J. Dionne, Jr., "Bush's Presidential Bid Is Shaky . . ., Poll Finds" *The New York Times,* July 26, 1987, pp. 1, 14.

43. Burt Solomon, "Color Gore a New Democrat Who May Yet Turn His Boss Into One," *National Journal* (November 1994).

44. Elaine Sciolino and Todd S. Purdum, "Gore Is No Typical Vice President in the Shadow," *The New York Times,* February 19, 1995, p. 16.

45. Paul Kengor, "The Foreign Policy Role of Vice President Al Gore," *Presidential Studies Quarterly* 27 (1) (Winter 1997).

46. Sciolino and Purdum, note 44, p. 1.

47. Richard Reeves, *Running in Place: How Bill Clinton Disappointed America* (Kansas City, Kan.: Andrews and McMeel, 1996), p. 97.

48. Richard L. Berke, "Al Gore: Good Scout," *New York Times Magazine,* February 20, 1994. Also see Bob Woodward, *The Agenda* (New York: Simon & Schuster, 1994), pp. 280–81.

49. Elizabeth Drew, *On The Edge: The Clinton Presidency* (New York: Simon & Schuster, 1994), p. 227.

50. Robert B. Reich, *Locked in the Cabinet* (New York: Knopf, 1997), p. 241.

51. David S. Broder and Bob Woodward, *The Man Who Would Be President: Dan Quayle* (New York: Simon & Schuster, 1992).

52. Quayle, note 17.

53. Jean Edward Smith, *George Bush's War* (New York: Henry Holt, 1992); and Burt Solomon, "War Bolsters Quayle's Visibility, but Hasn't Increased His Stature," *National Journal,* March 2, 1991.

54. Thomas M. Durbin, *Nomination and Election of the President and the Vice President of the United States* (Washington, D.C.: U.S. Government Printing Office, 1988).

55. Jules Witcover, *Crapshoot: Rolling the Dice on the Vice Presidency* (New York: Crown, 1992).

56. See *Official Report of the Vice Presidential Selection Commission of the Democratic Party,* December 19, 1974, mimeo. See also "Hearings of the Vice Presidential Selection Commission of the Democratic National Committee," *Congressional Record,* October 16, 1973, S19245 ff; and Allan P. Sindler, *Unchosen Presidents* (Berkeley: University of California Press, 1976).

57. Some of these proposals are discussed and assessed in Joel K. Goldstein, *The Modern American Vice Presidency* (Princeton, N.J.: Princeton University Press, 1982), p. 282ff.

58. Arthur M. Schlesinger, Jr., *The Imperial Presidency* (New York: Popular Library, 1974), p. 481.

59. Ibid., p. 493.

60. Report of the Study Group on Vice Presidential Selection, Institute of Politics, Kennedy School of Government, Harvard University, June 14, 1976, monograph, p. 7. This report contains several of the proposals summarized here.

61. Ibid.

62. Bernard Weinraub, "Mondale Outlines Job Qualities of Running Mate," *The New York Times,* June 13, 1984, p. B11.

63. Goldstein, note 57, p. 89.

64. John D. Feerick, "The Proposed Twenty-fifth Amendment to the Constitution," *Fordham Law Review* (December 1965), p. 197. See also Feerick's *The Twenty-fifth Amendment* (Bronx, N.Y.: Fordham University Press, 1976); and Herbert L. Abrams, "Shielding the President from the Constitution: Disability and the 25th Amendment," *Presidential Studies Quarterly* (Summer 1993), pp. 533–53.

65. J. W. Peltason, *Understanding the Constitution,* 8th ed. (New York: Holt, Rinehart and Winston, 1979), pp. 232–33.

66. Stephen E. Ambrose, *Eisenhower: The President* (New York: Simon & Schuster, 1984), p. 440.

67. Birch Bayh, "The White House Safety Net," *The New York Times,* April 8, 1995, editorial.

CHAPTER 11

1. Theodore C. Sorensen, *Watchmen in the Night* (Cambridge, Mass.: MIT Press, 1975).

2. James Madison, *The Federalist Papers,* No. 51. (Modern Library, 1937).

3. Sidney M. Milkis, *The President and the Parties* (New York: Oxford University Press, 1993).

4. Bruce Miroff, *Icons of Democracy* (New York: Basic Books, 1993), chap. 1.

5. James M. Burns, *Roosevelt: The Lion and the Fox* (San Diego: A Harvest/HJB Book, 1956, renewed 1984), p. 151.

6. Benjamin Barber, *Strong Democracy: Participatory Politics for a New Age* (Berkeley: University of California Press, 1984).

7. Benjamin R. Barber, "Neither Leaders Nor Followers: Citizenship under Strong Democracy," in Michael R. Beschloss and Thomas E. Cronin, eds., *Essays in Honor of James MacGregor Burns* (Englewood Cliffs, N.J.: Prentice-Hall, 1989), p. 117.

8. Bert Rockman, *The Leadership Question* (New York: Praeger, 1984), p. 221.

9. Quoted in Larry Berman, *The New American Presidency,* (Boston: Little, Brown, 1987), p. 344.

10. Attorney General Griffin B. Bell, address before the faculty and students, University of Kansas at Lawrence, January 25, 1979, mimeo, p. 5. The founders of

416

Notes

the southern Confederacy were also proponents of this idea and provided for a six-year single term for their president. A Washington-based foundation was organized in the 1970s to push for a six-year single presidential term as well as term limits for members of Congress. This Foundation for the Study of Presidential and Congressional Terms was established under the National Heritage Foundation as a nonpartisan, nonprofit organization.

11. Jack Valenti, "A Six-Year Presidency?," *Newsweek,* February 4, 1971, p. 11. See also Charles Bartlett, "That Six-Year Term," *Washington Star,* May 2, 1979, p. A19; and Tom Wicker, "Six Years for the Presidency?," *The New York Times Magazine,* June 26, 1983.

12. Mike Mansfield, *Statement in Support of Senate Joint Resolution 77,* before the Subcommittee on Constitutional Amendments of the U.S. Senate, Committee on the Judiciary, October 8, 1971, processed. For his extended views and those of several other witnesses, see U.S. Congress, Senate, Committee on the Judiciary, Subcommittee on Constitutional Amendments, *Single Six-Year Term for President,* 92nd Congress, 1st Session, 1972, p. 32.

13. Woodrow Wilson, letter placed in the *Congressional Record,* 64th Congress, 2d Session, August 15, 1916, 53. pt. 13:12620.

14. James MacGregor Burns, "Keeping the President in Line," *The New York Times,* April 1973, p. E15.

15. The possibility also exists that a six-year term, or "a term-and-a-half" as some call it, with reelection precluded, would intensify the presidential selection process. Certainly in such a winner-take-more situation, there is the likelihood that ideological competition would be more aggressive and perhaps more bitter than at present. Conflict would assuredly be heightened. How harmful this would be is difficult to assess, but judging from how corrupt the 1972 reelection campaign became, this factor must be considered.

16. Harry S. Truman, testimony before the Subcommittee on Constitutional Amendments of the U.S. Senate, Committee on the Judiciary Hearings on S.J. Resolution II: "Presidential Term of Office," 86th Congress, 1st session, 1959, pt. I, p. 7.

17. Thomas A. Baylis, *Governing By Committee: Collegial Leadership in Advanced Societies* (Albany, N.Y.: SUNY Press, 1989).

18. Herman Finer, *The Presidency: Crisis and Regeneration* (University of Chicago Press, 1960), p. viii. See also a more recent plea for a more collective presidency: Milton Eisenhower, *The President Is Calling* (Doubleday, 1974).

19. See Mark O. Hatfield, "Resurrecting Political Life in America . . ." *Congressional Record,* 93rd Congress, 1st session, October 12, 1973, 119, no. 153, S.19104-07.

20. Benjamin V. Cohen, "Presidential Responsibility and American Democracy," 1974 Royer Lecture, University of California, Berkeley, May 23, 1974, pp. 24–25.

21. Ibid.; see also John Orman, *Comparing Presidential Behavior: Carter, Reagan, and the Macho Presidential Style* (Westport, Conn.: Greenwood Press, 1987).

22. Cohen, ibid., p. 29. See also George Reedy, *The Twilight of the Presidency* (World, 1970).

23. Alexander Hamilton, *The Federalist Papers,* No. 70 (Modern Library, 1937), pp. 460–67.

24. The constitutional amendment proposed by Representative Henry S. Reuss can be found in Congressional Record, II 7158, July 21, 1975. The one place where it received attention and scrutiny was in the *George Washington Law Review,* which contained a symposium on the no-confidence proposal (January 1975), pp. 328–500.

25. James Sundquist, "Needed: A Workable Check on the Presidency," *Brookings Bulletin* 10 (4) (1973), p. 7. See also, Sundquist, *Constitutional Reform* (1986).

26. Samuel H. Beer puts it well from a comparative perspective: "The lesson of British experience, and indeed of comparative government generally, is that the vote of confidence device might unduly weaken the executive, or unduly strengthen the executive, or, possibly, bring about the nice adjustment that Representative Reuss desires. In short, the consequences of the proposed reform are incalculable. In view of the further fact that the impeachment process did work in the case of President Nixon, these prospects indicate to this author that Congress should leave things as they are." *George Washington Law Review* (January 1975), p. 371.

27. The best history and political analysis of third parties is Steven Rosenstore, Roy Behr, and Edward Lazarus, *Third Parties in America: Citizen Response to Major Party Failure* (Princeton, N.J.: Princeton University Press, 1984).

28. Committee on Federal Legislation, Bar Association of the City of New York, *The Law of Presidential Impeachment* (New York: Harrow Books, 1974), pp. 5–6.

29. John R. Labovitz, *Presidential Impeachment* (New Haven, Conn.: Yale University Press, 1978). See also Charles Black, *Impeachment: A Handbook* (New Haven, Conn.: Yale University Press, 1974); Raoul Berger, *Impeachment: The Constitutional Problems* (Cambridge, Mass.: Harvard University Press, 1973); and Stanley I. Kutler, *The Wars of Watergate* (New York: Knopf, 1990).

30. Labovitz, note 29.

31. Michael A. Genovese, "Presidential Corruption," paper presented at the annual meeting of the American Political Science Association, New York, September, 1994. Some would add President Clinton's problems with Whitewater to this list.

32. Michael A. Genovese, *The Nixon Presidency: Power and Politics in Turbulent Times* (Westport, Conn.: Greenwood Press, 1990).

33. Shelley Ross, *Fall From Grace* (New York: Ballantine Books, 1988), p. 269.

34. See, for example, James B. Stewart's provocative analysis in *Blood Sport: The President and His Adversaries* (New York: Simon & Schuster, 1996).

35. Peter M. Shane, "Independent Counsel," in Leonard Levy and Louis Fisher, eds., *Encyclopedia of the American Presidency* (New York: Simon & Schuster, 1994), p. 816.

36. Archibald Cox, "Curbing Special Counsels," *The New York Times,* December 12, 1996, p. A21.

37. Ibid.

38. Ibid. See also "Keep the Independent Counsel," *The New York Times,* December 21, 1996, p. 18, editorial.

39. *Morrison v. Olson* 487 U.S. 654 (1988).

40. Fred Wertheimer, "Stop Soft Money. Now," *The New York Times Magazine,* December 22, 1996, pp. 38–39.

41. Michael Duffy and Nancy Gibbs, "Money and Politics," *Time* magazine, November 11, 1996, p. 33.

42. Ibid.

43. Anthony Corrado, *Paying for Presidents: Public Financing in Presidential Elections* (New York: Twentieth Century Fund Press, 1993).

44. See Lance T. LeLoup and Steven A. Shull, *Congress and the President: The Policy Connection* (Belmont, Calif.: Wadsworth Publishing Company, 1993).

45. See the fresh interpretation in Philip Abbott, *Strong Presidents: A Theory of Leadership* (Knoxville: University of Tennessee Press, 1996).

Selected Bibliography

MAJOR REFERENCE WORKS ON THE PRESIDENCY

Goldsmith, William M. *The Growth of Presidential Power: A Documented History.* 3 vols. New York: Chelsea House, 1974.

Graff, Henry F., ed. *The Presidents: A Reference History.* New York: Scribner's, 1984.

Greenstein, Fred I., Larry Berman, and Alvin S. Felzenberg with Doris Lidtke. *Evolution of the Modern Presidency: A Bibliographical Survey.* Washington, D.C.: American Enterprise Institute, 1977.

Levy, Leonard W., and Louis Fisher, eds. *Encyclopedia of the American Presidency.* 4 vols. New York: Simon & Schuster, 1994.

Nelson, Michael, ed. *Guide to the Presidency.* Washington, D.C.: Congressional Quarterly Press, 1989.

————. *The Presidency, A to Z.* Washington, D.C.: Congressional Quarterly Press, 1992.

Shane, Peter M., and Harold H. Bruff, eds. *The Law of Presidential Power.* Durham, N.C., Carolina Academic Press, 1988.

Whitney, David C., and Robin Vaughn Whitney, eds. *The American Presidents,* 8th ed. Pleasantville, N.Y.: Reader's Digest, 1996.

OTHER IMPORTANT WORKS

Abbott, Philip. *Strong Presidents: A Theory of Leadership.* Knoxville: University of Tennessee Press, 1996.

Abraham, Henry J. *Justices and Presidents.* 2d ed. New York: Oxford University Press, 1985.

Abrams, Herbert L. *"The President Has Been Shot": Confusion, Disability and the 25th Amendment.* Stanford: Stanford University Press, 1994.

Adler, David Gray, and Larry N. George, eds. *The Constitution and the Conduct of Foreign Policy.* Lawrence: University Press of Kansas, 1996.

Arnold, Peri E. *Making The Managerial Presidency.* Princeton, N.J.: Princeton University Press, 1986.

Asher, Herbert. *Presidential Elections and American Politics.* 5th ed. New York: Harcourt Brace, 1992.

Baker, Nancy V. *Conflicting Loyalties: Law and Politics in the Attorney General's Office. 1789–1990.* Lawrence: University Press of Kansas, 1992.

Barber, James David. *The Presidential Character: Predicting Performance in the White House.* 4th ed. Englewood Cliffs, N.J.: Prentice-Hall, 1992.

Barilleaux, Ryan J., and Barbara Kellerman. *The President as World Leader.* New York: St. Martin's Press, 1991.

Bennett, Anthony. *The American President's Cabinet: From Kennedy to Bush.* New York: St. Martin's, 1996.

Bennis, Warren. *On Becoming a Leader.* Reading, Pa.: Addison-Wesley, 1989.

———, and Bert Nanus. *Leaders: Strategies for Taking Charge.* New York: HarperCollins, 1986.

Berman, Larry. *The New American Presidency.* Boston: Little, Brown, 1987.

Bessette, Joseph M., and Jeffrey Tulis, eds. *The Presidency in the Constitutional Order.* Baton Rouge: Louisiana State University Press, 1984.

Best, Judith A., et al. *The Choice of the People? Debating the Electoral College.* Lanham, Md.: Rowland & Littlefield, 1996.

Blakesley, Lance. *Presidential Leadership: From Eisenhower to Clinton.* Chicago: Nelson-Hall, 1995.

Bond, Jon R., and Richard Fleisher. *The President in the Legislative Arena.* Chicago: University of Chicago Press, 1990.

Brace, Paul, and Barbara Hinckley. *Follow the Leader: Opinion Polls and the Modern Presidents.* New York: Basic Books, 1992.

Brody, Richard A. *Assessing the President: The Media, Elite Opinion, and Public Support.* Stanford, Calif.: Stanford University Press, 1991.

Bryce, James. *The American Commonwealth.* New York: MacMillan, 1888.

Buchanan, Bruce. *The Presidential Experience: What the Office Does to the Man.* Englewood Cliffs, N.J.: Prentice-Hall, 1978.

Bunce, Valerie. *Do New Leaders Make a Difference?: Executive Successions and Public Policy Capitalism and Socialism.* Princeton, N.J.: Princeton University Press, 1981.

Burke, John P. *The Institutional Presidency.* Baltimore: Johns Hopkins University Press, 1992.

Burns, James MacGregor. *Leadership.* New York: HarperCollins, 1978.

———. *Deadlock of Democracy.* Englewood Cliffs, N.J.: Prentice-Hall, 1963.

———. *The Power to Lead: The Crisis of the American Presidency.* New York: Simon & Schuster, 1984.

———. *Presidential Government: The Crucible of Leadership.* Boston: Houghton Mifflin, 1965.

———. *Roosevelt: The Lion and the Fox.* San Diego: Harvest/HBJ, 1984.

———. *Roosevelt: Soldier of Freedom.* New York: Harcourt Brace, Jovanovich, 1970.

Campbell, Colin. *Managing the Presidency: Carter, Reagan and the Search for Executive Harmony.* Pittsburgh: University of Pittsburgh Press, 1986.

Cannon, Lou. *President Reagan: The Role of a Lifetime.* New York: Touchstone, 1991.

Carter, Jimmy. *Keeping Faith: Memoirs of a President.* New York: Bantam Books, 1982.

Caroli, Betty Boyd. *First Ladies.* New York: Oxford University Press, 1987.

Ceasar, James W. *Presidential Selection.* Princeton, N.J.: Princeton University Press, 1979.

Corrado, Anthony. *Creative Campaigning.* Boulder, Colo.: Westview, 1992.

Corvitz, L. Gordon, and Jeremy A. Rabkin, eds. *The Tethered Presidency.* Washington, D.C.: American Enterprise Institute, 1989.

Corwin, Edward S. *The President: Office and Powers, 1978–1984.* 5th ed. New York: New York University Press, 1984.

———. *Total War and the Constitution.* Westminister, Md.: Knopf, 1947.

Covington, Cary R., and Lester G. Seligman. *The Coalition Presidency.* Chicago: Dorsey Press, 1989.

Cox, Gary W., and Samuel Kernell, eds. *The Politics of Divided Government.* Boulder, Colo.: Westview, 1991.

Crabb, Cecil V., and Pat Holt. *Invitation to Struggle: Congress, the President, and Foreign Policy.* 3rd ed. Washington, D.C.: Congressional Quarterly Press, 1989.

———, and Kevin V. Mulcahy. *American National Security: A Presidential Perspective.* New York: Harcourt Brace, 1990.

———. *Presidents and Foreign Policymaking.* Baton Rouge: Louisiana State University Press, 1986.

Cronin, Thomas E., ed. *Inventing the American Presidency.* Lawrence: University Press of Kansas, 1989.

———, ed. *Rethinking the Presidency.* Boston: Little, Brown, 1982.

———. *The State of the Presidency.* 2nd ed. Boston: Little, Brown, 1980.

———, and Sanford Greenberg, eds. *The Presidential Advisory System.* New York: Harper & Row, 1969.

———, and Rexford Tugwell, eds. *The Presidency Reappraised.* New York: Praeger, 1977.

Dallek, Robert. *Hail to the Chief: The Making and Unmaking of American Presidents.* New York: Hyperion Books, 1996.

Darman, Richard. *Who's In Control? Polar Politics and the Sensible Center.* New York: Simon & Schuster, 1996.

Davis, James W. *The American Presidency.* 2d ed. Westport, Conn.: Praeger, 1995.

———. *The President as Party Leader.* New York: Praeger, 1992.

DeGrazia, Alfred. *Congress and the Presidency.* Washington, D.C.: American Enterprise Institute, 1967.

Donald, David H. *Lincoln.* New York: Simon & Schuster, 1995.

Draper, Theodore. *A Very Thin Line: The Iran-Contra Affair.* New York: Touchstone Books, 1991.

Eastland, Terry. *Energy in the Executive: The Case for the Strong Presidency.* New York: Free Press, 1992.

Edwards, George C. *At the Margins: Presidential Leadership in Congress.* New Haven, Conn.: Yale University Press, 1989.

———. *Presidential Influence in Congress.* San Francisco: W.H. Freeman, 1980.

———. *The Public Presidency.* New York: St. Martin's Press, 1983.

———, John H. Kessel, and Bert A. Rockman, eds. *Researching the Presidency: Vital Questions, New Approaches.* Pittsburgh: University of Pittsburgh Press, 1993.

———, and Stephen Wayne. *Presidential Leadership: Politics and Policy Making.* 3rd ed. New York: St. Martin's Press, 1994.

Ellis, Richard, and Aaron Wildavsky. *Dilemmas of Presidential Leadership.* New Brunswick, N.J.: Transaction Publishers, 1989.

Fishel, Jeff. *Presidents and Promises.* Washington, D.C.: Congressional Quarterly Press, 1985.

Fisher, Louis. *The Constitution between Friends.* New York: St. Martin's Press, 1978.

————. *Presidential War Power.* Lawrence: University Press of Kansas, 1995.

Franklin, Daniel P. *Extraordinary Measures.* Pittsburgh: University of Pittsburgh Press, 1991.

Gardner, John W. *On Leadership.* New York: Free Press, 1990.

Genovese, Michael A. *The Presidential Dilemma: Leadership in the American System.* New York: HarperCollins, 1995.

————. *The Nixon Presidency: Power and Politics in Turbulent Times.* Westport, Conn.: Greenwood Press, 1990.

————. *The Presidency in an Age of Limits.* Westport, Conn.: Greenwood Press, 1993.

————. *The Supreme Court, the Constitution, and Presidential Power.* Landham, Md.: University Press of America, 1980.

Greenstein, Fred. *The Hidden-Hand Presidency: Eisenhower as Leader.* New York: Basic Books, 1982.

Gregg, Gary L. *The Presidential Republic: Executive Representation and Deliberative Democracy.* Lanham, Md.: Rowland & Littlefield, 1997.

Grover, William F. *The President as Prisoner.* New York: State University of New York Press, 1989.

Hamilton, Alexander, James Madison, and John Jay. *Federalist Papers.* New York: New American Library, 1961.

Hargrove, Erwin C. *The Power of the Modern Presidency.* New York: Knopf, 1974.

————, and Michael Nelson. *Presidents, Politics and Policy.* Baltimore: Johns Hopkins University Press, 1984.

Harriger, Katy J. *Independent Justice: The Federal Special Prosecutor in American Politics.* Lawrence: University Press of Kansas, 1992.

Hart, Roderick. *The Sound of Leadership.* Chicago: University of Chicago Press, 1987.

————. *The Presidential Branch.* Chatham, N.J.: Chatham House, 1995.

Hess, Stephen. *Organizing the Presidency.* Washington, D.C.: Brookings, 1976.

Hughes, Emmet John. *The Living Presidency.* New York: Coward, McCann and Geoghegan, 1973.

Jamieson, Kathleen Hall. *Packaging the Presidency.* New York: Oxford University Press, 1996.

Johnson, Haynes, and David Broder. *The System.* Boston: Little, Brown, 1996.

Johnson, Loch K. *Secret Agencies: U.S. Intelligence Agencies in a Hostile World.* New Haven, Conn.: Yale University Press, 1996.

Johnson, Richard T. *Managing the White House.* New York: HarperCollins, 1974.

Jones, Charles O. *The Presidency in a Separated System.* Washington, D.C.: Brookings, 1994.

Kallenbach, Joseph E. *The American Chief Executive.* New York: Harper & Row, 1966.

Kernell, Samuel M. *Going Public: New Strategies of Presidential Leadership.* Washington, D.C.: Congressional Quarterly Press, 1993.

————, and Samuel Popkin, eds. *Chief of Staff.* Berkeley: University of California Press, 1986.

Kessel, John H. *Presidential Campaign Politics.* Chicago: Dorsey Press, 1988.

King, Gary, and Lyn Ragsdale. *The Elusive Executive: Discovering Statistical Pat-*

terns in the Presidency. Washington, D.C.: Congressional Quarterly Press, 1988.

Koenig, Louis W. *The Chief Executive.* 5th ed. San Diego: Harcourt Brace Jovanovich, 1986.

Koh, Harold Kongju. *The National Security Constitution: Sharing Power After the Iran-Contra Affair.* New Haven, Conn.: Yale University Press, 1990.

Kutler, Stanley K. *The Wars of Watergate.* New York: Knopf, 1990.

Lammers, William W. *Presidential Politics: Patterns and Prospects.* New York: HarperCollins, 1976.

Laski, Harold. *The American Presidency.* New York: Harper & Brothers, 1940.

LeLoup, Lance T., and Steven A. Shull. *Congress and the President: The Policy Connection.* Belmont, Calif.: Wadsworth, 1993.

Leuchtenburg, William E. *In the Shadow of FDR: From Harry Truman to Ronald Reagan.* Ithaca, N.Y.: Cornell University Press, 1983.

Levine, Myron A. *Presidential Campaigns and Elections.* Itasca, Ill.: F.E. Peacock, 1995.

Light, Paul C. *The President's Agenda.* Baltimore: Johns Hopkins University Press, 1982.

————. *Vice Presidential Power: Advice and Influence in the White House.* Baltimore: Johns Hopkins University Press, 1984.

Loevy, Robert D. *The Flawed Path to the Presidency.* Albany: State University of New York Press, 1995.

Lowi, Theodore J. *The Personal President: Power Invested Promise Unfulfilled.* Ithaca, N.Y.: Cornell University Press, 1985.

Maltese, John. *Spin Control: The White House Office of Communications and the Management of Presidential News.* Chapel Hill: University of North Carolina Press, 1992.

Mansfield, Harvey C., Jr. *Taming the Prince: The Ambivalence of Presidential Power.* New York: Free Press, 1989.

Maraniss, David. *First in His Class: The Biography of Bill Clinton.* New York: Touchstone, 1995.

Mayer, Jane, and Doyle McManus. *Landslide: The Unmaking of the President, 1984–1988.* Boston: Houghton Mifflin, 1988.

Mayer, William G., ed. *In Pursuit of the Presidency: How We Choose Our Presidential Nominees.* Chatham, N.J.: Chatham House, 1995.

Mayhew, David. *Divided We Govern.* New Haven, Conn.: Yale University Press, 1991.

Milkis, Sidney M. *The President and the Parties.* New York: Oxford University Press, 1993.

————, and Michael Nelson. *The American Presidency: Origins and Development.* 2nd ed. Washington, D.C.: Congressional Quarterly Press, 1994.

Miroff, Bruce. *Icons of Democracy.* New York: Basic Books, 1993.

————. *Pragmatic Illusions: The Presidential Politics of John F. Kennedy.* New York: David McKay, 1976.

Morris, Richard S. *Behind the Oval Office: Winning the Presidency in the 1990s.* New York: Random House, 1997.

Mosley, Zelma, Joseph Pika, and Richard A. Watson. *The Presidential Contest.* Washington, D.C.: Congressional Quarterly Press, 1992.

Mullen, William F. *Presidential Power and Politics.* New York: St. Martin's Press, 1976.

Murray, Robert K., and Tim H. Blessing. *Greatness in the White House: Rating the Presidents, Washington through Carter.* University Park: Pennsylvania State University Press, 1988.

Nathan, Richard. *The Administrative Presidency.* New York: John Wiley and Sons, 1983.

Neustadt, Richard E. *Presidential Power.* New York: Wiley, 1960.

———. *Presidential Power and the Modern Presidents.* New York: Free Press, 1990.

O'Toole, James. *Leading Change: The Argument for Values-based Leadership.* New York: Ballentine Books, 1996.

Patterson, Thomas E. *Out of Order.* New York: Vintage Books, 1994.

Peterson, Mark A. *Legislating Together.* Cambridge, Mass.: Harvard University Press, 1990.

Pfiffner, James P., ed. *The Managerial Presidency.* Pacific Grove, Calif.: Brooks/Cole, 1991.

———. *The Strategic Presidency: Hitting the Ground Running.* 2nd ed. Lawrence: University Press of Kansas, 1996.

Phelps, Glenn A. *George Washington and American Constitutionalism.* Lawrence: University Press of Kansas, 1994.

Pika, Joseph A., Norman C. Thomas, and Richard A. Watson. *The Politics of the Presidency.* Washington, D.C.: Congressional Quarterly Press, 1993.

Pious, Richard. *The American Presidency.* New York: Basic Books, 1979.

———. *The Presidency.* Boston: Allyn and Bacon, 1996.

Podhoretz, John. *Hell-of-a-Ride: Backstage at the White House Follies, 1989–1993.* New York: Simon & Schuster, 1993.

Polsby, Nelson W. *Political Innovation in America.* New Haven, Conn.: Yale University Press, 1985.

———, and Aaron Wildavsky. *Presidential Elections.* Chatham, N.J.: Chatham House, 1995.

Quayle, Dan. *Standing Firm.* New York: Harper Paperbacks, 1995.

Ragsdale, Lyn. *Presidential Politics.* Boston: Houghton Mifflin, 1993.

———. *Vital Statistics on the Presidency.* Washington, D.C.: Congressional Quarterly Press, 1995.

Reich, Robert. *Locked in the Cabinet.* New York: Knopf, 1997.

Reagan, Ronald. *An American Life: The Autobiography.* New York: Simon & Schuster, 1990.

Rimmerman, Craig A. *Presidency by Plebiscite.* Boulder, Colo.: Westview, 1993.

Robinson, Donald L. *"To the Best of My Ability": The Presidency and the Constitution.* New York: Norton, 1987.

Rockman, Bert. *The Leadership Question: The Presidency and the American System.* New York: Praeger, 1984.

Rose, Richard. *The Postmodern President: The White House Meets the World.* Chatham, N.J.: Chatham House, 1991.

Rosenstiel, Tom. *Strange Bedfellows: How Television and the Presidential Candidates Changed American Politics, 1992.* New York: Hyperion Books, 1993.

Rossiter, Clinton. *The American Presidency.* New York: Harcourt, Brace and World, 1956.

———. *Constitutional Dictatorship: Crisis Government in the Modern Democracy.* Princeton, N.J.: Princeton University Press, 1948.

Sabato, Larry. *Feeding Frenzy.* New York: Free Press, 1991.

Schlesinger, Arthur M., Jr. *The Cycles of American History.* Boston: Houghton Mifflin, 1986.

———. *The Imperial Presidency.* Boston: Houghton Mifflin, 1973.

Shogan, Robert. *The Riddle of Power.* New York: Dutton, 1991.

Shull, Steven A., ed. *The Two Presidencies: A Quarter Century Assessment.* Chicago: Nelson-Hall, 1991.

Shultz, George P. *Turmoil and Triumph: My Years as Secretary of State.* New York: Scribners, 1993.

Simonton, Dean Keith. *Why Presidents Succeed.* New Haven, Conn.: Yale University Press, 1987.

Sindler, Allan P. *Unchosen Presidents.* Berkeley: University of California Press, 1976.

Skowronek, Stephen. *The Politics Presidents Make.* Cambridge, Mass: Belknap, 1993.

Smith, Hedrick. *The Power Game: How Washington Works.* New York: Ballantine, 1988.

Sorensen, Theodore C. *Watchman in the Night.* Cambridge, Mass.: MIT Press, 1975.

———. *Why I Am A Democrat.* New York: Henry Holt, 1996.

Spitzer, Robert J. *President and Congress: Executive Hegemony at the Crossroads of American Government.* New York: McGraw-Hill, 1993.

———. *The Presidential Veto: Touchstone of the American Presidency.* Albany: State University of New York Press, 1988.

Stewart, James B. *Blood Sport: The President and His Adversaries.* New York: Simon & Schuster, 1996.

Stockman, David A. *The Triumph of Politics: The Inside Story of the Reagan Revolution.* New York: Avon, 1987.

Stuckey, Mary E. *The President as Interpreter-in-Chief.* Chatham, N.J.: Chatham House, 1991.

Sundquist, James L. *Constitutional Reform and Effective Government.* Washington, D.C.: Brookings, 1986.

Tatalovich, Raymond, and Bryon W. Daynes. *Presidential Power in the United States.* Belmont, Calif.: Brooks/Cole, 1984.

Thurber, James A., ed. *Divided Democracy.* Washington, D.C.: Congressional Quarterly Press, 1991.

———, ed. *Rivals for Power: Presidential-Congressional Relations.* Washington, D.C.: Congressional Quarterly Press, 1996.

Tulis, Jeffrey K. *The Rhetorical Presidency.* Princeton, N.J.: Princeton University Press, 1987.

Troy, Gil. *Affairs of State: The Rise and Rejection of the Presidential Couple since World War II.* New York: McGraw Hill, 1997.

Walsh, Kenneth T. *Feeding the Beast, the White House vs. Press.* New York: Random House, 1996.

Walsh, Lawrence E. *Firewall.* New York: Norton, 1997.

Warshaw, Shirley Anne. *Powersharing: White House–Cabinet Relations in the Modern Presidency.* Albany: State University of New York Press, 1996.

———. *The Domestic Presidency.* Boston: Allyn and Bacon, 1997.

Watson, Richard A. *Presidential Vetoes and Public Policy.* Lawrence: University Press of Kansas, 1993.

———. *The Presidential Contest.* New York: Wiley, 1984.

Wayne, Stephen J. *The Road to the White House, 1996.* New York: St. Martin's Press, 1995.

Weko, Thomas J. *The Politicizing Presidency: The White House Personnel Office, 1948–1994.* Lawrence: University Press of Kansas, 1995.

Whitcover, Jules. *Crapshoot: Rolling the Dice on the Vice Presidency.* New York: Crown, 1992.

Wildavsky, Aaron. *The Beleaguered Presidency.* New Brunswick, N.J.: Transaction Publishers, 1994.

Wills, Garry. *Certain Trumpets: The Call of Leaders.* New York: Simon & Schuster, 1994.

Wilson, Robert A., ed. *Character Above All: Ten Presidents from FDR to George Bush.* New York: Simon & Schuster, 1995.

Woodward, Bob. *The Agenda: Inside the Clinton White House.* New York: Simon & Schuster, 1994.

Index